Birkhäuser

Annals of the International Society of Dynamic Games
Volume 13

Series Editor
Tamer Başar

For further volumes:
http://www.springer.com/series/4919

Vlastimil Křivan • Georges Zaccour
Editors

Advances in Dynamic Games

Theory, Applications, and Numerical Methods

Birkhäuser

Editors
Vlastimil Křivan
Biology Centre AS CR
České Budějovice
Czech Republic

Georges Zaccour
GERAD
HEC Montréal
Montreal, QC, Canada

ISBN 978-3-319-02689-3 ISBN 978-3-319-02690-9 (eBook)
DOI 10.1007/978-3-319-02690-9
Springer Cham Heidelberg New York Dordrecht London

Library of Congress Control Number: 2013953331

Mathematics Subject Classification (2010): 49N70, 49N75, 49N90, 91A15, 91A22, 91A23, 91A24, 91A25, 91A80

Printed on acid-free paper

Springer is part of Springer Science+Business Media (www.birkhauser-science.com)

Preface

This edited volume focuses on various aspects of dynamic game theory providing state-of-the-art information on recent conceptual and methodological developments. It also includes novel applications in different areas such as economics, ecology, engineering, and management science. Most of the selected papers were presented at the 15th International Symposium on Dynamic Games and Applications held in Byšice, Czech Republic on July 19–22, 2012. The symposium is held every two years under the auspices of the International Society of Dynamic Games. The list of contributors consists of well-established and young researchers working in different countries. Every submitted paper has gone through a stringent reviewing process. This volume is made of 15 chapters that we classified into three parts. The first one regroups papers dealing with some theoretical and/or computational issues in dynamic games; the second part includes four chapters applying dynamic games in different areas; and finally, the third part has five chapters on pursuit–evasion games. As in any clustering of papers of this type, the one proposed here is far from being unique.

Part I. Dynamic Games: Theory and Computation

Arapostathis, Borkar, and Kumar study zero-sum stochastic differential games and establish the existence of a solution to the Isaac's equation for the ergodic game. They characterize the optimal stationary strategies without assuming the data and geometric ergodicity. The authors also study a relative value iteration scheme that takes the form of a parabolic Isaac's equation and show that it converges to the elliptic Isaac's equation as time goes to infinity under the hypothesis of geometric ergodicity. *Blueschke, Neck, and Behrens* present an algorithm (OPTGAME3) for the computation of Nash and Stackelberg equilibria, as well as Pareto-optimal solutions of dynamic games. The use of this algorithm is illustrated in the context of a stylized nonlinear two-country macroeconomic model of a monetary union for analyzing the interactions between fiscal (governments) and monetary (common central bank) policy makers, assuming different objective functions of these decision makers. *Botkin and Turova* consider the problem of aircraft control during take-off in a windshear. A four-dimensional dynamic programming method is proposed and stable numerical algorithms for solving Hamilton–Jacobi–Bellman–Isaacs equations arising from differential games with state constraints are utilized for the design of controls.

Carlson formulates and proves a general existence theorem for an optimal solution for the class of bi-level optimal control problems, when both upper and lower level problems are described by ordinary optimal control models of Lagrange type. An interesting feature of the result is that it does not require the lower level to have a unique best response to each admissible strategy of the upper level problem. *Pachter* studies a two-player linear-quadratic game, where the players have some private information. Each player is able to formulate an expression for his/her expected payoff without the need, à la Harsanyi, to provide a prior probability distribution function of the game's parameter and without recourse to the player Nature. The paper characterizes the conditions under which the principle of certainty equivalence holds. Finally, *Zusai* investigates a variety of conditions to establish the connection between an interior convergence in regular payoff monotone selections and versions of proper equilibrium and use the connection for equilibrium selection.

Part II. Dynamic Games: Applications

De Giovanni studies a differential game involving a manufacturer and a retailer interacting in a supply chain. The retailer, who is the leader of chain, sets the price and the advertising budget, whereas the manufacturer chooses the level of quality improvement. The solutions of two scenarios are characterized and compared, namely, a coordinated case, where the retailer supports the quality improvement program, and the uncoordinated case, where he/she does not. *Ramsey* studies evolution of parental care using evolutionary game theory. This research was motivated by various types of parental care observed in mouth brooding fishes. These types of behavior include paternal mouth brooding where the male only holds eggs, maternal mouth brooding where the female takes eggs, or biparental care where both males and females care for eggs. Ramsey surveys and extends existing models in several directions. In particular, he compares a model where partners take decisions simultaneously with a model where one sex decides first. These models describe a complex feedback between the sex ratio and patterns of parental care. *Scheffran* studies value-cost dynamic games, where multiple agents adjust the flow and allocation of investments to action pathways that affect the value of other agents. He determines conditions for cooperation and analyzes allocation priorities and the stability of equilibrium. The approach is applied to the trading between buyers and sellers of goods to determine conditions for mutually beneficial market exchange.

Staňková, Abate, and Sabelis model interactions between predatory and prey mites during a season. In particular, they focus on the time when these species begin to enter diapause (a physiological state of dormancy to survive winter). Although entering diapause is induced by environmental factor such as low temperature and short days, presence of predators can also induce diapause in prey mites as a survival strategy. Similarly, inability of finding dormant prey can induce diapause in predators, which leads to a game prey and predatory mites play. Authors argue that this is a Stackelberg game and they analyze the optimal behavior of prey and predators. *Troeva and Lukin* consider a differential game in which n players dump a pollutant, a by-product of their production process, in a water reservoir. The evolution of the pollution concentration level is described by a partial-differential equation. The authors prove the existence of an ε-Nash equilibrium for the class of piecewise-programmed strategies and illustrate their model with a series of numerical examples.

Part III. Pursuit–Evasion Games

Kamneva and Patsko deal with an open-loop solvability operator in two-person zero-sum differential games with simple motions. This operator, which takes a given terminal set to the set defined at the initial instant, possesses the semigroup property in the case of a convex terminal set. The authors provide sufficient conditions ensuring the semigroup property in the non-convex case and construct examples to illustrate the relevance of these conditions. *Kumkov, Patsko, and Le Ménec* deal with a zero-sum differential game, in which the first player controls two pursuing objects, whose aim is to minimize the minimum of the misses between each of them and the evader at some given instant. The authors consider the case where the pursuers have equal dynamic capabilities, but are less powerful than the evader, and provide some numerical results.

Le Ménec considers a team of autonomous vehicles, composed of a pursuing vehicle and of several unmanned aircraft vehicles (UAVs), using on-board sensors for tracking and intercepting a moving target. This situation is modeled as a zero-sum two-player pursuit–evasion differential game with costly information. The author solves the game for simple as well as complex kinematics and discusses the 4D guidance law and the coordination algorithm implemented for managing the UAVs. *Shinar, Glizer, and Turetsky* consider linear pursuit–evasion games with bounded controls. They analyze the cases of an ideal, a first-order, and a second-order pursuer against an ideal and a first-order evader and compare the values of these games. The authors show that replacing the second-order pursuer by a first-order approximation underestimates the value of the game.

České Budějovice, Czech Republic — Vlastimil Křivan
Montreal, QC, Canada — Georges Zaccour

Acknowledgments

We would like to extend our warmest thanks to the large number of reviewers who did an excellent job under very harsh time constraints. Our thanks to all the contributors and authors who submitted their work to this volume. We would also wish to thank Francine Benoît for her excellent editing work and GERAD for its financial and administrative support to ISDG activities.

Contents

Contributors

Alessandro Abate Delft Center for Systems & Control, Delft University of Technology, Delft, The Netherlands

Ari Arapostathis Department of Electrical and Computer Engineering, The University of Texas at Austin, Austin, TX, USA

Doris A. Behrens Department of Controlling and Strategic Management, Alpen-Adria-Universität Klagenfurt, Klagenfurt, Austria

Dmitri Blueschke Department of Economics, Alpen-Adria-Universität Klagenfurt, Klagenfurt, Austria

Vivek S. Borkar Department of Electrical Engineering, Indian Institute of Technology, Powai, Mumbai, India

Nikolai D. Botkin Technische Universität München, Garching bei München, Germany

Dean A. Carlson American Mathematical Society, Mathematical Reviews, Ann Arbor, MI, USA

Pietro De Giovanni Department of Information, Logistics and Innovation, VU Amsterdam University, Amsterdam, The Netherlands

Valery Y. Glizer Department of Applied Mathematics, Ort Braude College, Karmiel, Israel

Liudmila Kamneva Institute of Mathematics and Mechanics, Ekaterinburg, Russia

K. Suresh Kumar Department of Mathematics, Indian Institute of Technology, Powai, Mumbai, India

Sergey Kumkov Institute of Mathematics and Mechanics, Ekaterinburg, Russia

Stéphane Le Ménec EADS / MBDA France, Le Plessis-Robinson Cedex, France

Vassili Lukin Scientific Research Institute of Mathematics, NEFU, Yakutsk, Russia

Reinhard Neck Department of Economics, Alpen-Adria-Universität Klagenfurt, Klagenfurt, Austria

Meir Pachter Department of Electrical and Computer Engineering, Air Force Institute of Technology, Wright-Patterson AFB, OH, USA

Valerii Patsko Institute of Mathematics and Mechanics, Ekaterinburg, Russia

David M. Ramsey Department of Computer Science and Management, Wrocław University of Technology, Wrocław, Poland

Maurice W. Sabelis Institute for Biodiversity and Ecosystem Dynamics, University of Amsterdam, Amsterdam, The Netherlands

Jürgen Scheffran Institute of Geography, KlimaCampus, University of Hamburg, Hamburg, Germany

Josef Shinar Faculty of Aerospace Engineering, Technion - Israel Institute of Technology, Haifa, Israel

Kateřina Staňková Department of Knowledge Engineering, Maastricht University, Maastricht, The Netherlands

Marianna Troeva North-Eastern Federal University, Yakutsk, Russia

Vladimir Turetsky Department of Applied Mathematics, Ort Braude College, Karmiel, Israel

Varvara L. Turova Technische Universität München, Garching bei München, Germany

Dai Zusai Department of Economics, Temple University, Philadelphia, PA, USA

Part I
Dynamic Games: Theory and Computation

Chapter 1
Relative Value Iteration for Stochastic Differential Games

Ari Arapostathis, Vivek S. Borkar, and K. Suresh Kumar

Abstract We study zero-sum stochastic differential games with player dynamics governed by a nondegenerate controlled diffusion process. Under the assumption of uniform stability, we establish the existence of a solution to the Isaac's equation for the ergodic game and characterize the optimal stationary strategies. The data is not assumed to be bounded, nor do we assume geometric ergodicity. Thus our results extend previous work in the literature. We also study a relative value iteration scheme that takes the form of a parabolic Isaac's equation. Under the hypothesis of geometric ergodicity we show that the relative value iteration converges to the elliptic Isaac's equation as time goes to infinity. We use these results to establish convergence of the relative value iteration for risk-sensitive control problems under an asymptotic flatness assumption.

Keywords Stochastic differential games • Ergodic control • Relative value iteration • Risk-sensitive control

---

A. Arapostathis
Department of Electrical and Computer Engineering, The University of Texas at Austin, 1 University Station, Austin, TX 78712, USA
e-mail: ari@mail.utexas.edu

V.S. Borkar (✉)
Department of Electrical Engineering, Indian Institute of Technology, Powai, Mumbai 400076, India
e-mail: borkar.vs@gmail.com

K.S. Kumar
Department of Mathematics, Indian Institute of Technology, Powai, Mumbai 400076, India
e-mail: suresh@math.iitb.ac.in

V. Křivan and G. Zaccour (eds.), *Advances in Dynamic Games*, Annals of the International Society of Dynamic Games 13, DOI 10.1007/978-3-319-02690-9_1,

1.1 Introduction

In this paper we consider a relative value iteration for zero-sum stochastic differential games. This relative value iteration is introduced in Arapostathis and Borkar (2012) for stochastic control, and we follow the method introduced in this paper.

In Sect. 1.2, we prove the existence of a solution to the Isaac's equation corresponding to the ergodic zero-sum stochastic differential game. We do not assume that the data or the running payoff function is bounded, nor do we assume geometric ergodicity, so our results extend the work in Borkar and Ghosh (1992). In Sect. 1.3, we introduce a relative value iteration scheme for the zero-sum stochastic differential game and prove its convergence under a hypothesis of geometric ergodicity. In Sect. 1.4, we apply the results from Sect. 1.3 and study a value iteration scheme for risk-sensitive control under an asymptotic flatness assumption.

1.2 Problem Description

We consider zero-sum stochastic differential games with state dynamics modeled by a controlled non-degenerate diffusion process $X = \{X(t) : 0 \leq t < \infty\}$, and subject to a long-term average payoff criterion.

1.2.1 State Dynamics

Let U_i, $i = 1, 2$, be compact metric spaces and $V_i = \mathscr{P}(U_i)$ denote the space of all probability measures on U_i with Prohorov topology. Let

$$\bar{b} : \mathbb{R}^d \times U_1 \times U_2 \to \mathbb{R}^d \qquad \text{and} \qquad \sigma : \mathbb{R}^d \to \mathbb{R}^{d\times d}$$

be measurable functions. Assumptions on $\bar{b}$ and σ will be specified later. Define $b : \mathbb{R}^d \times V_1 \times V_2 \to \mathbb{R}^d$ as

$$b(x, v_1, v_2) := \int_{U_1}\int_{U_2} \bar{b}(x, u_1, u_2)\, v_1(\mathrm{d}u_1)\, v_2(\mathrm{d}u_2),$$

for $x \in \mathbb{R}^d$, $v_1 \in V_1$ and $v_2 \in V_2$. We model the controlled diffusion process X via the Itô s.d.e.

$$\mathrm{d}X(t) = b\big(X(t), v_1(t), v_2(t)\big)\,\mathrm{d}t + \sigma\big(X(t)\big)\,\mathrm{d}W(t). \tag{1.1}$$

All processes on (1.1) are defined in a common probability space $(\Omega, \mathscr{F}, P)$ which is assumed to be complete. The process $W = \{W(t) : 0 \leq t < \infty\}$ is an $\mathbb{R}^d$-valued standard Wiener process which is independent of the initial condition X_0

of (1.1). Player i, with $i = 1, 2$, controls the dynamics X through her strategy $v_i(\cdot)$, a V_i-valued process which is jointly measurable in $(t, \omega) \in [0, \infty) \times \Omega$ and non-anticipative, i.e., for $s < t$, $W(t) - W(s)$ is independent of

$$\mathscr{F}_s := \text{the completion of } \sigma(X_0, v_1(r), v_2(r), W(r), r \le s).$$

We denote the set of all such controls (admissible controls) for player i by $\mathscr{U}_i$, $i = 1, 2$.

Assumptions on the Data: We assume the following conditions on the coefficients $\bar{b}$ and σ to ensure existence of a unique solution to (1.1).

(A1) The functions $\bar{b}$ and σ are locally Lipschitz continuous in $x \in \mathbb{R}^d$, uniformly over $(u_1, u_2) \in U_1 \times U_2$, and have at most a linear growth rate in $x \in \mathbb{R}^d$, i.e., for some constant κ,

$$\|\bar{b}(x, u_1, u_2)\|^2 + \|\sigma(x)\|^2 \le \kappa\big(1 + \|x\|^2\big) \qquad \forall (x, u_1, u_2) \in \mathbb{R}^d \times U_1 \times U_2,$$

where $\|\sigma\|^2 := \operatorname{trace}\big(\sigma\sigma^{\mathsf{T}}\big)$, with ${}^{\mathsf{T}}$ denoting the transpose. Also $\bar{b}$ is continuous.

(A2) For each $R > 0$ there exists a constant $\kappa(R) > 0$ such that

$$z^{\mathsf{T}} a(x) z \ge \kappa(R)\|z\|^2 \qquad \text{for all } \|x\| \le R \text{ and } z \in \mathbb{R}^d,$$

where $a := \sigma\sigma^{\mathsf{T}}$.

Definition 1.1. For $f \in C^2(\mathbb{R}^d)$ define

$$\bar{L} f(x, u_1, u_2) := \bar{b}(x, u_1, u_2) \cdot \nabla f(x) + \frac{1}{2} \operatorname{tr}\big(a(x)\nabla^2 f(x)\big)$$

for $x \in \mathbb{R}^d$ and $(u_1, u_2) \in U_1 \times U_2$. Also define the *relaxed extended controlled generator* L by

$$Lf(x, v_1, v_2) := \int_{U_1}\int_{U_2} Lf(x, u_1, u_2)\, v_1(\mathrm{d}u_1)\, v_2(\mathrm{d}u_2)\,, \quad f \in C^2(\mathbb{R}^d)\,,$$

for $x \in \mathbb{R}^d$ and $(v_1, v_2) \in V_1 \times V_2$.

We denote the set of all stationary Markov strategies of player i by $\mathscr{M}_i$, $i = 1, 2$.

1.2.2 Zero-Sum Ergodic Game

Let $\bar{h} : \mathbb{R}^d \times U_1 \times U_2 \to [0, \infty)$ be a continuous function, which is also locally Lipschitz continuous in its first argument. We define the *relaxed running payoff function* $h : \mathbb{R}^d \times V_1 \times V_2 \to [0, \infty)$ by

$$h(x, v_1, v_2) := \int_{U_1} \int_{U_2} \bar{h}(x, u_1, u_2)\, v_1(\mathrm{d}u_1)\, v_2(\mathrm{d}u_2)\,.$$

Player 1 seeks to maximize the average payoff given by

$$\liminf_{T \to \infty} \frac{1}{T}\, E_x\left[\int_0^T h(X(t), v_1(t), v_2(t))\, \mathrm{d}t\right] \tag{1.2}$$

over all admissible controls $v_1 \in \mathscr{U}_1$, while Player 2 seeks to minimize (1.2) over all $v_2 \in \mathscr{U}_2$. Here E_x is the expectation operator corresponding to the probability measure on the canonical space of the process starting at $X(0) = x$.

Since we shall analyze the average payoff as a limiting case of the discounted payoff in the "vanishing discount" limit, we shall also consider the infinite horizon discounted payoff

$$E_x\left[\int_0^\infty e^{-\alpha t} h(X(t), v_1(t), v_2(t))\, \mathrm{d}t\right],$$

where $\alpha > 0$ is the discount factor.

Assumptions on Ergodicity: We consider the following ergodicity assumptions:

(A3) There exist a positive inf-compact function $\mathscr{V} \in C^2(\mathbb{R}^d)$ and positive constants k_0, k_1, and k_2 such that

$$\bar{L}\mathscr{V}(x, u_1, u_2) \le k_0 - 2k_1 \mathscr{V}(x),$$

$$\max_{u_1 \in U_1, u_2 \in U_2} \bar{h}(x, u_1, u_2) \le k_2\, \mathscr{V}(x)$$

for all $(u_1, u_2) \in U_1 \times U_2$, and $x \in \mathbb{R}^d$. Without loss of generality we assume $\mathscr{V} \ge 1$.

(A3$'$) There exist nonnegative inf-compact functions $\mathscr{V} \in C^2(\mathbb{R}^d)$ and $g \in C(\mathbb{R}^d)$, and positive constants k_0 and k_2 such that

$$\bar{L}\mathscr{V}(x, u_1, u_2) \le k_0 - g(x),$$

$$\max_{u_1 \in U_1, u_2 \in U_2} \bar{h}(x, u_1, u_2) \le k_2\, g(x)$$

for all $(u_1, u_2) \in U_1 \times U_2$, and $x \in \mathbb{R}^d$. Also,

$$\frac{\max_{u_1 \in U_1, u_2 \in U_2} \bar{h}(x, u_1, u_2)}{g(x)} \xrightarrow[\|x\| \to \infty]{} 0\,.$$

Without loss of generality we assume $\mathscr{V} \ge 1$ and $g \ge 1$.

In this section we use assumption (A3$'$), while in Sect. 1.3 we employ (A3) which is stronger and equivalent to geometric ergodicity in the time-homogeneous Markov case.

For the uncontrolled (i.e., Markov) case, (A3$'$) is the so-called g-norm ergodicity in the terminology of Meyn and Tweedie (1993) which implies, in addition to convergence of laws to a unique stationary distribution, convergence of $\frac{1}{t}\int_0^t E[f(X(s))]\,\mathrm{d}s$ to the corresponding stationary expectation as $t \uparrow \infty$ for all f with growth rate at most that of g and vice versa. Assumption (A3) corresponds to the same with $h = V$ and implies in particular exponential convergence to stationary averages (and vice versa). This is the so-called geometric ergodicity. When (A3$'$) holds in the controlled case, it implies in particular tightness of stationary distributions attainable under stationary Markov controls. In fact this condition is necessary and sufficient. See Arapostathis et al. (2011, Lemma 3.3.4) for this and other equivalent characterizations. Thus (A3$'$) is the best possible condition for uniform stability in this sense. While the results of Arapostathis and Borkar (2012) can be extended to control problems when instability is possible but is penalized by the cost structure, this does not extend naturally to the zero-sum game, because what is penalty for one agent is a reward for the other.

We start with a theorem which characterizes the value of the game under a discounted infinite horizon criterion. For this we need the following notation: For a continuous function $\mathscr{V}\colon\mathbb{R}^d \to (0,\infty)$, $C_{\mathscr{V}}(\mathbb{R}^d)$ denotes the space of functions in $C(\mathbb{R}^d)$ satisfying $\sup_{x\in\mathbb{R}^d}\left|\frac{f(x)}{\mathscr{V}(x)}\right| < \infty$. This is a Banach space under the norm

$$\|f\|_{\mathscr{V}} := \sup_{x\in\mathbb{R}^d}\left|\frac{f(x)}{\mathscr{V}(x)}\right| .$$

Theorem 1.1. *Assume* (A1), (A2), *and* (A3$'$). *For* $\alpha > 0$, *there exists a solution* $\psi_\alpha \in C_{\mathscr{V}}(\mathbb{R}^d) \cap C^2(\mathbb{R}^d)$ *to the p.d.e.*

$$\begin{aligned}\alpha\psi_\alpha(x) &= \min_{v_2\in V_2}\max_{v_1\in V_1}\left[L\psi_\alpha(x,v_1,v_2) + h(x,v_1,v_2)\right]\\ &= \max_{v_1\in V_1}\min_{v_2\in V_2}\left[L\psi_\alpha(x,v_1,v_2) + h(x,v_1,v_2)\right]\end{aligned} \tag{1.3}$$

and is characterized by

$$\begin{aligned}\psi_\alpha(x) &= \sup_{v_1\in\mathscr{U}_1}\inf_{v_2\in\mathscr{U}_2}\mathrm{E}_x\left[\int_0^\infty e^{-\alpha t}h\big(X(t),v_1(t),v_2(t)\big)\,\mathrm{d}t\right]\\ &= \inf_{v_2\in\mathscr{U}_2}\sup_{v_1\in\mathscr{U}_1}\mathrm{E}_x\left[\int_0^\infty e^{-\alpha t}h\big(X(t),v_1(t),v_2(t)\big)\,\mathrm{d}t\right].\end{aligned}$$

Proof. Let B_R denote the open ball of radius R centered at the origin in $\mathbb{R}^d$. The p.d.e.

$$\begin{aligned} \alpha\varphi_\alpha^R(x) &= \min_{v_2\in V_2}\max_{v_1\in V_1}\left[L\varphi_\alpha^R(x,v_1,v_2)+h(x,v_1,v_2)\right], \\ \varphi_\alpha^R &= 0 \qquad \text{on } \partial B_R \end{aligned} \tag{1.4}$$

has a unique solution φ_α^R in $C^2(B_R)\cap C(\overline{B_R})$, see Gilbarg and Trudinger (1983, Theorem 15.12, p. 382). Since

$$\begin{aligned} &\min_{v_2\in V_2}\max_{v_1\in V_1}\left[L\varphi_\alpha^R(x,v_1,v_2)+h(x,v_1,v_2)\right] \\ &\qquad\qquad = \max_{v_1\in V_1}\min_{v_2\in V_2}\left[L\varphi_\alpha^R(x,v_1,v_2)+h(x,v_1,v_2)\right], \end{aligned}$$

it follows that $\varphi_\alpha^R\in C^2(B_R)\cap C(\overline{B_R})$ is also a solution to

$$\begin{aligned} \alpha\varphi_\alpha^R(x) &= \max_{v_1\in V_1}\min_{v_2\in V_2}\left[L\varphi_\alpha^R(x,v_1,v_2)+h(x,v_1,v_2)\right], \\ \varphi_\alpha^R &= 0 \qquad \text{on } \partial B_R. \end{aligned} \tag{1.5}$$

Let $v_{1\alpha}^R\colon B_R\to V_1$ be a measurable selector for the maximizer in (1.5) and $v_{2\alpha}^R\colon B_R\to V_2$ be a measurable selector for the minimizer in (1.4). If we let

$$F\big(x,v_1;\varphi_\alpha^R\big) := \min_{v_2\in V_2}\left[L\varphi_\alpha^R(x,v_1,v_2)+h(x,v_1,v_2)\right],$$

then $(x,v_1)\mapsto F\big(x,v_1;\varphi_\alpha^R\big)$ is continuous and also Lipschitz in x, and φ_α^R satisfies

$$\begin{aligned} \alpha\varphi_\alpha^R(x) &= F\big(x,v_{1\alpha}^R(x);\varphi_\alpha^R\big) \\ &= \min_{v_2\in V_2}\left[L\varphi_\alpha^R(x,v_{1\alpha}^R(x),v_2)+h(x,v_{1\alpha}^R(x),v_2)\right], \\ \varphi_\alpha^R &= 0 \qquad \text{on } \partial B_R. \end{aligned}$$

By a routine application of Itô's formula, it follows that

$$\varphi_\alpha^R(x) = \inf_{v_2\in\mathcal{U}_2} \mathrm{E}_x\left[\int_0^{\tau_R} e^{-\alpha t}h\big(X(t),v_{1\alpha}^R(X(t)),v_2(t)\big)\,\mathrm{d}t\right], \tag{1.6}$$

where

$$\tau_R := \inf\{t\ge 0 : \|X(t)\|\ge R\}$$

and X is the solution to (1.1) corresponding to the control pair $(v_{1\alpha}^R, v_2)$, with $v_2\in\mathcal{U}_2$.

Repeating the above argument with the outer minimizer $v_{2\alpha}^R$ of (1.4), we similarly obtain

$$\varphi_\alpha^R(x) = \sup_{v_1 \in \mathscr{U}_1} \mathrm{E}_x\left[\int_0^{\tau_R} e^{-\alpha t} h\big(X(t), v_1(t), v_{2\alpha}^R(X(t))\big)\, \mathrm{d}t\right]. \tag{1.7}$$

Combining (1.6) and (1.7), we obtain

$$\begin{aligned}\inf_{v_2 \in \mathscr{U}_2} \sup_{v_1 \in \mathscr{U}_1} \mathrm{E}_x\left[\int_0^{\tau_R} e^{-\alpha t} h\big(X(t), v_1(t), v_2(t)\big)\, \mathrm{d}t\right] &\le \varphi_\alpha^R(x) \\ &\le \sup_{v_1 \in \mathscr{U}_1} \inf_{v_2 \in \mathscr{U}_2} \mathrm{E}_x\left[\int_0^{\tau_R} e^{-\alpha t} h\big(X(t), v_1(t), v_2(t)\big)\, \mathrm{d}t\right],\end{aligned}$$

which implies that

$$\begin{aligned}\varphi_\alpha^R(x) &= \sup_{v_1 \in \mathscr{U}_1} \inf_{v_2 \in \mathscr{U}_2} \mathrm{E}_x\left[\int_0^{\tau_R} e^{-\alpha t} h\big(X(t), v_1(t), v_2(t)\big)\, \mathrm{d}t\right] \\ &= \inf_{v_2 \in \mathscr{U}_2} \sup_{v_1 \in \mathscr{U}_1} \mathrm{E}_x\left[\int_0^{\tau_R} e^{-\alpha t} h\big(X(t), v_1(t), v_2(t)\big)\, \mathrm{d}t\right].\end{aligned}$$

It is evident that $\varphi_\alpha^R(x) \le \tilde{\psi}_\alpha(x)$, $x \in \mathbb{R}^d$, where

$$\tilde{\psi}_\alpha(x) := \sup_{v_1 \in \mathscr{U}_1} \inf_{v_2 \in \mathscr{U}_2} \mathrm{E}_x\left[\int_0^{\infty} e^{-\alpha t} h\big(X(t), v_1(t), v_2(t)\big)\, \mathrm{d}t\right], \quad x \in \mathbb{R}^d.$$

Also φ_α^R is nondecreasing in R. By Assumption (A3′), it follows that

$$\tilde{\psi}_\alpha(x) \le k_2\, \mathrm{E}_x\left[\int_0^{\infty} e^{-\alpha t} g\big(X(t)\big)\, \mathrm{d}t\right],$$

where X is a solution to (1.1) corresponding to some stationary Markov control pair. Since the function $x \mapsto \mathrm{E}_x\big[\int_0^\infty e^{-\alpha t} g\big(X(t)\big)\, \mathrm{d}t\big]$ is continuous, it follows that $\tilde{\psi}_\alpha \in L^p_{loc}(\mathbb{R}^d)$ for $1 < p < \infty$.

Beneš' measurable selection theorem (Beneš 1970) asserts that there exists a pair of controls $(v_{1\alpha}^R, v_{2\alpha}^R) \in \mathscr{M}_1 \times \mathscr{M}_2$ which realizes the minimax in (1.4)–(1.5), i.e., for all $x \in B_R$ the following holds:

$$\begin{aligned}&\max_{v_1 \in V_1} \min_{v_2 \in V_2} \big[L\varphi_\alpha^R(x, v_1, v_2) + h(x, v_1, v_2)\big] \\ &\qquad = L\varphi_\alpha^R\big(x, v_{1\alpha}^R(x), v_{2\alpha}^R(x)\big) + h\big(x, v_{1\alpha}^R(x), v_{2\alpha}^R(x)\big).\end{aligned}$$

Hence $\varphi_\alpha^R \in C^2(B_R) \cap C(\overline{B_R})$ is a solution to

$$\alpha\varphi_\alpha^R(x) = L\varphi_\alpha^R\big(x, v_{1\alpha}^R(x), v_{2\alpha}^R(x)\big) + h\big(x, v_{1\alpha}^R(x), v_{2\alpha}^R(x)\big)\,, \quad x \in B_R\,.$$

Hence by Arapostathis et al. (2011, Lemma A.2.5, p. 305), for each $1 < p < \infty$ and $R' > 2R$, we have

$$\begin{aligned}\big\|\varphi_\alpha^{R'}\big\|_{W^{2,p}(B_R)} &\le K_1\Big(\big\|\varphi_\alpha^{R'}\big\|_{L^p(B_{2R})} + \big\|L\varphi_\alpha^{R'} - \alpha\varphi_\alpha^{R'}\big\|_{L^p(B_{2R})}\Big)\\ &\le K_1\Big(\big\|\tilde{\psi}_\alpha\big\|_{L^p(B_{2R})} + \big\|h\big(\cdot\,, v_{1\alpha}^{R'}(\cdot), v_{2\alpha}^{R'}(\cdot)\big)\big\|_{L^p(B_{2R})}\Big)\\ &\le K_1\Big(\big\|\tilde{\psi}_\alpha\big\|_{L^p(B_{2R})} + K_2(R)|B_{2R}|^{1/p}\Big)\,,\end{aligned}$$

where $K_1 > 0$ is a constant independent of R' and $K_2(R)$ is a constant depending only on the bound of h on B_{2R}. Using standard approximation arguments involving Sobolev imbedding theorems, see Arapostathis et al. (2011, p. 111), it follows that there exists $\psi_\alpha \in W^{2,p}_{loc}(\mathbb{R}^d)$ such that $\varphi_\alpha^R \uparrow \psi_\alpha$ as $R \uparrow \infty$ and ψ_α is a solution to

$$\alpha\psi_\alpha(x) = \max_{v_1\in V_1}\min_{v_2\in V_2}\big[L\psi_\alpha(x, v_1, v_2) + h(x, v_1, v_2)\big].$$

By standard regularity arguments, see Arapostathis et al. (2011, p. 109), one can show that $\psi_\alpha \in C^{2,r}(\mathbb{R}^d)$, $0 < r < 1$. Also using the minimax condition, it follows that $\psi_\alpha \in C^{2,r}(\mathbb{R}^d)$, $0 < r < 1$, is a solution to

$$\begin{aligned}\alpha\psi_\alpha(x) &= \min_{v_2\in V_2}\max_{v_1\in V_1}\big[L\psi_\alpha(x, v_1, v_2) + h(x, v_1, v_2)\big]\\ &= \max_{v_1\in V_1}\min_{v_2\in V_2}\big[L\psi_\alpha(x, v_1, v_2) + h(x, v_1, v_2)\big].\end{aligned}$$

Let $v_1^\alpha \in \mathscr{M}_1$ and $v_2^\alpha \in \mathscr{M}_2$ be an outer maximizing and an outer minimizing selector for (1.3), respectively, corresponding to ψ_α given above. Then ψ_α satisfies the p.d.e.

$$\alpha\psi_\alpha(x) = \max_{v_1\in V_1}\big[L\psi_\alpha\big(x, v_1, v_2^\alpha(x)\big) + h\big(x, v_1, v_2^\alpha(x)\big)\big]\,.$$

For $v_1 \in \mathscr{U}_1$, let X be the solution to (1.1) corresponding to (v_1, v_2^α) and the initial condition $x \in \mathbb{R}^d$. Applying the Itô–Dynkin formula, we obtain

$$\mathrm{E}_x\big[e^{-\alpha\tau_R}\psi_\alpha(X(\tau_R))\big] - \psi_\alpha(x) \le -\mathrm{E}_x\left[\int_0^{\tau_R} e^{-\alpha t}h\big(X(t), v_1(t), v_2^\alpha(X(t))\big)\,\mathrm{d}t\right].$$

Since $\psi_\alpha \geq 0$, we have

$$\psi_\alpha(x) \;\geq\; \mathrm{E}_x\biggl[\int_0^{\tau_R} e^{-\alpha t} h\bigl(X(t), v_1(t), v_2^\alpha(X(t))\bigr)\,\mathrm{d}t\biggr].$$

Using Fatou's lemma we obtain

$$\psi_\alpha(x) \;\geq\; \mathrm{E}_x\biggl[\int_0^{\infty} e^{-\alpha t} h\bigl(X(t), v_1(t), v_2^\alpha(X(t))\bigr)\,\mathrm{d}t\biggr]. \tag{1.8}$$

Therefore

$$\psi_\alpha(x) \;\geq\; \sup_{v_1\in\mathscr{U}_1} \mathrm{E}_x\biggl[\int_0^{\infty} e^{-\alpha t} h\bigl(X(t), v_1(t), v_2^\alpha(X(t))\bigr)\,\mathrm{d}t\biggr]. \tag{1.9}$$

Similarly, for $v_2 \in \mathscr{U}_2$, let X be the solution to (1.1) corresponding to (v_1^α, v_2) and the initial condition $x \in \mathbb{R}^d$. By applying the Itô–Dynkin formula, we obtain

$$\mathrm{E}_x\bigl[e^{-\alpha\tau_R}\psi_\alpha(X(\tau_R))\bigr] - \psi_\alpha(x) \;\geq\; -\mathrm{E}_x\biggl[\int_0^{\tau_R} e^{-\alpha t} h\bigl(X(t), v_1^\alpha(X(t)), v_2(t)\bigr)\,\mathrm{d}t\biggr].$$

Hence

$$\psi_\alpha(x) \;\leq\; \mathrm{E}_x\biggl[\int_0^{\infty} e^{-\alpha t} h\bigl(X(t), v_1^\alpha(X(t)), v_2(t)\bigr)\,\mathrm{d}t\biggr] + \mathrm{E}_x\bigl[e^{-\alpha\tau_R}\psi_\alpha(X(\tau_R))\bigr].$$

By Arapostathis et al. (2011, Remark A.3.8, p. 310), it follows that

$$\lim_{R\uparrow\infty} \mathrm{E}_x\bigl[e^{-\alpha\tau_R}\psi_\alpha(X(\tau_R))\bigr] \;=\; 0\,.$$

Hence, we have

$$\psi_\alpha(x) \;\leq\; \mathrm{E}_x\biggl[\int_0^{\infty} e^{-\alpha t} h\bigl(X(t), v_1^\alpha(X(t)), v_2(t)\bigr)\,\mathrm{d}t\biggr]. \tag{1.10}$$

Therefore

$$\psi_\alpha(x) \;\leq\; \inf_{v_2\in\mathscr{U}_2} \mathrm{E}_x\biggl[\int_0^{\infty} e^{-\alpha t} h\bigl(X(t), v_1^\alpha(X(t)), v_2(t)\bigr)\,\mathrm{d}t\biggr]. \tag{1.11}$$

By (1.9) and (1.11), we obtain

$$\psi_\alpha(x) \;=\; \mathrm{E}_x\biggl[\int_0^{\infty} e^{-\alpha t} h\bigl(X(t), v_1^\alpha(X(t)), v_2^\alpha(X(t))\bigr)\,\mathrm{d}t\biggr]. \tag{1.12}$$

Also by (1.8) and (1.10) we have

$$\inf_{v_2\in\mathscr{U}_2}\sup_{v_1\in\mathscr{U}_1}\mathrm{E}_x\left[\int_0^\infty e^{-\alpha t}h\big(X(t),v_1(t),v_2(t)\big)\,\mathrm{d}t\right] \le \psi_\alpha(x)$$

$$\le \sup_{v_1\in\mathscr{U}_1}\inf_{v_2\in\mathscr{U}_2}\mathrm{E}_x\left[\int_0^\infty e^{-\alpha t}h\big(X(t),v_1(t),v_2(t)\big)\,\mathrm{d}t\right].$$

This implies the desired characterization. □

Remark 1.1. Using Theorem 1.1, one can easily show that any pair of measurable outer maximizing and outer minimizing selectors of (1.3) is a saddle point equilibrium for the stochastic differential game with state dynamics given by (1.1) and with a discounted criterion under the running payoff function h.

Theorem 1.2. *Assume* (A1), (A2) *and* (A3′). *Then there exists a solution* $(\beta,\varphi^*)\in\mathbb{R}\times C_{\mathscr{V}}(\mathbb{R}^d)\cap C^2(\mathbb{R}^d)$ *to the Isaac's equation*

$$\begin{aligned}\beta &= \min_{v_2\in V_2}\max_{v_1\in V_1}\big[L\varphi^*(x,v_1,v_2)+h(x,v_1,v_2)\big]\\ &= \max_{v_1\in V_1}\min_{v_2\in V_2}\big[L\varphi^*(x,v_1,v_2)+h(x,v_1,v_2)\big], \qquad (1.13)\\ \varphi^*(0) &= 0\end{aligned}$$

such that β *is the value of the game.*

Proof. For $(v_1,v_2)\in\mathscr{M}_1\times\mathscr{M}_2$, define

$$J_\alpha(x,v_1,v_2) := \mathrm{E}_x\left[\int_0^\infty e^{-\alpha t}h\big(X(t),v_1(X(t)),v_2(X(t))\big)\,\mathrm{d}t\right],\quad x\in\mathbb{R}^d,$$

where X is a solution to (1.1) corresponding to $(v_1,v_2)\in\mathscr{M}_1\times\mathscr{M}_2$. Hence from (1.12), we have

$$\psi_\alpha(x) = J_\alpha(x,v_1^\alpha,v_2^\alpha),$$

where $(v_1^\alpha,v_2^\alpha)\in\mathscr{M}_1\times\mathscr{M}_2$ is a pair of measurable outer maximizing and outer minimizing selectors of (1.3). Using (A3′), it is easy to see that (v_1^α,v_2^α) is a pair of stable stationary Markov controls. Hence by the arguments in the proof of Arapostathis et al. (2011, Theorem 3.7.4, pp. 128–131), we have the following estimates:

$$\|\psi_\alpha-\psi_\alpha(0)\|_{W^{2,p}(B_R)} \le \frac{K_3}{\eta[v_1^\alpha,v_2^\alpha](B_R)}\left(\frac{\beta[v_1^\alpha,v_2^\alpha]}{\eta[v_1^\alpha,v_2^\alpha](B_R)} + \max_{(x,v_1,v_2)\in B_{4R}\times V_1\times V_2}h(x,v_1,v_2)\right), \qquad (1.14)$$

$$\sup_{x \in B_R} \alpha \psi_\alpha(x) \;\le\; K_3 \left(\frac{\beta[v_1^\alpha, v_2^\alpha]}{\eta[v_1^\alpha, v_2^\alpha](B_R)} + \max_{(x, v_1, v_2) \in B_{4R} \times V_1 \times V_2} h(x, v_1, v_2) \right), \qquad (1.15)$$

where $\eta[v_1^\alpha, v_2^\alpha]$ is the unique invariant probability measure of the process (1.1) corresponding to (v_1^α, v_2^α) and

$$\beta[v_1^\alpha, v_2^\alpha] \;:=\; \int_{\mathbb{R}^d} h\bigl(x, v_1^\alpha(x), v_2^\alpha(x)\bigr)\, \eta[v_1^\alpha, v_2^\alpha](\mathrm{d}x).$$

It follows from Arapostathis et al. (2011, Corollary 3.3.2, p. 97) that

$$\sup_{\alpha > 0} \beta[v_1^\alpha, v_2^\alpha] \;<\; \infty. \qquad (1.16)$$

Also from Arapostathis et al. (2011, (2.6.9a); p. 69 and (3.3.9); p. 97) it follows that

$$\inf_{\alpha > 0} \eta[v_1^\alpha, v_2^\alpha](B_R) \;>\; 0. \qquad (1.17)$$

Combining (1.14)–(1.17), we have

$$\begin{aligned} \|\psi_\alpha - \psi_\alpha(0)\|_{W^{2,p}(B_R)} \;&\le\; K_4\,, \\ \sup_{x \in B_R} \alpha \psi_\alpha(x) \;&\le\; K_4\,, \end{aligned} \qquad (1.18)$$

where $K_4 > 0$ is a constant independent of $\alpha > 0$.

Define

$$\bar{\psi}_\alpha(x) \;:=\; \psi_\alpha(x) - \psi_\alpha(0)\,, \qquad x \in \mathbb{R}^d\,.$$

In view of (1.18), one can use the arguments in Arapostathis et al. (2011, Lemma 3.5.4, pp. 108–109) to show that along some sequence $\alpha_n \downarrow 0$, $\alpha_n \psi_\alpha(0)$ converges to a constant ϱ and $\bar{\psi}_{\alpha_n}$ converges uniformly on compact sets to a function $\varphi^* \in C^2(\mathbb{R}^d)$, where the pair (ϱ, φ^*) is a solution to the p.d.e.

$$\begin{aligned} \varrho \;&=\; \min_{v_2 \in V_2} \max_{v_1 \in V_1} \bigl[L\varphi^*(x, v_1, v_2) + h(x, v_1, v_2) \bigr]\,, \\ \varphi^*(0) \;&=\; 0. \end{aligned}$$

Moreover, using the Isaac's condition, it follows that $(\varrho, \varphi^*) \in \mathbb{R} \times C^2(\mathbb{R}^d)$ satisfies (1.13).

We claim that $\varphi^* \in \mathfrak{o}(\mathscr{V})$, i.e., $\frac{\varphi^*(x)}{\mathscr{V}(x)} \to 0$ as $\|x\| \to \infty$. To prove the claim let $(v_1^*, v_2^*) \in \mathscr{M}_1 \times \mathscr{M}_2$ be a pair of measurable outer maximizing and outer minimizing selectors of (1.13) corresponding to φ^*. Let X be the solution to (1.1) under the control (v_1^*, v_2^*). Then by an application of the Itô–Dynkin formula and the help of Fatou's lemma, we can show that for all $x \in \mathbb{R}^d$

$$\varphi^*(x) \;\geq\; \mathrm{E}_x\left[\int_0^{\breve{\tau}_r} \Big(h\big(X(t), v_1^*(X(t)), v_2^*(X(t))\big) - \varrho\Big)\, \mathrm{d}t\right] + \min_{\|y\|=r} \varphi^*(y)\,, \quad (1.19)$$

where

$$\breve{\tau}_r \;=\; \inf\,\{t \geq 0 : \|X(t)\| \leq r\}.$$

Let $v_1^\alpha \in \mathscr{M}_1$ be a measurable outer maximizing selector in (1.3). Then the function $\psi_\alpha \in C^{2,r}(\mathbb{R}^d)$ given in Theorem 1.1 satisfies the p.d.e.

$$\alpha\psi_\alpha \;=\; \min_{v_2 \in V_2} \big[L\psi_\alpha(x, v_1^\alpha(x), v_2) + h(x, v_1^\alpha(x), v_2)\big]\,. \quad (1.20)$$

Let X be the solution to (1.1) under the control (v_1^α, v_2), with $v_2 \in \mathscr{U}_2$, and initial condition $x \in \mathbb{R}^d$. Then by applying the Itô–Dynkin formula to $e^{-\alpha t}\psi_\alpha(X(t))$ and using (1.20), we obtain

$$\mathrm{E}_x\big[e^{-\alpha(\breve{\tau}_r \wedge \tau_R)}\psi_\alpha(X(\breve{\tau}_r \wedge \tau_R))\big] - \psi_\alpha(x) \;\geq\; -\mathrm{E}_x\left[\int_0^{\breve{\tau}_r \wedge \tau_R} h\big(X(t), v_1^\alpha(X(t)), v_2(t)\big)\, \mathrm{d}t\right],$$

which we write as

$$\psi_\alpha(x) \;\leq\; \mathrm{E}_x\left[\int_0^{\breve{\tau}_r} h\big(X(t), v_1^\alpha(X(t)), v_2(t)\big)\, \mathrm{d}t\right] + \mathrm{E}_x\big[e^{-\alpha(\breve{\tau}_r \wedge \tau_R)}\psi_\alpha(X(\breve{\tau}_r \wedge \tau_R))\big]\,. \quad (1.21)$$

Using Arapostathis et al. (2011, Remark A.3.8, p. 310), it follows that

$$\mathrm{E}_x\big[e^{-\alpha\tau_R}\psi_\alpha(X(\tau_R)) I\{\breve{\tau}_r \geq \tau_R\}\big] \;\leq\; \mathrm{E}_x\big[e^{-\alpha\tau_R}\psi_\alpha(X(\tau_R))\big] \xrightarrow[R\to\infty]{} 0\,. \quad (1.22)$$

Hence from (1.21) and (1.22), we obtain

$$\psi_\alpha(x) \;\leq\; \mathrm{E}_x\left[\int_0^{\breve{\tau}_r} h\big(X(t), v_1^\alpha(X(t)), v_2(t)\big)\, \mathrm{d}t\right] + \mathrm{E}_x\big[e^{-\alpha\breve{\tau}_r}\psi_\alpha(X(\breve{\tau}_r))\big]\,.$$

Therefore,

$$\begin{aligned}
\bar{\psi}_\alpha(x) &\le \mathrm{E}_x\biggl[\int_0^{\breve{\tau}_r} h\bigl(X(t), v_1^\alpha(X(t)), v_2(t)\bigr)\,\mathrm{d}t\biggr] \\
&\qquad\qquad + \mathrm{E}_x\bigl[e^{-\alpha\breve{\tau}_r}\psi_\alpha(X(\breve{\tau}_r)) - \psi_\alpha(0)\bigr] \\
&= \mathrm{E}_x\biggl[\int_0^{\breve{\tau}_r} \Bigl(h\bigl(X(t), v_1^\alpha(X(t)), v_2(t)\bigr) - \varrho\Bigr)\,\mathrm{d}t\biggr] \\
&\qquad + \mathrm{E}_x\bigl[\psi_\alpha(X(\breve{\tau}_r)) - \psi_\alpha(0)\bigr] \\
&\qquad\qquad + \mathrm{E}_x\bigl[\alpha^{-1}(1 - e^{-\alpha\breve{\tau}_r})\bigl(\varrho - \alpha\psi_\alpha(X(\breve{\tau}_r))\bigr)\bigr] \\
&\le \mathrm{E}_x\biggl[\int_0^{\breve{\tau}_r} \Bigl(h\bigl(X(t), v_1^\alpha(X(t)), v_2(t)\bigr) - \varrho\Bigr)\,\mathrm{d}t\biggr] \\
&\qquad + M(r) + \mathrm{E}_x\bigl[\breve{\tau}_r\bigr] \sup_{\|y\|=r} \bigl|\varrho - \alpha\psi_\alpha(y)\bigr| \\
&\le \sup_{v_1\in\mathcal{M}_1} \mathrm{E}_x\biggl[\int_0^{\breve{\tau}_r} \Bigl(h\bigl(X(t), v_1(X(t)), v_2(t)\bigr) - \varrho\Bigr)\,\mathrm{d}t\biggr] \\
&\qquad + M(r) + \sup_{\|y\|=r} \bigl|\varrho - \alpha\psi_\alpha(y)\bigr| \sup_{v_1\in\mathcal{M}_1} \mathrm{E}_x[\breve{\tau}_r]
\end{aligned}$$

for some nonnegative constant $M(r)$ such that $M(r) \to 0$ as $r \downarrow 0$. Next from the definition of φ^*, by letting $\alpha \downarrow 0$ along the sequence given in the proof of Theorem 1.2, we obtain

$$\varphi^*(x) \le \sup_{v_1\in\mathcal{M}_1} \mathrm{E}_x\biggl[\int_0^{\breve{\tau}_r} \Bigl(h\bigl(X(t), v_1(X(t)), v_2(t)\bigr) - \varrho\Bigr)\,\mathrm{d}t\biggr] + M(r). \quad (1.23)$$

By combining (1.19) and (1.23), the result follows by Arapostathis et al. (2011, Lemma 3.7.2, p. 125). This completes the proof of the claim.

Let $(v_1^*, v_2^*) \in \mathcal{M}_1 \times \mathcal{M}_2$ be a pair of measurable outer maximizing and minimizing selectors in (1.13) corresponding to φ^*. Then (ϱ, φ^*) satisfies the p.d.e.

$$\varrho = \max_{v_1\in V_1} \bigl[L\varphi^*\bigl(x, v_1, v_2^*(x)\bigr) + h\bigl(x, v_1, v_2^*(x)\bigr)\bigr].$$

Let $v_1 \in \mathcal{U}_1$ and X be the process in (1.1) under the control (v_1, v_2^*) and initial condition $x \in \mathbb{R}^d$. By applying the Itô–Dynkin formula, we obtain

$$\mathrm{E}_x\bigl[\varphi^*(X(t\wedge\tau_R))\bigr] - \varphi^*(x) \le -\mathrm{E}_x\biggl[\int_0^{t\wedge\tau_R} \Bigl(h\bigl(X(t), v_1(t), v_2^*(X(t))\bigr) - \varrho\Bigr)\,\mathrm{d}t\biggr].$$

Hence

$$\varrho t \;\geq\; \mathrm{E}_x\biggl[\int_0^{t\wedge\tau_R} h\bigl(X(t), v_1(t), v_2^*(X(t))\bigr)\,\mathrm{d}t\biggr] + \mathrm{E}_x\bigl[\varphi^*(X(t\wedge\tau_R))\bigr] - \varphi^*(x)$$

for all $t \geq 0$. Using Fatou's lemma and Arapostathis et al. (2011, Lemma 3.7.2, p. 125), we obtain

$$\varrho t \geq \mathrm{E}_x\biggl[\int_0^{t} h\bigl(X(t), v_1(t), v_2^*(X(t))\bigr)\,\mathrm{d}t\biggr] + \mathrm{E}_x\bigl[\varphi^*(X(t))\bigr] - \varphi^*(x)\,, \qquad t \geq 0\,.$$

Dividing by t and taking limits again using Arapostathis et al. (2011, Lemma 3.7.2, p. 125), we obtain

$$\varrho \;\geq\; \liminf_{t\to\infty}\;\frac{1}{t}\;\mathrm{E}_x\biggl[\int_0^{t} h\bigl(X(t), v_1(t), v_2^*(X(t))\bigr)\,\mathrm{d}t\biggr].$$

Since $v_1 \in \mathscr{U}_1$ was arbitrary, we have

$$\begin{aligned}\varrho \;&\geq\; \sup_{v_1\in\mathscr{U}_1}\;\liminf_{t\to\infty}\;\frac{1}{t}\;\mathrm{E}_x\biggl[\int_0^{t} h\bigl(X(t), v_1(t), v_2^*(X(t))\bigr)\,\mathrm{d}t\biggr]\\ &\geq\; \inf_{v_2\in\mathscr{U}_2}\;\sup_{v_1\in\mathscr{U}_1}\;\liminf_{t\to\infty}\;\frac{1}{t}\;\mathrm{E}_x\biggl[\int_0^{t} h\bigl(X(t), v_1(t), v_2(t)\bigr)\,\mathrm{d}t\biggr].\end{aligned} \tag{1.24}$$

The pair (ϱ, φ^*) also satisfies the p.d.e.

$$\varrho \;=\; \min_{v_2\in V_2}\;\bigl[L\varphi^*(x, v_1^*(x), v_2) + h(x, v_1^*(x), v_2)\bigr]\,.$$

Let $v_2 \in \mathscr{U}_2$ and X be the process in (1.1) corresponding to (v_1^*, v_2) and initial condition $x \in \mathbb{R}^d$. By applying the Itô–Dynkin formula, we obtain

$$\mathrm{E}_x\bigl[\varphi^*(X(t\wedge\tau_R))\bigr] - \varphi^*(x) \;\geq\; -\mathrm{E}_x\biggl[\int_0^{t\wedge\tau_R}\Bigl(h\bigl(X(t), v_1^*(X(t)), v_2(t)\bigr) - \varrho\Bigr)\,\mathrm{d}t\biggr].$$

Hence

$$\varrho\,\mathrm{E}_x[t\wedge\tau_R] \;\leq\; \mathrm{E}_x\biggl[\int_0^{t} h\bigl(X(t), v_1^*(X(t)), v_2(t)\bigr)\,\mathrm{d}t + \varphi^*(X(t\wedge\tau_R))\biggr] - \varphi^*(x).$$

Next, by letting $R \to \infty$ and using the dominated convergence theorem for the l.h.s. and Arapostathis et al. (2011, Lemma 3.7.2, p. 125) for the r.h.s., we obtain

$$\varrho t \;\leq\; \mathrm{E}_x\biggl[\int_0^{t} h\bigl(X(t), v_1^*(X(t)), v_2(t)\bigr)\,\mathrm{d}t\biggr] + \mathrm{E}_x\bigl[\varphi^*(X(t))\bigr] - \varphi^*(x).$$

Also by Arapostathis et al. (2011, Lemma 3.7.2, p. 125), we obtain

$$\varrho \leq \liminf_{t\to\infty} \frac{1}{t} \mathrm{E}_x\left[\int_0^t h\big(X(t), v_1^*(X(t)), v_2(t)\big)\, \mathrm{d}t\right].$$

Since $v_2 \in \mathscr{U}_2$ was arbitrary, we have

$$\begin{aligned} \varrho &\leq \inf_{v_2\in\mathscr{U}_2} \liminf_{t\to\infty} \frac{1}{t} \mathrm{E}_x\left[\int_0^t h\big(X(t), v_1^*(X(t)), v_2(t)\big)\, \mathrm{d}t\right] \\ &\leq \sup_{v_1\in\mathscr{U}_1} \inf_{v_2\in\mathscr{U}_2} \liminf_{t\to\infty} \frac{1}{t} \mathrm{E}_x\left[\int_0^t h\big(X(t), v_1(t), v_2(t)\big)\, \mathrm{d}t\right]. \end{aligned} \tag{1.25}$$

Combining (1.24) and (1.25), we obtain

$$\begin{aligned} \varrho &= \inf_{v_2\in\mathscr{U}_2} \sup_{v_1\in\mathscr{U}_1} \liminf_{t\to\infty} \frac{1}{t} \mathrm{E}_x\left[\int_0^t h\big(X(t), v_1(t), v_2(t)\big)\, \mathrm{d}t\right] \\ &= \sup_{v_1\in\mathscr{U}_1} \inf_{v_2\in\mathscr{U}_2} \liminf_{t\to\infty} \frac{1}{t} \mathrm{E}_x\left[\int_0^t h\big(X(t), v_1(t), v_2(t)\big)\, \mathrm{d}t\right], \end{aligned}$$

i.e. $\varrho = \beta$, the value of the game. This completes the proof. □

Remark 1.2. Using Theorem 1.2, one can easily prove that any pair of measurable outer maximizing and outer minimizing selectors of (1.3) is a saddle point equilibrium for the stochastic differential game with state dynamics given by (1.1) and with the ergodic criterion under the running payoff function h.

The following corollary, stated here without proof, follows along the lines of the proof of Arapostathis et al. (2011, Theorem 3.7.12).

Corollary 1.1. *The solution φ^* has the stochastic representation*

$$\begin{aligned} \varphi^*(x) &= \lim_{r\downarrow 0} \sup_{v_1\in\mathscr{M}_1} \inf_{v_2\in\mathscr{M}_2} \mathrm{E}_x\left[\int_0^{\breve{\tau}_r} \Big(h\big(X(t), v_1(X(t)), v_2(X(t))\big) - \beta\Big)\, \mathrm{d}t\right] \\ &= \lim_{r\downarrow 0} \inf_{v_2\in\mathscr{M}_2} \sup_{v_1\in\mathscr{M}_1} \mathrm{E}_x\left[\int_0^{\breve{\tau}_r} \Big(h\big(X(t), v_1(X(t)), v_2(X(t))\big) - \beta\Big)\, \mathrm{d}t\right] \\ &= \lim_{r\downarrow 0} \mathrm{E}_x\left[\int_0^{\breve{\tau}_r} \Big(h\big(X(t), v_1^*(X(t)), v_2^*(X(t))\big) - \beta\Big)\, \mathrm{d}t\right] \end{aligned}$$

and is unique in the class of functions that do not grow faster than $\mathscr{V}$ and vanish at $x = 0$.

1.3 Relative Value Iteration

We consider the following relative value iteration equation.

$$\begin{aligned} \frac{\partial \varphi}{\partial t}(t,x) &= \min_{v_2 \in V_2} \max_{v_1 \in V_1} \left[L\varphi(t,x,v_1,v_2) + h(x,v_1,v_2) \right] - \varphi(t,0) , \\ \varphi(0,x) &= \varphi_0(x), \end{aligned} \tag{1.26}$$

where $\varphi_0 \in C_{\mathcal{V}}(\mathbb{R}^d) \cap C^2(\mathbb{R}^d)$. This can be viewed as a continuous time continuous state space variant of the relative value iteration algorithm for Markov decision processes (White 1963).

Convergence of this relative value iteration scheme is obtained through the study of the value iteration equation which takes the form

$$\begin{aligned} \frac{\partial \overline{\varphi}}{\partial t}(t,x) &= \min_{v_2 \in V_2} \max_{v_1 \in V_1} \left[L\overline{\varphi}(t,x,v_1,v_2) + h(x,v_1,v_2) \right] - \beta , \\ \overline{\varphi}(0,x) &= \varphi_0(x), \end{aligned} \tag{1.27}$$

where β is the value of the average payoff game in Theorem 1.2.

Under Assumption (A3), it is straightforward to show that for each $T > 0$ there exists a unique solution $\overline{\varphi}$ in $C_{\mathcal{V}}([0,T] \times \mathbb{R}^d) \cap C^{1,2}((0,T) \times \mathbb{R}^d)$ to the p.d.e. (1.27).

First, we prove the following important estimate which is crucial for the proof of convergence.

Lemma 1.1. *Assume* (A1)–(A3). *Then for each* $T > 0$, *the p.d.e. in* (1.26) *has a unique solution* $\varphi \in C_{\mathcal{V}}((0,T) \times \mathbb{R}^d) \cap C^{1,2}((0,T) \times \mathbb{R}^d)$.

Proof. The proof follows by mimicking the arguments in Arapostathis and Borkar (2012, Lemma 4.1), using the following estimate

$$\mathrm{E}_x\left[\mathcal{V}(X(t))\right] \le \frac{k_0}{2k_1} + \mathcal{V}(x)e^{-2k_1 t} , \tag{1.28}$$

where X is the solution to (1.1) corresponding to any admissible controls v_1 and v_2 and initial condition $x \in \mathbb{R}^d$. The estimate for φ follows from the arguments in Arapostathis et al. (2011, Lemma 2.5.5, pp. 63–64), noting that for all $v_i \in \mathcal{U}_i$, $i = 1, 2$, we have

$$\begin{aligned} \int_0^t \mathrm{E}_x\left[h^n\big(X(s), v_1(s), v_2(s)\big)\right] \mathrm{d}s &\le k_2 \int_0^t \mathrm{E}_x\left[\mathcal{V}(X(s))\right] \mathrm{d}s \\ &\le \frac{k_2}{2k_1}\big(k_0 t + \mathcal{V}(x)\big) , \end{aligned}$$

where $h^n(x, v_1, v_2) := n \wedge h(x, v_1, v_2)$ is the truncation of h at $n \geq 0$. □

Next, we turn our attention to the p.d.e. in (1.27). It is straightforward to show that the solution $\overline{\varphi}$ to (1.27) also satisfies

$$\begin{aligned} \frac{\partial \overline{\varphi}}{\partial t}(t,x) &= \max_{v_1 \in V_1} \min_{v_2 \in V_2} \left[L\,\overline{\varphi}(t,x,v_1,v_2) + h(x,v_1,v_2) \right] - \beta, \\ \overline{\varphi}(0,x) &= \varphi_0(x), \end{aligned} \tag{1.29}$$

Definition 1.2. We let $\bar{v}_i : \mathbb{R}_+ \times \mathbb{R}^d \to V_i$ for $i = 1, 2$ be an outer maximizing and an outer minimizing selector of (1.29) and (1.27), respectively. For each $t \geq 0$ we define the (nonstationary) Markov control

$$\bar{v}_i^t := \left\{ \bar{v}_i^t(s, \cdot) = \bar{v}_i(t-s, \cdot), \; s \in [0, t] \right\}.$$

We also let $\mathrm{P}_x^{v_1,v_2}$ denote the probability measure and $\mathrm{E}_x^{v_1,v_2}$ the expectation operator on the canonical space of the process under the control $v_i \in \mathscr{U}_i$, $i = 1, 2$, conditioned on the process X starting from $x \in \mathbb{R}^d$ at $t = 0$.

It is straightforward to show that the solution $\overline{\varphi}$ of (1.27) satisfies,

$$\begin{aligned} \overline{\varphi}(t,x) &= \mathrm{E}_x^{\bar{v}_1^t, \bar{v}_2^t} \left[\int_0^{t-s} \Big(h\big(X(\tau), \bar{v}_1(t-\tau, X(\tau)), \bar{v}_2(t-\tau, X(\tau))\big) - \beta \Big) \, \mathrm{d}\tau \right. \\ &\qquad \left. + \overline{\varphi}\big(s, X(t-s)\big) \right] \\ &= \inf_{v_2 \in \mathscr{U}_2} \sup_{v_1 \in \mathscr{U}_1} \mathrm{E}_x^{v_1,v_2} \left[\int_0^{t-s} \Big(h\big(X(\tau), v_1(\tau), v_2(\tau)\big) - \beta \Big) \, \mathrm{d}\tau \right. \\ &\qquad \left. + \overline{\varphi}\big(s, X(t-s)\big) \right] \\ &= \sup_{v_1 \in \mathscr{U}_1} \inf_{v_2 \in \mathscr{U}_2} \mathrm{E}_x^{v_1,v_2} \left[\int_0^{t-s} \Big(h\big(X(\tau), v_1(\tau), v_2(\tau)\big) - \beta \Big) \, \mathrm{d}\tau \right. \\ &\qquad \left. + \overline{\varphi}\big(s, X(t-s)\big) \right] \end{aligned} \tag{1.30}$$

for all $t \geq s \geq 0$.

Lemma 1.2. *Assume* (A1)–(A3). *For each $\varphi_0 \in C_{\mathscr{V}}(\mathbb{R}^d) \cap C^2(\mathbb{R}^d)$, the solution $\overline{\varphi}$ of the p.d.e.* (1.27) *satisfies the following estimate*

$$\left| \overline{\varphi}(t,x) - \varphi^*(x) \right| \leq \| \overline{\varphi}(s, \cdot) - \varphi^* \|_{\mathscr{V}} \left(\frac{k_0}{2k_1} + \mathscr{V}(x)\, e^{-2k_1(t-s)} \right) \quad \forall x \in \mathbb{R}^d,$$

and for all $t \geq s \geq 0$, where φ^* is as in Theorem 1.2.

Proof. Let $v_1^* \in \mathscr{M}_1$ and $v_2^* \in \mathscr{M}_2$ be an outer maximizing and outer minimizing selector of (1.13), respectively. By (1.30) we obtain

$$\overline{\varphi}(t,x) - \varphi^*(x) \;\leq\; \mathrm{E}_x^{\bar{v}_1^t, v_2^*}\bigl[\overline{\varphi}\bigl(s, X(t-s)\bigr) - \varphi^*\bigl(X(t-s)\bigr)\bigr] \tag{1.31}$$

and

$$\varphi^*(x) - \overline{\varphi}(t,x) \;\leq\; \mathrm{E}_x^{v_1^*, \bar{v}_2^t}\bigl[\varphi^*\bigl(X(t-s)\bigr) - \overline{\varphi}\bigl(s, X(t-s)\bigr)\bigr] \tag{1.32}$$

for all $t \geq s \geq 0$. By (1.31)–(1.32) we obtain

$$\bigl|\overline{\varphi}(t,x) \;-\; \varphi^*(x)\bigr| \;\leq\; \sup_{(v_1,v_2)\,\in\,\mathscr{U}_1\times\mathscr{U}_2} \mathrm{E}_x^{v_1,v_2}\Bigl[\bigl|\overline{\varphi}\bigl(s, X(t-s)\bigr) - \varphi^*\bigl(X(t-s)\bigr)\bigr|\Bigr],$$

and an application of (1.28) completes the proof. □

Arguing as in the proof of Arapostathis and Borkar (2012, Lemma 4.4), we can show the following:

Lemma 1.3. *Assume* (A1)–(A3). *If* $\overline{\varphi}(0,x) = \varphi(0,x) = \varphi_0(x)$ *for some* $\varphi_0 \in C_{\mathscr{V}}(\mathbb{R}^d) \cap C^2(\mathbb{R}^d)$, *then*

$$\varphi(t,x) - \varphi(t,0) \;=\; \overline{\varphi}(t,x) - \overline{\varphi}(t,0),$$

and

$$\varphi(t,x) \;=\; \overline{\varphi}(t,x) - e^{-t}\int_0^t e^s\,\overline{\varphi}(s,0)\,\mathrm{d}s + \beta(1 - e^{-t})$$

for all $x \in \mathbb{R}^d$ *and* $t \geq 0$.

Convergence of the relative value iteration is asserted in the following theorem.

Theorem 1.3. *Assume* (A1)–(A3). *For each* $\varphi_0 \in C_{\mathscr{V}}(\mathbb{R}^d) \cap C^2(\mathbb{R}^d)$, $\overline{\varphi}(t,x)$ *converges to* $\varphi^*(x) +$ *constant and* $\varphi(t,x)$ *converges to* $\varphi^*(x) + \beta$ *as* $t \to \infty$.

Proof. By Lemma 1.2 the map $x \mapsto \overline{\varphi}(t,x)$ is locally bounded, uniformly in $t \geq 0$. It then follows that $\bigl\{\frac{\partial^2\overline{\varphi}(t,x)}{\partial x_i \partial x_j},\ t \geq 1\bigr\}$ are locally Hölder equicontinuous (see Ladyženskaja et al. 1967, Theorem 5.1). Therefore the ω-limit set $\omega(\varphi_0)$ of any initial condition $\varphi_0 \in C_{\mathscr{V}}(\mathbb{R}^d) \cap C^2(\mathbb{R}^d)$ is a nonempty compact subset of $C_{\mathscr{V}}(\mathbb{R}^d) \cap C^2_{loc}(\mathbb{R}^d)$.

To simplify the notation we define

$$\Phi_t(x) \;:=\; \overline{\varphi}(t,x) - \varphi^*(x)\,, \quad (t,x) \in \mathbb{R}_+ \times \mathbb{R}^d.$$

By Lemma 1.2, we have

$$\limsup_{t\to\infty} |\Phi_t(x)| \le \frac{k_0}{2k_1} \|\varphi_0 - \varphi^*\|_{\mathscr{V}}.$$

Let $\{t_n\ n \in \mathbb{N}\} \subset \mathbb{R}_+$ be any increasing sequence such that $t_n \uparrow \infty$ and

$$\Phi_{t_n} \to f \in C_{\mathscr{V}}(\mathbb{R}^d) \cap C^2(\mathbb{R}^d) \qquad \text{as } n \to \infty.$$

Dropping to a subsequence we assume that $t_{n+1} - t_n \uparrow \infty$ as $n \to \infty$. By construction $f + \varphi^* \in \omega(\varphi_0)$.

We first show that f is a constant. We define

$$\bar{f} := \sup_{x\in\mathbb{R}^d} f(x),$$

and a subsequence $\{k_n\} \subset \mathbb{N}$ by

$$k_n := \sup\left\{k \in \mathbb{N} : \sup_{x\in B_k} \Phi_{t_n}(x) \le \bar{f} + \frac{1}{k}\right\}. \tag{1.33}$$

Since Φ_{t_n} converges to f uniformly on compact sets as $n \to \infty$, it follows that $k_n \uparrow \infty$ as $n \to \infty$. Let D be any fixed closed ball centered at the origin such that

$$\inf_{x\in D^c} \mathscr{V}(x) \ge \frac{2k_0}{k_1}.$$

It is straightforward to verify using (1.28) that if X is the solution to (1.1) corresponding to any admissible controls v_1 and v_2 and initial condition $x \in \mathbb{R}^d$ then there exists $T_0 < \infty$ depending only on x, such that

$$\mathrm{P}_x\big(X_t \in D\big) \ge \frac{1}{2} \qquad \forall x \in \mathbb{R}^d, \quad \forall t \ge T_0(x). \tag{1.34}$$

By the standard estimates of hitting probabilities for diffusions (see Gruber 1984, Lemma 1.1) for any $r > 0$ there exists a constant $\gamma > 0$ depending only on r and D, such that with $B_r(y)$ denoting the open ball of radius r centered at $y \in \mathbb{R}^d$ we have

$$\mathrm{P}_x\big(X_t \in B_r(y)\big) \ge \gamma \qquad \forall t \in [0,1], \quad \forall x, y \in D. \tag{1.35}$$

Let $\mathrm{I}_A(\cdot)$ denote the indicator function of a set $A \subset \mathbb{R}^d$. An equivalent statement to (1.35) is that if $g : D \to \mathbb{R}_+$ is a Hölder continuous function then there exists a continuous function $\Gamma : \mathbb{R}_+ \to \mathbb{R}_+$, satisfying $\Gamma(z) > 0$ for $z > 0$ and depending only on D and the Hölder constant of g, such that

$$\mathrm{E}_x\big(g(X_t)\,\mathrm{I}_D(X_t)\big) \ge \Gamma\Big(\max_{y\in D} g(y)\Big) \qquad \forall t \in [0,1], \quad \forall x \in D. \tag{1.36}$$

Combining (1.34) and (1.36) and using the Markov property, we obtain

$$\begin{aligned}
\mathrm{E}_x\bigl[g(X_t)\,\mathrm{I}_D(X_t)\bigr] &\geq \mathrm{E}_x\Bigl[\mathrm{E}_{X_{t-1}}\bigl[g(X_1)\,\mathrm{I}_D(X_1)\bigr]\,\mathrm{I}_D(X_{t-1})\Bigr] \\
&\geq \Gamma\Bigl(\max_{y\in D}\ g(y)\Bigr)\,\mathrm{P}_x\bigl(X_{t-1}\in D\bigr) \\
&\geq \frac{1}{2}\,\Gamma\Bigl(\max_{y\in D}\ g(y)\Bigr) \qquad \forall t \geq T_0(x)+1. \qquad (1.37)
\end{aligned}$$

and for all $x \in \mathbb{R}^d$. Note that if n is sufficiently large, then $D \subset B_{k_n}$ and therefore the function $x \mapsto \bar{f} + \frac{1}{k_n} - \Phi_{t_n}(x)$ is nonnegative on D. Thus the local Hölder equicontinuity of $\{\Phi_t\,,\ t > 0\}$ (this collection of functions locally share a common Hölder exponent) allows us to apply (1.37) for any fixed $x \in \mathbb{R}^d$ to obtain

$$\begin{aligned}
&\mathrm{E}_x^{\bar{v}_1^{t_n+1},v_2^*}\Bigl[\Bigl(\bar{f} + \frac{1}{k_n} - \Phi_{t_n}\bigl(X(t_{n+1}-t_n)\bigr)\Bigr)\,\mathrm{I}_D\bigl(X(t_{n+1}-t_n)\bigr)\Bigr] \\
&\qquad\qquad \geq \frac{1}{2}\,\Gamma\Bigl(\bar{f} + \frac{1}{k_n} - \min_{y\in D}\Phi_{t_n}(y)\Bigr), \qquad (1.38)
\end{aligned}$$

for all n large enough. For $A \subset \mathbb{R}^d$ and $x \in \mathbb{R}^d$ we define

$$\Psi_n(x;A) := \mathrm{E}_x^{\bar{v}_1^{t_n+1},v_2^*}\Bigl[\Phi_{t_n}\bigl(X(t_{n+1}-t_n)\bigr)\,\mathrm{I}_A\bigl(X(t_{n+1}-t_n)\bigr)\Bigr]$$

By (1.31), (1.33), and (1.38) we have

$$\begin{aligned}
\Phi_{t_{n+1}}(x) &\leq \mathrm{E}_x^{\bar{v}_1^{t_n+1},v_2^*}\bigl[\Phi_{t_n}\bigl(X(t_{n+1}-t_n)\bigr)\bigr] \\
&= \Psi_n(x;D) + \Psi_n(x;B_{k_n}\setminus D) + \Psi_n(x;B_{k_n}^c) \\
&\leq \Bigl(\bar{f} + \frac{1}{k_n}\Bigr)\,\mathrm{E}_x^{\bar{v}_1^{t_n+1},v_2^*}\Bigl[\mathrm{I}_{B_{k_n}}\bigl(X(t_{n+1}-t_n)\bigr)\Bigr] \\
&\qquad\qquad - \frac{1}{2}\,\Gamma\Bigl(\bar{f} + \frac{1}{k_n} - \min_{y\in D}\Phi_{t_n}(y)\Bigr) + \Psi_n(x;B_{k_n}^c) \\
&\leq \bar{f} + \frac{1}{k_n} - \frac{1}{2}\,\Gamma\Bigl(\bar{f} + \frac{1}{k_n} - \min_{y\in D}\Phi_{t_n}(y)\Bigr) + \Psi_n(x;B_{k_n}^c)\,. \qquad (1.39)
\end{aligned}$$

We claim that $\Psi_n(x;B_{k_n}^c) \to 0$ as $n \to \infty$. Indeed if X is the solution to (1.1) corresponding to any admissible controls v_1 and v_2 and initial condition $x \in \mathbb{R}^d$, then by (1.28) we have

$$\mathrm{E}_x\bigl[\Phi_t\bigl(X(s)\bigr)\,\mathrm{I}_{B_R^c}\bigl(X(s)\bigr)\bigr] \;\le\; \bigl\|\Phi_t\,\mathrm{I}_{B_R^c}\bigr\|_{\mathscr{V}}\left(\frac{k_0}{2k_1} + \mathscr{V}(x)e^{-2k_1 s}\right), \tag{1.40}$$

By Lemma 1.2 we have

$$\bigl\|\Phi_t\,\mathrm{I}_{B_R^c}\bigr\|_{\mathscr{V}} \;\le\; \|\Phi_0\|_{\mathscr{V}}\left(\frac{k_0}{2k_1\,\inf_{x\in B_R^c}\,\mathscr{V}(x)} + e^{-2k_1 t}\right). \tag{1.41}$$

It follows by (1.40) and (1.41) that

$$\mathrm{E}_x\bigl[\Phi_t\bigl(X(s)\bigr)\,\mathrm{I}_{B_R^c}\bigl(X(s)\bigr)\bigr] \xrightarrow[\min\{t,R\}\to\infty]{} 0$$

uniformly in $s \ge 0$, which proves that $\Psi_n(x; B_{k_n}^c) \to 0$ as $n \to \infty$. Thus, taking limits as $n \to \infty$ in (1.39), we obtain

$$f(x) \;\le\; \bar{f} - \frac{1}{2}\,\Gamma\Bigl(\bar{f} - \min_{y\in D} f(y)\Bigr) \qquad \forall x \in \mathbb{R}^d. \tag{1.42}$$

Taking the supremum over $x \in \mathbb{R}^d$ of the left-hand side of (1.42) it follows that $\Gamma\Bigl(\bar{f} - \min_{y\in D} f(y)\Bigr) = 0$ which implies that f is constant on D. Since D was arbitrary if follows that f must be a constant.

We next show that f is unique. We argue by contradiction. Suppose that $\Phi_{t_n'} \to f'$ over some increasing sequence $\{t_n'\}$ with $t_n' \uparrow \infty$ as $n \to \infty$. Without loss of generality we assume $t_n < t_n' < t_{n+1}$ for each n. By (1.31) we have

$$\Phi_{t_{n+1}}(x) \;\le\; \mathrm{E}_x^{\bar{v}_1^{t_{n+1}},v_2^*}\bigl[\Phi_{t_n'}\bigl(X(t_{n+1} - t_n')\bigr)\bigr], \tag{1.43}$$

and taking limits as $n \to \infty$ in (1.43) we obtain $f \le f'$. Reversing the roles of f and f' shows that $f = f'$.

By Lemma 1.3 we have

$$\varphi(t,x) \;=\; \overline{\varphi}(t,x) + \int_0^t e^{s-t}\bigl(\beta - \overline{\varphi}(s,0)\bigr)\,\mathrm{d}s\,.$$

Hence, since $\overline{\varphi}(t,x)$ converges to $\varphi^*(x) + f$, we obtain that $\varphi(t,x) \to \varphi^*(x) + \beta$ as $t \to \infty$. □

1.4 Risk-Sensitive Control

In this section, we apply the results from Sect. 1.3 to study the convergence of a relative value iteration scheme for the risk-sensitive control problem which is described as follows. Let U be a compact metric space and $V = \mathscr{P}(U)$ denote the space of all probability measures on U with Prohorov topology. We consider the risk-sensitive control problem with state equation given by the controlled s.d.e. (in relaxed form)

$$\mathrm{d}X(t) = b(X(t), v(t))\,\mathrm{d}t + \sigma(X(t))\,\mathrm{d}W(t)\,, \tag{1.44}$$

and payoff criterion

$$J(x, v) := \liminf_{T\to\infty} \frac{1}{T} \ln \mathrm{E}_x\left[\exp\left(\int_0^T h(X(t), v(t))\,\mathrm{d}t\right) \;\Big|\; X(0) = x\right].$$

This is called the risk-sensitive payoff because in some sense it is sensitive to higher moments of the running cost and not merely its mean, thus capturing 'risk' in the sense understood in economics Whittle (1990).

All processes in (1.44) are defined in a common probability space $(\Omega, \mathscr{F}, P)$ which is assumed to be complete. The process W is an $\mathbb{R}^d$-valued standard Wiener process which is independent of the initial condition X_0 of (1.1). The control v is a V-valued process which is jointly measurable in $(t, \omega) \in [0, \infty) \times \Omega$ and non-anticipative, i.e., for $s < t$, $W(t) - W(s)$ is independent of $\mathscr{F}_s :=$ the completion of $\sigma(X_0, v(r), W(r), r \le s)$. We denote the set of all such controls (admissible controls) by $\mathscr{U}$.

Assumptions on the Data: We assume the following properties for the coefficients b and σ:

(B1) The functions b and σ are continuous and bounded, and also Lipschitz continuous in $x \in \mathbb{R}^d$ uniformly over $v \in V$. Also $(\sigma\sigma^{\mathsf{T}})^{-1}$ is Lipschitz continuous.

(B2) For each $R > 0$ there exists a constant $\kappa(R) > 0$ such that

$$z^{\mathsf{T}} a(x) z \ge \kappa(R)\|z\|^2 \qquad \text{for all } \|x\| \le R \text{ and } z \in \mathbb{R}^d\,,$$

where $a := \sigma\sigma^{\mathsf{T}}$.

Asymptotic Flatness Hypothesis: We assume the following property:

(B3) (i) There exists a $c > 0$ and a positive definite matrix Q such that for all x, $y \in \mathbb{R}^d$ with $x \ne y$, we have

$$2\big(b(x,v)-b(y,v)\big)^{\mathsf{T}} Q(x-y) + \mathrm{tr}\Big(\big(\sigma(x)-\sigma(y)\big)\big(\sigma(x)-\sigma(y)\big)^{\mathsf{T}} Q\Big) - \frac{\big\|\big(\sigma(x)-\sigma(y)\big)^{\mathsf{T}} Q(x-y)\big\|^2}{(x-y)^{\mathsf{T}} Q(x-y)} \le -c\,\|x-y\|^2\,.$$

(ii) Let $\mathrm{Lip}(f)$ denote the Lipschitz constant of a Lipschitz continuous function f. Then

$$2\,\|\sigma\sigma^{\mathsf{T}}\|_{\infty}^{2}\,\mathrm{Lip}(h)\,\mathrm{Lip}\big((\sigma\sigma^{\mathsf{T}})^{-1}\big) \;\le\; c^{2}.$$

The asymptotic flatness hypothesis was first introduced by Basak and Bhattacharya (1992) for the study of ergodicity in degenerate diffusions and is a little more general than the condition introduced by Fleming and McEneaney (1995) in risk-sensitive control to facilitate the analysis of the corresponding HJB equation, which is our motivation as well. An important consequence of this condition is that if we fix a non-anticipative control process and consider two diffusion processes with this control differing only in their initial conditions, they approach each other in mean at an exponential rate (Arapostathis et al. 2011, Lemma 7.3.4). This ensures a bounded gradient for the solution of the HJB equation, a key step in the analysis of its well-posedness.

We quote the following result from Borkar and Suresh Kumar (2010, Theorems 2.2 and 2.3):

Theorem 1.4. *Assume* (B1)–(B3). *The p.d.e.*

$$\begin{aligned}\beta &= \min_{v\in V}\,\max_{w\in\mathbb{R}^d}\Big[\tilde{L}\varphi^*(x,w,v)+h(x,v)-\tfrac{1}{2}\,w^{\mathsf{T}}\big(a^{-1}(x)\big)w\Big]\\ &= \max_{w\in\mathbb{R}^d}\,\min_{v\in V}\Big[\tilde{L}\varphi^*(x,w,v)+h(x,v)-\tfrac{1}{2}\,w^{\mathsf{T}}\big(a^{-1}(x)\big)w\Big]\,, \qquad (1.45)\\ \varphi^*(0) &= 0\,,\end{aligned}$$

where

$$\tilde{L}f(x,w,v) \;:=\; (b(x,v)+w)\cdot\nabla f(x)+\frac{1}{2}\,\mathrm{tr}\big(a(x)\nabla^2 f(x)\big)\,,\quad f\in C^2(\mathbb{R}^d),$$

has a unique solution $(\beta,\varphi^*)\in\mathbb{R}\times C^2(\mathbb{R}^d)\cap o(\|x\|)$. *Moreover,* β *is the value of the risk-sensitive control problem and any measurable outer minimizing selector in* (1.45) *is risk-sensitive optimal. Also in* (1.45), *the supremum can be restricted to a closed ball* $\tilde{V}=\overline{B_R}$ *for*

$$R \;:=\; \frac{\mathrm{Lip}(h)}{c}+\frac{\mathrm{Lip}\big((\sigma\sigma^{\mathsf{T}})^{-1}\big)K^2}{2\sqrt{c}}\,,$$

where K *is the smallest positive root (using* (B3) (ii)*) of*

$$\frac{\sqrt{c}}{2}\,\|\sigma\sigma^{\mathsf{T}}\|_{\infty}\,\mathrm{Lip}\big((\sigma\sigma^{\mathsf{T}})^{-1}\big)\,x^2-c^{5/4}x+\mathrm{Lip}(h)\|\sigma\sigma^{\mathsf{T}}\|_{\infty} \;=\; 0\,.$$

For the stochastic differential game in (1.45) we consider the following relative value iteration equation:

$$\begin{aligned}\frac{\partial \varphi}{\partial t}(t,x) &= \min_{v\in V}\max_{w\in\tilde V}\left[\tilde L\varphi(t,x,w,v) + h(x,v) - \tfrac{1}{2}w^{\mathsf T}\big(a^{-1}(x)\big)w\right] - \varphi(t,0)\,,\\ \varphi(0,x) &= \varphi_0(x),\end{aligned}$$

where $\varphi_0 \in C_{\mathcal V}(\mathbb R^d)\cap C^2(\mathbb R^d)$ with

$$\mathcal V(x) = \frac{(x^{\mathsf T}Qx)^{1+\alpha}}{\varepsilon + (x^{\mathsf T}Qx)^{1/2}}\,,$$

for some positive constants ε and α. Here note that Assumption (B3) implies Assumption (A3) of Sect. 1.2 for the Lyapunov function $\mathcal V$ given above, see Arapostathis et al. (2011, (7.3.6), p. 257).

By Theorems 1.3 and 1.4 the following holds.

Theorem 1.5. *Assume* (B1)–(B3). *For each* $\varphi_0 \in C_{\mathcal V}(\mathbb R^d)\cap C^2(\mathbb R^d)$, $\varphi(t,x)$ *converges to* $\varphi^*(x)+\beta$ *as* $t\to\infty$.

The relative value iteration equation for the risk-sensitive control problem is given by

$$\begin{aligned}\frac{\partial \psi}{\partial t}(t,x) &= \min_{v\in V}\left[L\psi(t,x,v) + (h(x,v) - \ln\psi(t,0))\psi(t,x)\right],\\ \psi(0,x) &= \psi_0(x),\end{aligned} \tag{1.46}$$

where

$$Lf(x,v) := b(x,v)\cdot\nabla f(x) + \frac{1}{2}\operatorname{tr}\big(a(x)\nabla^2 f(x)\big)\,,\quad f\in C^2(\mathbb R^d).$$

That one has $\ln\psi(t,0)$ instead of $\psi(t,0)$ as the "offset" is only natural, because we are trying to approximate the *logarithmic* growth rate of the cost. We have the following theorem:

Theorem 1.6. *Let* ψ^* *be the unique solution in the class of functions which grow no faster than* $e^{\|x\|^2}$ *of the HJB equation for the risk-sensitive control problem given by*

$$\beta\psi^* = \min_{v\in V}\left[L\psi^*(x,v) + h(x,v)\psi^*\right],\quad \psi^*(0) = 1\,.$$

Under assumptions (B1)–(B3) *the solution* $\psi(t,x)$ *of the relative value iteration in* (1.46) *converges as* $t\to\infty$ *to* $e^{\beta}\psi^*(x)$ *where* β *is the value of the risk-sensitive control problem given in Theorem 1.4.*

Proof. A straightforward calculation shows that $\psi^* = e^{\varphi^*}$, where φ^* is given in Theorem 1.4. Then it easily follows that $\psi(t,x) = e^{\varphi(t,x)}$, where φ is the solution of the relative value iteration for the stochastic differential game in (1.45). From Theorem 1.5, it follows that $\psi(t,x) \to e^{\beta}\psi^*(x)$ as $t \to \infty$, which establishes the claim. □

Acknowledgments The work of Ari Arapostathis was supported in part by the Office of Naval Research under the Electric Ship Research and Development Consortium. The work of Vivek Borkar was supported in part by Grant #11IRCCSG014 from IRCC, IIT, Mumbai.

References

Arapostathis A, Borkar VS (2012) A relative value iteration algorithm for nondegenerate controlled diffusions. SIAM J Control Optim 50(4):1886–1902

Arapostathis A, Borkar VS, Ghosh MK (2011) Ergodic control of diffusion processes. Encyclopedia of mathematics and its applications, vol 143. Cambridge University Press, Cambridge

Basak GK, Bhattacharya RN (1992) Stability in distribution for a class of singular diffusions. Ann Probab 20(1):312–321

Beneš VE (1970) Existence of optimal strategies based on specified information, for a class of stochastic decision problems. SIAM J Control 8:179–188

Borkar VS, Ghosh MK (1992) Stochastic differential games: occupation measure based approach. J Optim Theory Appl 73(2):359–385

Borkar VS, Suresh Kumar K (2010) Singular perturbations in risk-sensitive stochastic control. SIAM J Control Optim 48(6):3675–3697

Fleming WH, McEneaney WM (1995) Risk-sensitive control on an infinite time horizon. SIAM J Control Optim 33(6):1881–1915

Gilbarg D, Trudinger NS (1983) Elliptic partial differential equations of second order, 2nd edn. Grundlehren der Mathematischen Wissenschaften, vol 224. Springer, Berlin

Gruber M (1984) Harnack inequalities for solutions of general second order parabolic equations and estimates of their Hölder constants. Math Z 185(1):23–43

Ladyženskaja OA, Solonnikov VA, Ural′ceva NN (1967) Linear and quasilinear equations of parabolic type. Translated from the Russian by S. Smith. Translations of Mathematical Monographs, Vol. 23. American Mathematical Society, Providence, R.I.

Meyn SP, Tweedie RL (1993) Stability of Markovian processes. III. Foster-Lyapunov criteria for continuous-time processes. Adv Appl Probab 25(3):518–548

White DJ (1963) Dynamic programming, Markov chains, and the method of successive approximations. J Math Anal Appl 6:373–376

Whittle P (1990) Risk-sensitive optimal control. Wiley-Interscience Series in Systems and Optimization. Wiley, Chichester

Chapter 2
OPTGAME3: A Dynamic Game Solver and an Economic Example

Dmitri Blueschke, Reinhard Neck, and Doris A. Behrens

Abstract In this paper we present the OPTGAME3 algorithm, which can be used to calculate equilibrium and optimum control solutions of dynamic games. The algorithm was programmed in C# and MATLAB[1] and allows the calculation of approximate cooperative Pareto-optimal solutions and non-cooperative Nash and Stackelberg equilibrium solutions. In addition we present an application of the OPTGAME3 algorithm where we use a small stylized nonlinear two-country macroeconomic model of a monetary union for analysing the interactions between fiscal (governments) and monetary (common central bank) policy makers, assuming different objective functions of these decision makers. Several dynamic game experiments are run for different information patterns and solution concepts. We show how the policy makers react optimally to demand and supply shocks. Some comments are given about possible applications to the recent sovereign debt crisis in Europe.

Keywords Numerical methods for control and dynamic games • Economic dynamics • Monetary union

---

[1]The source code of the OPTGAME3 algorithm is available from the authors on request.

D. Blueschke • R. Neck (✉)
Department of Economics, Alpen-Adria-Universität Klagenfurt,
Universitätsstrasse 65–67, 9020 Klagenfurt, Austria
e-mail: dmitri.blueschke@aau.at; reinhard.neck@uni-klu.ac.at

D.A. Behrens
Department of Controlling and Strategic Management, Alpen-Adria-Universität Klagenfurt,
Universitätsstrasse 65–67, 9020 Klagenfurt, Austria
e-mail: doris.behrens@aau.at

V. Křivan and G. Zaccour (eds.), *Advances in Dynamic Games*, Annals of the International Society of Dynamic Games 13, DOI 10.1007/978-3-319-02690-9_2,

2.1 Introduction

When we think about economic policy making in one single country, it is preferable to consider the government controlling fiscal policy and the central bank controlling monetary policy as independent players. When considering a country inside a monetary union where monetary policy is no longer an instrument of national institutions, it is essential to look at the government and the central bank of the monetary union separately. Moreover, the interests of other countries inside the union, which primarily pursue their own national interests and do not necessarily care about the spillovers of their actions to other countries should also be taken into account by the decision makers when determining the best policy actions. Such problems can best be modelled by using the concepts and methods of dynamic game theory, which has been developed mostly by engineers and mathematicians but which has proved to be a valuable analytical tool for economists, too (see, e.g., Başar and Olsder 1999; Dockner et al. 2000; Petit 1990).

The theory of dynamic games is well developed for linear-quadratic games. It is also well known that considering linear problems alone is a very strong limitation, thus a lot of research is required to extend the theory for nonlinear games. This paper follows this line of research and presents an algorithm which is designed for the solution of nonlinear-quadratic dynamic tracking games. The algorithm is called OPTGAME3 and is programmed in C# and MATLAB. Due to their nonlinearity, the problems cannot be solved analytically but only numerically. The algorithm allows the calculation of approximate cooperative Pareto-optimal solutions and non-cooperative Nash and Stackelberg equilibrium solutions.

In addition we present an application of the OPTGAME3 algorithm for a monetary union. Dynamic games have been used by several authors (Hager et al. 2001; Pohjola 1986) for modelling conflicts between monetary and fiscal policies. There is also a large body of literature on dynamic conflicts between policy makers from different countries on issues of international stabilization (Hughes Hallett 1986; Levine and Brociner 1994; Miller and Salmon 1985). Both types of conflict are present in a monetary union, because a supranational central bank interacts strategically with sovereign governments as national fiscal policy makers in the member states. Such conflicts can be analysed using either large empirical macroeconomic models (Engwerda et al. 2012; Haber et al. 2002; Plasmans et al. 2006) or small stylized models (van Aarle et al. 2002; Neck and Behrens 2004, 2009). We follow the latter line of research and use a small stylized nonlinear two-country macroeconomic model of a monetary union for analysing the interactions between fiscal (governments) and monetary (common central bank) policy makers, assuming different objective functions of these decision makers. We show how the policy makers react optimally to demand and supply shocks. Some comments are given about possible applications to the recent sovereign debt crisis in Europe.

2.2 The Dynamic Game Problem

We consider intertemporal nonlinear game-theoretic problems which are given in tracking form. The players aim at minimizing quadratic deviations of the equilibrium values from given target (desired) values. Thus each player minimizes an objective function J^i:

$$\min_{u_1^i,\ldots,u_T^i} J^i = \min_{u_1^i,\ldots,u_T^i} \sum_{t=1}^{T} L_t^i(x_t, u_t^1, \ldots, u_t^N), \qquad i = 1, \ldots, N, \tag{2.1}$$

with

$$L_t^i(x_t, u_t^1, \ldots, u_t^N) = \frac{1}{2}[X_t - \tilde{X}_t^i]' \Omega_t^i [X_t - \tilde{X}_t^i], \qquad i = 1, \ldots, N, \tag{2.2}$$

The parameter N denotes the number of players (decision makers). T is the terminal period of the finite planning horizon, i.e. the duration of the game. X_t is an aggregated vector

$$X_t := [x_t \;\; u_t^1 \;\; u_t^2 \;\; \ldots \;\; u_t^N]', \tag{2.3}$$

which consists of an $(n_x \times 1)$ vector of state variables

$$x_t := [x_t^1 \;\; x_t^2 \;\; \ldots \;\; x_t^{n_x}]', \tag{2.4}$$

and N $(n_i \times 1)$ vectors of control variables determined by the players $i = 1, \ldots, N$:

$$\begin{aligned} u_t^1 &:= [u_t^{11} \;\; u_t^{12} \;\; \ldots \;\; u_t^{1n_1}]', \\ u_t^2 &:= [u_t^{21} \;\; u_t^{22} \;\; \ldots \;\; u_t^{2n_2}]', \\ &\vdots \\ u_t^N &:= [u_t^{N1} \;\; u_t^{N2} \;\; \ldots \;\; u_t^{Nn_N}]'. \end{aligned} \tag{2.5}$$

Thus X_t (for all $t = 1, \ldots, T$) is an r-dimensional vector, where

$$r := n_x + n_1 + n_2 + \cdots + n_N. \tag{2.6}$$

The desired levels of the state variables and the control variables of each player enter the quadratic objective functions (as given by (2.1) and (2.2)) via the terms

$$\tilde{X}_t^i := [\tilde{x}_t^i \;\; \tilde{u}_t^{i1} \;\; \tilde{u}_t^{i2} \;\; \ldots \;\; \tilde{u}_t^{iN}]'. \tag{2.7}$$

It must be pointed out that each player $i = 1, \ldots, N$ may be allowed to observe and monitor the control variables of the other players, i.e. deviations of other control variables can be punished in one's own objective function.[2]

Finally, (2.2) contains an $(r \times r)$ penalty matrix Ω_t^i $(i = 1, \ldots, N)$, weighting the deviations of states and controls from their desired levels in any time period t $(t = 1, \ldots, T)$. Thus the matrices

$$\Omega_t^i = \begin{bmatrix} Q_t^i & 0 & \cdots & 0 \\ 0 & R_t^{i1} & 0 & \vdots \\ \vdots & 0 & \ddots & 0 \\ 0 & \cdots & 0 & R_t^{iN} \end{bmatrix}, \; i = 1, \ldots, N, \; t = 1, \ldots, T, \tag{2.8}$$

are of block-diagonal form, where the blocks Q_t^i and R_t^{ij} $(i, j = 1, \ldots, N)$ are symmetric. These blocks Q_t^i and R_t^{ij} correspond to penalty matrices for the states and the controls, respectively. The matrices $Q_t^i \geq 0$ are positive semi-definite for all $i = 1, \ldots, N$; the matrices R_t^{ij} are positive semi-definite for $i \neq j$ but positive definite for $i = j$. This guarantees that the matrices $R_t^{ii} > 0$ are invertible, a necessary prerequisite for the analytical tractability of the algorithm.

In a frequent special case, a discount factor α is used to calculate the penalty matrix Ω_t^i in time period t:

$$\Omega_t^i = \alpha^{t-1} \Omega_0^i, \qquad i = 1, \ldots, N, \; t = 1, \ldots, T, \tag{2.9}$$

where the initial penalty matrix Ω_0^i of player i is given.

The dynamic system, which constrains the choices of the decision makers, is given in state-space form by a first-order system of nonlinear difference equations:

$$x_t = f(x_{t-1}, x_t, u_t^1, \ldots, u_t^N, z_t), \qquad x_0 = \bar{x}_0. \tag{2.10}$$

$\bar{x}_0$ contains the initial values of the state variables. The vector z_t contains non-controlled exogenous variables. f is a vector-valued function where f^k $(k = 1, \ldots, n_x)$ denotes the kth component of f. For the algorithm, we require that the first and second derivatives of the system function f with respect to x_t, x_{t-1} and $u_t^1, \ldots, u_t^N$ exist and are continuous. The assumption of a first-order system of difference equations as stated in (2.10) is not really restrictive as higher-order difference equations can be reduced to systems of first-order difference equations by suitably redefining variables as new state variables and augmenting the state vector.

[2]For example, the central bank in a monetary union, which controls monetary policy, can also penalize "bad" fiscal policies of member countries.

Equations (2.1), (2.2), and (2.10) define a nonlinear dynamic tracking game problem to be solved. That means, we try to find N trajectories of control variables u_t^i, $i = 1, \ldots, N$ which minimize the postulated objective functions subject to the dynamic system. In the next section, the OPTGAME3 algorithm, which is designed to solve such types of problems, is presented.

2.3 OPTGAME3

This section describes the OPTGAME algorithm in its third version (OPTGAME3), which was programmed in C# and MATLAB. For a better understanding, first a very simplified structure of the OPTGAME algorithm will be presented.

Algorithm 1 Rough structure of the OPTGAME algorithm

1: iteration step $k = 0$
2: initialize input parameters x_0, $(\overset{\circ}{u}{}_t^i)_{t=1}^T$, $(\tilde{x}_t^i)_{t=1}^T$, $(\tilde{u}_t^{ij})_{t=1}^T$, $(z_t)_{t=1}^T$ and $f(...)$
3: calculate tentative path for states $x_t = f(x_{t-1}, x_t, \overset{\circ}{u}{}_t^1, \ldots, \overset{\circ}{u}{}_t^N, z_t)$, $t = 1, ..., T$
4: **while** the stopping criterion is not met (*nonlinearity loop*) **do**
5:     **for** T to 1 (*backward loop*) **do**
6:         linearise the system of equations: $x_t = A_t x_{t-1} + \sum_{i=1}^N B_t^i u_t^i + c_t$
7:         min J^i, get feedback matrices: G_t^i and g_t^i
8:     **end for**
9:     **for** 1 to T (*forward loop*) **do**
10:         calculate the solution: $u_t^{i*} = G_t^i x_{t-1}^* + g_t^i$ and $x_t^* = f(x_{t-1}^*, x_t, u_t^{1*}, \ldots, u_t^{N*}, z_t)$
11:     **end for**
12:     at the end of the forward loop the solution for the current iteration of the nonlinearity loop is calculated: $(u_t^{i*}, x_t^*)_{t=1}^T$
13:     set new tentative control paths: $u_t^{i*} \rightarrow \overset{\circ}{u}{}_t^i \quad \forall\ t, i$
14:     calculate new tentative path for states $x_t = f(x_{t-1}, x_t, \overset{\circ}{u}{}_t^1, \ldots, \overset{\circ}{u}{}_t^N, z_t)$, $t = 1, ..., T$
15:     $k \rightarrow k + 1$
16: **end while**
17: final solution is calculated: $(u_t^{i*})_{t=1}^T$, $(x_t^*)_{t=1}^T$, J^{i*}, J^*

The algorithm starts with the input of all required data. As indicated in step 2, for all players ($i = 1, \ldots, N$) the initial tentative paths of the control variables $(\overset{\circ}{u}{}_t^i)_{t=1}^T$ are given as inputs. In order to find an initial tentative path for the state variables we apply an appropriate system solver like Newton–Raphson, Gauss–Seidel, Levenberg–Marquardt or Trust region to $x_t - f(x_{t-1}, x_t, \overset{\circ}{u}{}_t^1, \ldots, \overset{\circ}{u}{}_t^N, z_t) = 0$ in step 3. After that the nonlinearity loop is started where we iteratively approximate the final solution of the nonlinear dynamic tracking game. To this end, following a procedure introduced by Chow (1975) for optimum control problems, we linearise the nonlinear system f along the tentative path determined in the previous iteration steps. Note that we do not linearise the system only once prior to launching the optimization procedure (cf. step 4) but repeatedly linearise the entire system during the iterative optimization process along the current tentative paths (for both controls and states). This allows for replacing the autonomous nonlinear sytem by a non-autonomous linear system evaluated along a tentative path that changes with each iteration step. Accordingly, for each time period t we compute the reduced form of the linearised structure of (2.10) and approximate the nonlinear system by a linear system with time-dependent parameters in step 6.

The dynamic tracking game can then be solved for the linearised system using known optimization techniques, which results in feedback matrices G_t^i and g_t^i (see step 7). These feedback matrices allow us to calculate in a forward loop the solutions (u_t^{i*} and x_t^*) of the current iteration of the nonlinearity loop and, at the end of the nonlinearity loop, the final solutions. If the new tentative path falls into an ϵ-tube around the old tentative path, no variable differs by more than a value of ϵ between two successive iterations, and, for ϵ small enough, the consecutive paths are (more or less) identical. In particular, then, the state path calculated according to the nonlinear system dynamics (2.10) by using $x_t - f(x_{t-1}, x_t, u_t^1, \ldots, u_t^N, z_t) = 0$ equals the state path calculated according to the linearized system representation evaluated along the current tentative path (step 5). In other words, the algorithm has converged,[3] and the paths obtained indeed solve the original problem, i.e. (2.1) and (2.2) subject to (2.10).

The core of the OPTGAME algorithm appears in step 7, where the linearised system has to be optimized by each player. The optimization technique for minimizing the objective functions depends on the type of the game or solution concept. The OPTGAME3 algorithm determines four game strategies: one cooperative (Pareto optimal) and three non-cooperative game types: the Nash game for the open-loop information pattern, the Nash game for the feedback information pattern, and the Stackelberg game for the feedback information pattern.

Generally, open-loop Nash equilibrium solutions of affine-quadratic games are determined using Pontryagin's maximum principle. Feedback Nash and Stackelberg equilibrium solutions of affine-quadratic games are calculated using the dynamic programming (Hamilton–Jacobi–Bellman) technique. How to calculate the dynamic game solutions depending on the type of the game will be discussed separately in the next subsections.[4]

2.3.1 The Pareto-Optimal Solution

To determine a cooperative solution of the dynamic game, we have to define a joint objective function of all the players. This joint objective function J corresponds to

---

[3]Note that if convergence has not been obtained before k has reached its terminal value, then the iteration process terminates without succeeding in finding an equilibrium feedback solution.

[4]The mathematical details are based on Behrens and Neck (2007) and reflect the calculations and proofs in Başar and Olsder (1999).

the solution concept of Pareto-optimality and is given by a convex combination of the individual cost functions,

$$J = \sum_{t=1}^{T} \sum_{i=1}^{N} \mu^i L_t^i(x_t, u_t^1, \ldots, u_t^N), \quad \sum_{i=1}^{N} \mu^i = 1. \tag{2.11}$$

L_t^i is defined by (2.2). The parameters μ^i ($i = 1, \ldots, N$) reflect player i's "power" or "importance" in the joint objective function. Therefore the solution of the cooperative Pareto-optimal game with N players can be determined by solving a classical optimum control problem. First, define the following matrices:

$$Q_t := \sum_{i=1}^{N} \mu^i Q_t^i, \tag{2.12}$$

$$q_t := \sum_{i=1}^{N} \mu^i Q_t^i \tilde{x}_t^i, \tag{2.13}$$

$$R_t^j := \sum_{i=1}^{N} \mu^i R_t^{ij}, \quad j = 1, \ldots, N, \tag{2.14}$$

$$r_t^j := \sum_{i=1}^{N} \mu^i R_t^{ij} \tilde{u}_t^{ij}, \quad j = 1, \ldots, N. \tag{2.15}$$

The Riccati matrices H_t and h_t for all players and for all time periods $t \in 1, \ldots, T$ are derived by backward iteration according to the following system of recursive matrix equations:

$$H_{t-1} = Q_{t-1} + K_t' H_t K_t + \sum_{j=1}^{N} G_t^{j'} R_t^j G_t^j, \quad H_T = Q_T, \tag{2.16}$$

$$h_{t-1} = q_{t-1} - K_t'[H_t k_t - h_t] + \sum_{j=1}^{N} G_t^{j'} [r_t^j - R_t^j g_t^j], \quad h_T = q_T, \tag{2.17}$$

where

$$K_t := A_t + \sum_{j=1}^{N} B_t^j G_t^j, \tag{2.18}$$

$$k_t := s_t + \sum_{j=1}^{N} B_t^j g_t^j. \tag{2.19}$$

The feedback matrices G_t^i and g_t^i for $i = 1, \ldots, N$ are determined as solutions of the following set of linear matrix equations:

$$B_t^{j'} H_t A_t + [R_t^j + B_t^{j'} H_t B_t^j] G_t^j + B_t^{j'} H_t \sum_{\substack{k=1, \\ k \neq j}}^{N} B_t^k G_t^k = 0, \tag{2.20}$$

$$[R_t^j + B_t^{j'} H_t B_t^j] g_t^j + B_t^{j'} H_t \sum_{\substack{k=1, \\ k \neq j}}^{N} B_t^k g_t^k + B_t^{j'} H_t s_t - B_t^{j'} h_t - r_t^j = 0. \tag{2.21}$$

In each time period t the auxiliary matrices are determined in the following order:

1. H_t^i, h_t^i: according to (2.16) and (2.17),
2. G_t^i, g_t^i: according to (2.20) and (2.21),[5]
3. K_t, k_t: according to (2.18) and (2.19).

Using the Riccati matrices H_t^i and h_t^i for $i = 1, \ldots, N$ and the feedback matrices G_t^i and g_t^i for $i = 1, \ldots, N$ for all time periods, we can compute the matrices K_t and k_t as defined by (2.18) and (2.19). Then the states and controls forming a cooperative Pareto-optimal solution of the game can be determined by forward iteration according to:

$$x_t^* = K_t x_{t-1}^* + k_t, \quad x_0^* = x_0, \tag{2.22}$$

$$u_t^{i*} = G_t^i x_{t-1}^* + g_t^i, \quad i = 1, \ldots, N. \tag{2.23}$$

2.3.2 The Feedback Nash Equilibrium Solution

To approximate the feedback Nash equilibrium solution of the game, the algorithm proceeds as follows: Riccati matrices H_t^i and h_t^i for all players $i = 1, \ldots, N$ and for all time periods $t \in \{1, \ldots, T\}$ are derived by backward iteration according to the following system of recursive matrix equations:

$$H_{t-1}^i = Q_{t-1}^i + K_t' H_t^i K_t + \sum_{j=1}^{N} G_t^{j'} R_t^{ij} G_t^j, \quad H_T^i = Q_T^i, \tag{2.24}$$

$$h_{t-1}^i = Q_{t-1}^i \tilde{x}_{t-1}^i - K_t' [H_t^i k_t - h_t^i] + \sum_{j=1}^{N} G_t^{j'} R_t^{ij} [\tilde{u}_t^{ij} - g_t^j], \quad h_T^i = Q_T^i \tilde{x}_T^i, \tag{2.25}$$

[5] It is important to mention that we have a system with two simultaneous equations and two unknown parameters in step 2 ((2.20) and (2.21)). Therefore a system solver must be applied. In OPTGAME3, the Gauss–Seidel method is applied for this purpose.

where K_t is defined by (2.18) and k_t is given by (2.19). The feedback matrices G_t^i and g_t^i for $i = 1, \ldots, N$ are determined as solutions of the following set of linear matrix equations:

$$D_t^i G_t^i + B_t^{i'} H_t^i \sum_{\substack{j=1,\\ j \neq i}}^{N} B_t^j G_t^j + B_t^{i'} H_t^i A_t = 0, \tag{2.26}$$

$$D_t^i g_t^i + B_t^{i'} H_t^i \sum_{\substack{j=1,\\ j \neq i}}^{N} B_t^j g_t^j + v_t^i = 0, \tag{2.27}$$

where

$$D_t^i := R_t^{ii} + B_t^{i'} H_t^i B_t^i, \tag{2.28}$$

$$v_t^i := B_t^{i'} [H_t^i s_t - h_t^i] - R_t^{ii} \tilde{u}_t^{ii}. \tag{2.29}$$

In each time period t the auxiliary matrices are determined in the following order:

1. H_t^i, h_t^i: according to (2.16) and (2.17)
2. G_t^i, g_t^i: according to (2.26) and (2.27)
3. K_t, k_t: according to (2.18) and (2.19).

Using the Riccati matrices H_t^i and h_t^i for $i = 1, \ldots, N$ and the feedback matrices G_t^i and g_t^i for $i = 1, \ldots, N$ for all time periods, we can compute the matrices K_t and k_t as defined by (2.18) and (2.19). Then approximate feedback Nash equilibrium values of the states and controls forming a solution of the game can be determined by a forward loop in accordance with:

$$x_t^* = K_t x_{t-1}^* + k_t, \quad x_0^* = x_0, \tag{2.30}$$

$$u_t^{i*} = G_t^i x_{t-1}^* + g_t^i, \quad i = 1, \ldots, N. \tag{2.31}$$

2.3.3 The Open-Loop Nash Equilibrium Solution

At the beginning of an open-loop Nash game, each of the N simultaneously acting players makes a binding commitment to stick to a chosen policy rule for the entire time horizon $t = 1, \ldots, T$. As long as these commitments hold, the solution is an equilibrium in the sense that none of the players can improve their individual welfare by one-sided deviations from the open-loop Nash equilibrium path. Although the open-loop Nash equilibrium solution is not time consistent, for certain situations this solution concept could be the right choice. Furthermore, even if this kind of policy is not very realistic, its analysis can help compare the quality of other solutions.

In the following, the procedure of finding the open-loop Nash equilibrium solution is described. For an invertible matrix

$$\Lambda_t := I + \sum_{j=1}^{N} B_t^j [R_t^{jj}]^{-1} B_t^{j'} H_t^j, \tag{2.32}$$

the Riccati matrices for player i ($i = 1, \ldots, N$) are determined by backward iteration according to the following recursive system of matrix equations:

$$H_{t-1}^i = Q_{t-1}^i + A_t' H_t^i [\Lambda_t]^{-1} A_t, \quad H_T^i = Q_T^i \tag{2.33}$$

$$h_{t-1}^i = -Q_{t-1}^i \tilde{x}_{t-1}^i + A_t' [H_t^i [\Lambda_t]^{-1} \eta_t + h_t^i], \quad h_T^i = -Q_T^i \tilde{x}_T^i, \tag{2.34}$$

where

$$\eta_t := s_t + \sum_{j=1}^{N} B_t^j [\tilde{u}_t^{jj} - [R_t^{jj}]^{-1} B_t^{j'} h_t^j]. \tag{2.35}$$

In each time period t the auxiliary matrices are determined as follows:

1. H_t^i, h_t^i: according to (2.33) and (2.34)
2. Λ_t: according to (2.32).

With the Riccati matrices H_t^i and h_t^i, stored for all time periods $t \in \{1, \ldots, T\}$, the approximate open-loop Nash equilibrium values of the state and the control variables for all players ($i = 1, \ldots, N$) are determined by forward loop according to

$$x_t^* = [\Lambda_t]^{-1} [A_t x_{t-1}^* + \eta_t] \tag{2.36}$$

and

$$u_t^{i*} = \tilde{u}_t^{ii} - [R_t^{ii}]^{-1} B_t^{i'} [H_t^i x_t^* + h_t^i], \tag{2.37}$$

starting with the initial condition $x_0^* = x_0$.

2.3.4 The Feedback Stackelberg Equilibrium Solution

The feedback Stackelberg equilibrium solution is asymmetric: The Stackelberg leader (player 1) announces his decision rule, $u_t^1 = \varphi^1(x_{t-1})$, to all other players while the actions of the other players (players $i = 2, \ldots, N$, the Stackelberg followers) are based on the current state x_{t-1} and on the decision of the leader according to the reaction function $u_t^i = \varphi^i(x_{t-1}, u_t^1)$. At the time of optimizing his performance, the leader considers the reaction coefficients,

$$\Psi_t^i := \frac{\partial u_t^i}{\partial u_t^1}, \quad i = 2, \ldots, N, \tag{2.38}$$

as rational reactions of the followers $i = 2, \ldots, N$. These reaction coefficients Ψ_t^i ($i = 2, \ldots, N$) are determined as solutions of the following set of $N - 1$ linear matrix equations:

$$B_t^{i'} H_t^i B_t^1 + D_t^i \Psi_t^i + B_t^{i'} H_t^i \sum_{\substack{j=2, \\ j \neq i}}^{N} B_t^j \Psi_t^j = 0, \quad i = 2, \ldots, N, \tag{2.39}$$

where H_t^i denotes the Riccati matrices of the feedback Stackelberg game calculated as

$$H_{t-1}^i = Q_{t-1}^i + K_t' H_t^i K_t + \sum_{j=1}^{N} G_t^{j'} R_t^{ij} G_t^j, \quad H_T^i = Q_T^i, \; i = 1, \ldots, N, \tag{2.40}$$

and the matrix D_t^i is given by

$$D_t^i := R_t^{ii} + B_t^{i'} H_t^i B_t^i. \tag{2.41}$$

The matrices W_t^i and w_t^i (for $i = 2, \ldots, N$), which are required for determining the feedback matrices G_t^i and g_t^i (for $i = 1, \ldots, N$), are calculated as solutions of the following set of $N - 1$ linear matrix equations:

$$D_t^i W_t^i + B_t^{i'} H_t^i \left[A_t + \sum_{\substack{j=2, \\ j \neq i}}^{N} B_t^j W_t^j \right] = 0, \tag{2.42}$$

$$D_t^i w_t^i + B_t^{i'} (H_t^i s_t - h_t^i) - R_t^{ii} \tilde{u}_t^{ii} + B_t^{i'} H_t^i \sum_{\substack{j=2, \\ j \neq i}}^{N} B_t^j w_t^j = 0. \tag{2.43}$$

Using

$$h_{t-1}^i = Q_{t-1}^i \tilde{x}_{t-1}^i - K_t' [H_t^i k_t - h_t^i] + \sum_{j=1}^{N} G_t^{j'} R_t^{ij} [\tilde{u}_t^{ij} - g_t^j], \quad h_T^i = Q_T^i \tilde{x}_T^i, \tag{2.44}$$

and given that the matrix

$$\bar{\Lambda}_t := R_t^{11} + \bar{B}_t' H_t^1 \bar{B}_t + \sum_{j=2}^{N} \Psi_t^{j'} R_t^{1j} \Psi_t^j \tag{2.45}$$

is invertible, for

$$\bar{B}_t := B_t^1 + \sum_{j=2}^{N} B_t^j \Psi_t^j \tag{2.46}$$

we can derive the Riccati matrices, H_{t-1}^i and h_{t-1}^i for $i = 1, \ldots, N$, by backward iteration according to the Riccati equations (2.40) and (2.44). The feedback matrices are determined by

$$G_t^1 := -\left[\bar{\Lambda}_t\right]^{-1} \left[\bar{B}_t' H_t^1 A_t + \sum_{j=2}^{N} \bar{D}_t^j W_t^j \right], \tag{2.47}$$

$$g_t^1 := -\left[\bar{\Lambda}_t\right]^{-1} \left[\upsilon_t^1 + \bar{\upsilon}_t + \sum_{j=2}^{N} \bar{D}_t^j w_t^j \right], \tag{2.48}$$

$$G_t^i := W_t^i + \Psi_t^i G_t^1, \quad i = 2, \ldots, N, \tag{2.49}$$

$$g_t^i := w_t^i + \Psi_t^i g_t^1, \quad i = 2, \ldots, N, \tag{2.50}$$

where

$$\bar{D}_t^i := \Psi_t^{i'} R_t^{1i} + \bar{B}_t' H_t^1 B_t^i, \quad i = 2, \ldots, N, \tag{2.51}$$

$$\bar{\upsilon}_t := \sum_{j=2}^{N} \Psi_t^{j'} \left[B_t^{j'} H_t^1 s_t - B_t^{j'} h_t^1 - R_t^{1j} \tilde{u}_t^{1j} \right]. \tag{2.52}$$

Using the Riccati matrices H_t^i and h_t^i for $i = 1, \ldots, N$ and the feedback matrices G_t^i and g_t^i for $i = 1, \ldots, N$ for all time periods, we can compute the matrices K_t and k_t by:

$$K_t := A_t + \sum_{j=1}^{N} B_t^j G_t^j, \tag{2.53}$$

$$k_t := s_t + \sum_{j=1}^{N} B_t^j g_t^j. \tag{2.54}$$

In each time period t the auxiliary matrices are determined in the following order:

1. H_t^i, h_t^i: according to (2.40) and (2.44)
2. Ψ_t^i: according to (2.39)
3. W_t^i, w_t^i: according to (2.42) and (2.43)

4. G_t^i, g_t^i: according to (2.47), (2.48), (2.49), and (2.50)
5. K_t, k_t: according to (2.53) and (2.54).

Finally, the approximate feedback Stackelberg equilibrium values of the states and controls for the game for all players ($i = 1, \ldots, N$) are determined by forward iteration according to the following functional relationships:

$$x_t^* = K_t x_{t-1}^* + k_t, \quad x_0^* = x_0, \tag{2.55}$$

$$u_t^{i*} = G_t^i x_{t-1}^* + g_t^i, \quad i = 1, \ldots, N. \tag{2.56}$$

2.4 An Application

2.4.1 *The MUMOD1 Model*

In order to show the applicability of the OPTGAME3 algorithm we use a simplified macroeconomic model of a monetary union consisting of two countries (or two blocs of countries) with a common central bank. This model is called MUMOD1 and slightly improves on the one introduced in Blueschke and Neck (2011). For a similar framework in continuous time, see van Aarle et al. (2002). The model is calibrated so as to deal with the problem of public debt targeting in a situation that resembles the one currently prevailing in the European Union, but no attempt is made to describe a monetary union in general or the EMU in every detail.

In the following, capital letters indicate nominal values, while lowercase letters correspond to real values. Variables are denoted by Roman letters and model parameters are denoted by Greek letters. Three active policy makers are considered: the governments of the two countries responsible for decisions about fiscal policy and the common central bank of the monetary union controlling monetary policy. The two countries are labelled 1 and 2 or core and periphery, respectively. The idea is to create a stylized model of a monetary union consisting of two homogeneous blocs of countries, which in the current European context might be identified with the stability-oriented bloc (core) and the PIIGS bloc (countries with problems due to high public debt).

The model is formulated in terms of deviations from a long-run growth path. The goods markets are modelled for each country by a short-run income-expenditure equilibrium relation (IS curve). The two countries under consideration are linked through their goods markets, namely exports and imports of goods and services. The common central bank decides on the prime rate, that is, a nominal rate of interest under its direct control (for instance, the rate at which it lends money to private banks).

Real output (or the deviation of short-run output from a long-run growth path) in country i ($i = 1, 2$) at time t ($t = 1, \ldots, T$) is determined by a reduced form demand-side equilibrium equation:

$$y_{it} = \delta_i(\pi_{jt} - \pi_{it}) - \gamma_i(r_{it} - \theta) + \rho_i y_{jt} - \beta_i \pi_{it} + \kappa_i y_{i(t-1)} - \eta_i g_{it} + zd_{it}, \quad (2.57)$$

for $i \neq j$ $(i, j = 1, 2)$. The variable π_{it} denotes the rate of inflation in country i, r_{it} represents country i's real rate of interest and g_{it} denotes country i's real fiscal surplus (or, if negative, its fiscal deficit), measured in relation to real GDP. g_{it} in (2.57) is assumed to be country i's fiscal policy instrument or control variable. The natural real rate of output growth, $\theta \in [0, 1]$, is assumed to be equal to the natural real rate of interest. The parameters $\delta_i, \gamma_i, \rho_i, \beta_i, \kappa_i, \eta_i$, in (2.57) are assumed to be positive. The variables zd_{1t} and zd_{2t} are non-controlled exogenous variables and represent exogenous demand-side shocks in the goods market.

For $t = 1, \ldots, T$, the current real rate of interest for country i $(i = 1, 2)$ is given by:

$$r_{it} = I_{it} - \pi^e_{it}, \quad (2.58)$$

where π^e_{it} denotes the expected rate of inflation in country i and I_{it} denotes the nominal interest rate for country i, which is given by:

$$I_{it} = R_{Et} - \lambda_i g_{it} + \chi_i D_{it} + zhp_{it}, \quad (2.59)$$

where R_{Et} denotes the prime rate determined by the central bank of the monetary union (its control variable); $-\lambda_i$ and χ_i (λ_i and χ_i are assumed to be positive) are risk premiums for country i's fiscal deficit and public debt level. This allows for different nominal (and a fortiori also real) rates of interest in the union in spite of a common monetary policy due to the possibility of default or similar risk of a country (a bloc of countries) with high government deficit and debt. zhp_{it} allows for exogenous shocks on the nominal rate of interest, e.g. negative after-effects of a haircut or a default.

The inflation rates for each country $i = 1, 2$ and $t = 1, \ldots, T$ are determined according to an expectations-augmented Phillips curve, i.e. the actual rate of inflation depends positively on the expected rate of inflation and on the goods market excess demand (a demand-pull relation):

$$\pi_{it} = \pi^e_{it} + \xi_i y_{it} + zs_{it}, \quad (2.60)$$

where ξ_1 and ξ_2 are positive parameters; zs_{1t} and zs_{2t} denote non-controlled exogenous variables and represent exogenous supply-side shocks, such as oil price increases, introducing the possibility of cost-push inflation; π^e_{it} denotes the rate of inflation in country i expected to prevail during time period t, which is formed at (the end of) time period $t - 1$. Inflationary expectations are formed according to the hypothesis of adaptive expectations:

$$\pi^e_{it} = \varepsilon_i \pi_{i(t-1)} + (1 - \varepsilon_i)\pi^e_{i(t-1)}, \quad (2.61)$$

where $\varepsilon_i \in [0, 1]$ are positive parameters determining the speed of adjustment of expected to actual inflation.

The average values of output and inflation in the monetary union are given by:

$$y_{Et} = \omega y_{1t} + (1-\omega) y_{2t}, \quad \omega \in [0,1], \tag{2.62}$$

$$\pi_{Et} = \omega \pi_{1t} + (1-\omega) \pi_{2t}, \quad \omega \in [0,1]. \tag{2.63}$$

The parameter ω expresses the weight of country 1 in the economy of the whole monetary union as defined by its output level. The same weight ω is used for calculating union-wide inflation in (2.63).

The government budget constraint is given as an equation for government debt of country i $(i = 1, 2)$:

$$D_{it} = (1 + r_{i(t-1)}) D_{i(t-1)} - g_{it} + zh_{it}, \tag{2.64}$$

where D_i denotes real public debt of country i measured in relation to (real) GDP. No seigniorage effects on governments' debt are assumed to be present. zh_{it} allows us to model an exogenous shock on public debt; for instance, if negative it may express default or debt relief (a haircut).

Both national fiscal authorities are assumed to care about stabilizing inflation (π), output (y), debt (D), and fiscal deficits of their own countries (g) at each time t. This is a policy setting which seems plausible for the real EMU as well, with full employment (output at its potential level) and price level stability relating to country (or bloc) i's primary domestic goals, and government debt and deficit relating to its obligations according to the Maastricht Treaty of the European Union. The common central bank is interested in stabilizing inflation and output in the entire monetary union, also taking into account a goal of low and stable interest rates in the union.

Equations (2.57)–(2.64) constitute a dynamic game with three players, each of them having one control variable. The model contains 14 endogenous variables and four exogenous variables and is assumed to be played over a finite time horizon. The objective functions are quadratic in the paths of deviations of state and control variables from their desired values. The game is nonlinear-quadratic and hence cannot be solved analytically but only numerically. To this end, we have to specify the parameters of the model.

The parameters of the model are specified for a slightly asymmetric monetary union; see Table 2.1. Here an attempt has been made to calibrate the model parameters so as to fit for the EMU. The data used for calibration include average economic indicators for the 16 EMU countries from EUROSTAT up to the year 2007. Mainly based on the public finance situation, the EMU is divided into two blocs: a core (country or bloc 1) and a periphery (country or bloc 2). The first bloc has a weight of 60% in the entire economy of the monetary union (i.e. the parameter ω is equal to 0.6). The second bloc has a weight of 40% in the economy of the union; it consists of countries with higher public debt and deficits and higher interest and inflation rates on average. The weights correspond to the respective shares in EMU real GDP. For the other parameters of the model, we use values in accordance with econometric studies and plausibility considerations.

Table 2.1 Parameter values for an asymmetric monetary union, $i = 1, 2$

T	θ	ω	$\delta_i, \beta_i, \eta_i, \varepsilon_i$	$\gamma_i, \rho_i, \kappa_i, \xi_i, \lambda_i$	χ_i
30	3	0.6	0.5	0.25	0.0125

Table 2.2 Initial values of the two-country monetary union

$y_{i,0}$	$\pi_{i,0}$	$\pi^e_{i,0}$	$D_{1,0}$	$D_{2,0}$	$R_{E,0}$	$g_{1,0}$	$g_{2,0}$
0	2	2	60	80	3	0	0

Table 2.3 Target values for an asymmetric monetary union

$\tilde{y}_{it}$	$\tilde{D}_{1t}$	$\tilde{D}_{2t}$	$\tilde{\pi}_{it}$	$\tilde{\pi}_{Et}$	$\tilde{y}_{Et}$	$\tilde{g}_{it}$	$\tilde{R}_{Et}$
0	60↘50	80↘60	1.8	1.8	0	0	3

The initial values of the macroeconomic variables, which are the state variables of the dynamic game model, are presented in Table 2.2. The desired or ideal values assumed for the objective variables of the players are given in Table 2.3. Country 1 (the core bloc) has an initial debt level of 60% of GDP and aims to decrease this level in a linear way over time to arrive at a public debt of 50% at the end of the planning horizon. Country 2 (the periphery bloc) has an initial debt level of 80% of GDP and aims to decrease its level to 60% at the end of the planning horizon, which means that it is going to fulfil the Maastricht criterion for this economic indicator. The ideal rate of inflation is calibrated at 1.8%, which corresponds to the Eurosystem's aim of keeping inflation below, but close to, 2%. The initial values of the two blocs' government debts correspond to those at the beginning of the Great Recession, the recent financial and economic crisis. Otherwise, the initial situation is assumed to be close to equilibrium, with parameter values calibrated accordingly.

2.4.2 Equilibrium Fiscal and Monetary Policies

The MUMOD1 model can be used to simulate the effects of different shocks acting on the monetary union, which are reflected in the paths of the exogenous non-controlled variables, and of policy reactions towards these shocks. In this paper we show the applicability of the OPTGAME3 algorithm. To this end we assume a mixed asymmetric shock which occurs both on demand (zd_i) and supply side (zs_i) as given in Table 2.4.

In the first three periods, both countries experience the same negative demand shock (zd_i) which reflects a financial and economic crisis like the Great Recession. After three periods the economic environment of country 1 stabilizes again, but for country 2 the crisis continues for two more periods.

Starting with time period 3 both countries also experience adverse supply side shocks, which lead to increases in inflation rates. These shocks last three periods for both countries (or blocs) but vary in their strength. The core bloc experiences

Table 2.4 Negative asymmetric shock on demand and supply side

t	1	2	3	4	5	6	⋯	30
zd_1	−2	−4	−2	0	0	0	⋯	0
zd_2	−2	−4	−2	−2	−1	0	⋯	0
zs_1	0	0	2	2	2	0	⋯	0
zs_2	0	0	4	4	4	0	⋯	0

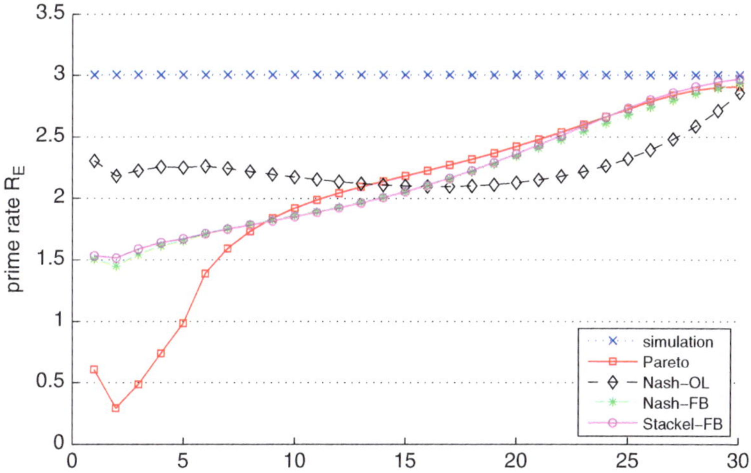

Fig. 2.1 Prime rate R_{Et} controlled by the central bank

an increase in inflation of 2 percentage points, the periphery bloc an increase of 4 percentage points.

In this section, we investigate how the dynamics of the model and the results of the policy game (2.57)–(2.64) depend on the strategy choice of the decision makers. For this game, we calculate five different solutions: a baseline solution with the shock but with policy instruments held at pre-shock levels (zero for the fiscal balance, 3 for the central bank's interest rate), three non-cooperative game solutions and one cooperative game solution. The baseline solution does not include any policy intervention and describes a simple simulation of the dynamic system. It can be interpreted as resulting from a policy ideology of market fundamentalism prescribing non-intervention in the case of a recession.

Figures 2.1–2.5 show the simulation and optimization results of our experiment. Figures 2.1 and 2.2 show the results for the control variables of the players and Figs. 2.3–2.5 show the results of selected state variables: output, inflation and public debt.

Without policy intervention (baseline scenario, denoted by "simulation"), both countries suffer dramatically from the economic downturn modelled by the demand side shock in the first periods. The output of country 1 drops by 6% and that of country 2 by more than 7%, which for several European countries is a fairly good approximation of what happened in reality. This economic crisis lowers the inflation

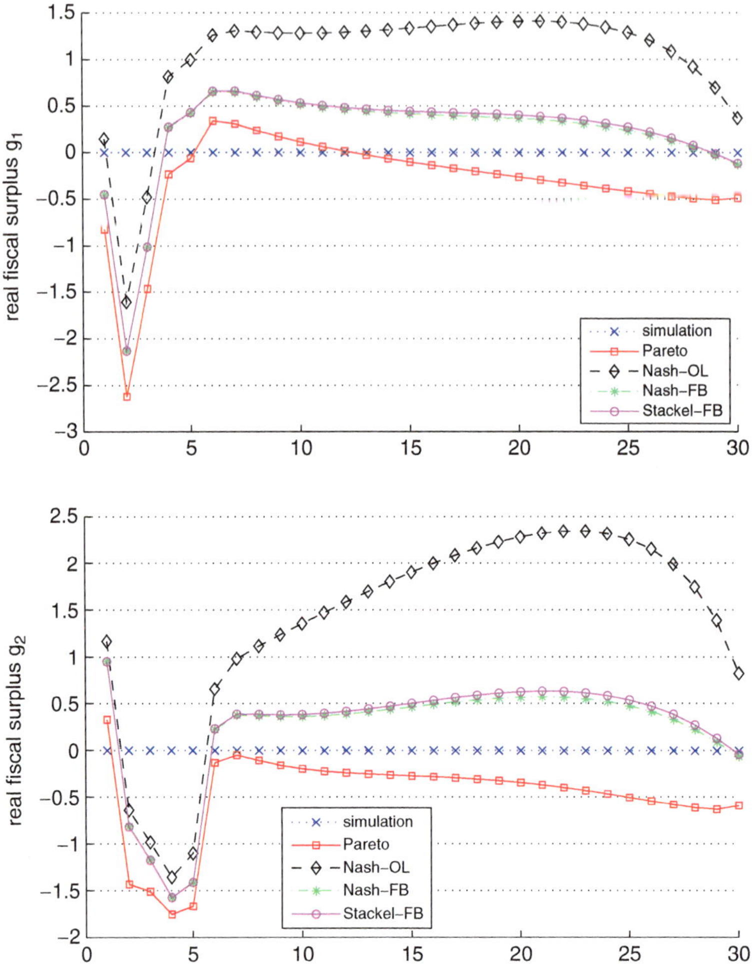

Fig. 2.2 Country i's fiscal surplus g_{it} (control variable) for $i = 1$ (core; *top*) and $i = 2$ (periphery; *bottom*)

rates to values very close to zero, but with the appearance of the supply side shock, inflation rates go up and reach 3% for country 1 and nearly 6% for country 2 in the non-controlled baseline scenario. Even more dramatic is the development of public debt. Without policy intervention it increases during the whole planning horizon and arrives at levels of 145% of GDP for country 1 (or core bloc) and 180% for country 2 (or periphery bloc), which shows a need for policy actions to stabilize the economies of the monetary union.

If the players want to react optimally to the demand and supply side shocks, their actions and their intensity depend on the presence or absence of cooperation. For example, optimal monetary policy has to be expansionary in all strategies, but

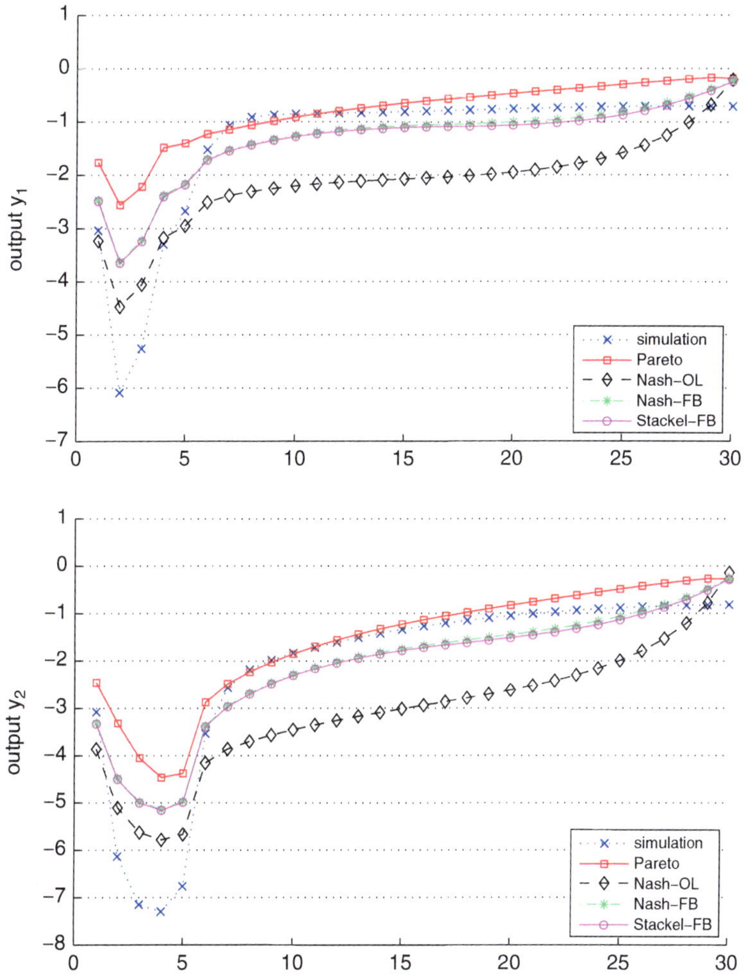

Fig. 2.3 Country i's output y_{it} for $i = 1$ (core; *top*) and $i = 2$ (periphery; *bottom*)

in the cooperative Pareto solution it is more active during the first eight periods. The open-loop Nash equilibrium solution, in contrast, is more or less constant during the whole optimization period, which causes the central bank to be less active at the beginning and relatively more active at the end of the optimization horizon.

With respect to fiscal policy, both countries are required to set expansionary actions and to create deficits in the first four periods in order to absorb the demand side shock. After that a trade-off occurs and the governments need to take care of the financial situation and to produce primary surpluses. The only exception is the cooperative Pareto solution: cooperation between the countries and the central bank

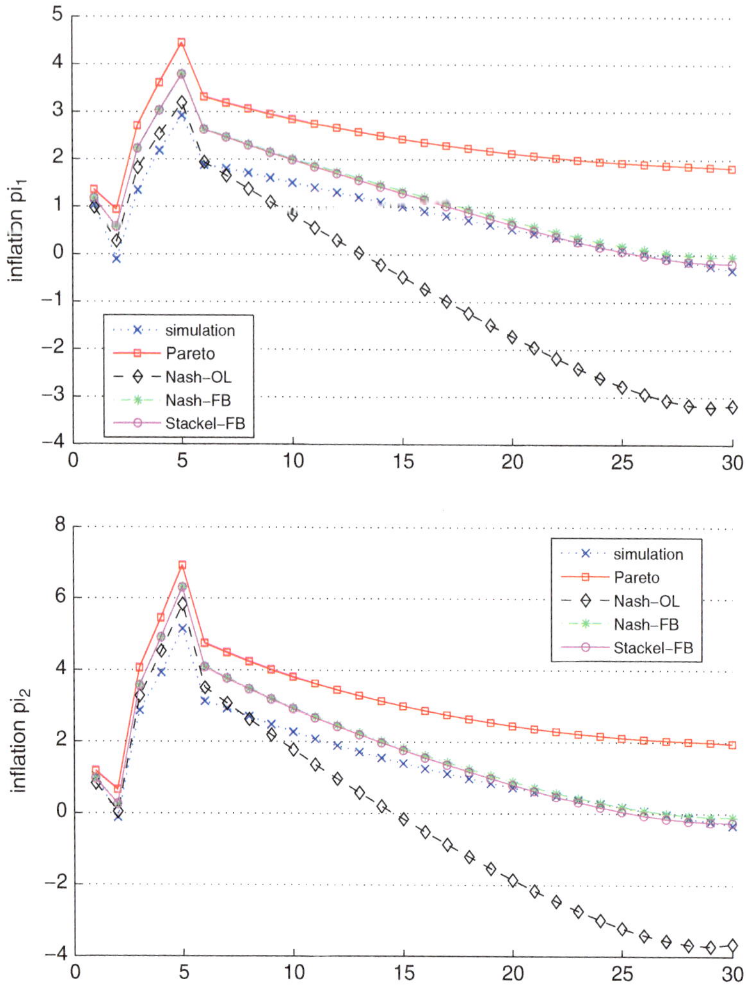

Fig. 2.4 Country i's inflation rate π_{it} for $i = 1$ (core; *top*) and $i = 2$ (periphery; *bottom*)

(which runs an expansionary monetary policy) and the high inflation means that the balance of public finances can be held close below zero. Even so the countries are able to stabilize and to bring down their public debts to the targeted values.

The non-cooperative Nash feedback and Stackelberg feedback solutions give very similar results. In comparison with a Pareto-optimal solution, the central bank acts less actively and the countries run more active fiscal policies (except during the negative demand shock). As a result, output and inflation are slightly below the values achieved in a cooperative solution, and public debt is slightly above.

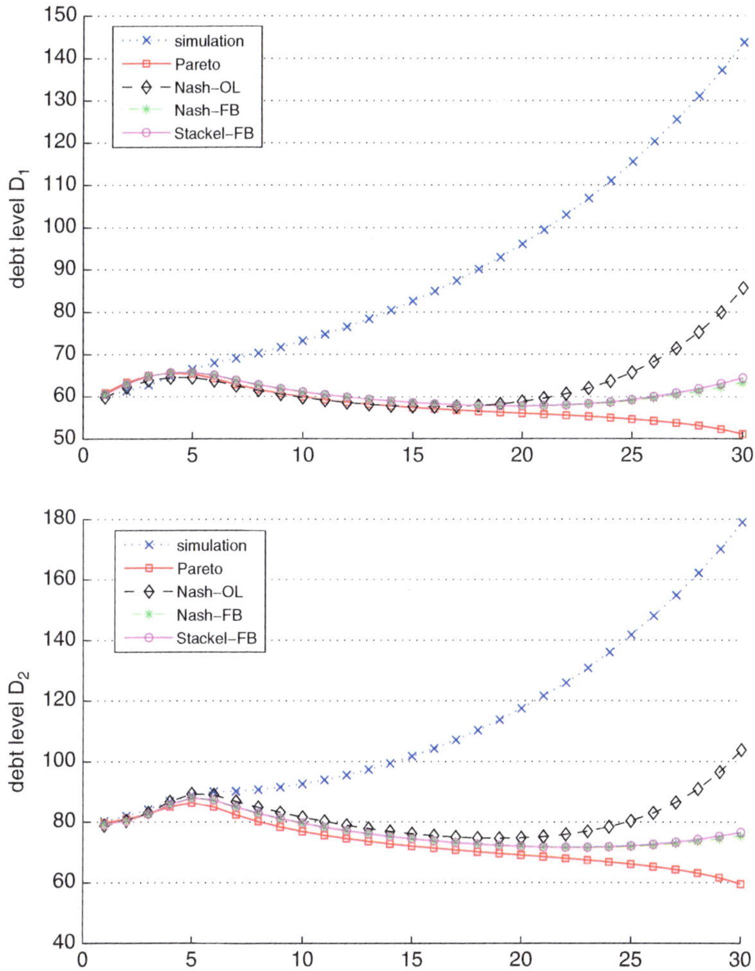

Fig. 2.5 Country i's debt level D_{it} for $i = 1$ (core; *top*) and $i = 2$ (periphery; *bottom*)

2.5 Concluding Remarks

In this paper we show the framework of the OPTGAME3 algorithm which allows us to find approximate solutions for nonlinear-quadratic dynamic tracking games. The algorithm was programmed in C# and MATLAB and includes the cooperative Pareto-optimal solution and non-cooperative Nash and Stackelberg equilibrium solutions. The applicability of the algorithm was shown using the MUMOD1 model, a small stylized nonlinear two-country macroeconomic model of a monetary union. We analyse the interaction between fiscal (governments) and monetary (common central bank) policy makers. By applying a dynamic game approach to a simple macroeconomic model of a two-country monetary union, we obtain some insights into the design of economic policies facing negative asymmetric shocks on the demand and the supply side. The monetary union is assumed to be asymmetric in the sense of consisting of a core with less initial public debt and a periphery with higher initial public debt, which is meant to reflect the situation in the EMU.

Our model implies that optimal and equilibrium policies of both the governments and the common central bank are counter-cyclical during the immediate influence of the demand shock but not afterwards. The later occurrence of a negative supply side shock increases the inflation rates and supports the countries in reducing their public debts. In the case of the Pareto solution, this leads to the situation that the countries can reduce their debts even without strongly restrictive fiscal policies. Taken together, the two negative shocks worsen the economic situation in the monetary union and produce growth rates of real output below the natural or long-run growth rate. We also show that the cooperative Pareto solution gives the best response to these shocks especially regarding output and public debt results.

Acknowledgements An earlier version of this paper was presented at the 15th ISDG Symposium on Dynamic Games and Applications. Thanks are due to very helpful comments and suggestions by participants at this symposium and by an anonymous referee.

References

van Aarle B, Bartolomeo GD, Engwerda J, Plasmans J (2002) Monetary and fiscal policy design in the EMU: An overview. Open Econ Rev 13(4):321–340

Başar T, Olsder G (1999) Dynamic noncooperative game theory, 2nd edn. SIAM, Philadelphia

Behrens D, Neck R (2007) OPTGAME: an algorithm approximating solutions for multi-player difference games. In: Elleithy K (ed) Advances and innovations in systems, computing sciences and software engineering. Springer, Dordrecht, pp 93–98

Blueschke D, Neck R (2011) "Core" and "periphery" in a monetary union: A macroeconomic policy game. Int Adv Econ Res 17:334–346

Chow GC (1975) Analysis and control of dynamic economic systems. Wiley, New York

Dockner E, Jorgensen S, Long N, Sorger G (2000) Differential games in economics and management science. Cambridge University Press, Cambridge

Engwerda J, Boldea O, Michalak T, Plasmans J, Salmah S (2012) A simulation study of an ASEAN monetary union. Econ Model 29(5):1870–1890

Haber G, Neck R, McKibbin W (2002) Global implications of monetary and fiscal policy rules in the EMU. Open Econ Rev 13(4):363–379

Hager M, Neck R, Behrens D (2001) Solving dynamic macroeconomic policy games using the algorithm OPTGAME 1.0. Optim Control Appl Methods 22(5–6):301–332

Hughes Hallett AJ (1986) Autonomy and the choice of policy in asymmetrically dependent economies: An investigation of the gains from international policy co-ordination. Oxf Econ Pap 38(3):516–544

Levine PL, Brociner A (1994) Fiscal policy coordination and EMU: A dynamic game approach. J Econ Dyn Control 18(3–4):699–729

Miller M, Salmon M (1985) Policy coordination and dynamic games. In: Buiter WH, Marston RC (eds) International economic policy coordination. Cambridge University Press, Cambridge, pp 184–213

Neck R, Behrens D (2004) Macroeconomic policies in a monetary union: A dynamic game. Cent Europ J Oper Res 12(2):171–186

Neck R, Behrens D (2009) A macroeconomic policy game for a monetary union with adaptive expectations. Atl Econ J 37(4):335–349

Petit M (1990) Control theory and dynamic games in economic policy analysis. Cambridge University Press, Cambridge

Plasmans J, Engwerda J, van Aarle B, Bartolomeo GD, Michalak T (2006) Dynamic modeling of monetary and fiscal cooperation among nations. Springer, New York

Pohjola M (1986) Applications of dynamic game theory to macroeconomics. In: Başar T (ed) Dynamic games and applications in economics. Springer, Berlin, pp 103–133

Chapter 3
Dynamic Programming Approach to Aircraft Control in a Windshear

Nikolai D. Botkin and Varvara L. Turova

Abstract Application of a dynamic programming method to the problem of aircraft control during take-off in a windshear is considered. A simplified four-dimensional model of the aircraft dynamics is used, and stable numerical algorithms for solving Hamilton–Jacobi–Bellman–Isaacs equations arising from differential games with state constraints are utilized for the design of feedback controls.

Keywords Takeoff control • Conflict control problems • State constraints • Hamilton–Jacobi–Bellman–Isaacs equations • Viscosity solutions • Numerical approximations

3.1 Introduction

Many aircraft accidents occur due to severe windshears such as microbursts. The microburst appears when a descending air flow hits the earth surface. This phenomenon is especially dangerous for aircrafts passing the microburst zone during the landing or take-off, because quick changes of the wind velocity occur at relatively low altitudes. Therefore, the development of automatic controls that can counteract the sudden microburst attack is of great importance.

Papers of Miele et al. (1986a,b), Chen and Pandey (1989), Leitmann and Pandey (1990, 1991), and Leitmann et al. (1993) address aircraft control during take-off in the presence of windshears. In works of Miele et al. (1986a,b), the wind velocity field is assumed to be known. Open-loop controls are constructed in Miele et al. (1986b) by solving appropriate optimization problems. In Miele et al. (1986a), local information about the wind is used for the construction of feedback controls. More realistic situation assumes that the wind velocity field is not known at all. Having

N.D. Botkin (✉) • V.L. Turova
Technische Universität München, Boltzmannstr. 3, 85747 Garching bei München, Germany
e-mail: botkin@ma.tum.de; turova@ma.tum.de

V. Křivan and G. Zaccour (eds.), *Advances in Dynamic Games*, Annals of the International Society of Dynamic Games 13, DOI 10.1007/978-3-319-02690-9_3,

this in mind, different variants of feedback controls are proposed in papers of Chen and Pandey (1989), Leitmann and Pandey (1990, 1991), and Leitmann et al. (1993). In Chen and Pandey (1989), the design of a feedback robust control strategy is based on the construction of an appropriate Lyapunov function. The robust control theory is used in Leitmann and Pandey (1990) to develop feedback controls stabilizing the relative path inclination, and in Leitmann and Pandey (1991) and Leitmann et al. (1993) for the design of feedback controls stabilizing the climb rate. The approach based on differential game theory advanced in books of Krasovskii and Subbotin (1974, 1988) (the second book is a revised translation of the first, original, one) is applied in papers of Botkin et al. (1993) and Turova (1991) to counteract to unknown wind disturbances. Reasonable feedback control algorithms based on solving appropriate linear differential games are suggested.

In this paper, feedback controls that are effective against microbursts are designed using dynamic programming techniques. Both the case of known wind velocity fields and the case of unknown wind disturbances are considered. Our method is based on numerical solving Hamilton–Jacobi equations arising from suitable nonlinear differential games of fixed time duration with state constraints.

3.2 Model Equations

We use a simplified model of the aircraft dynamics assuming the motion in the vertical plane (see Miele et al. 1986a,b). The following four ordinary differential equations governing the aircraft relative velocity V, the relative path inclination γ, the horizontal distance x, and the altitude h are considered:

$$\begin{aligned} m\dot{V} &= T\cos(\alpha+\delta) - D - mg\sin\gamma - m\dot{W}_x\cos\gamma - m\dot{W}_h\sin\gamma, \\ mV\dot{\gamma} &= T\sin(\alpha+\delta) + L - mg\cos\gamma + m\dot{W}_x\sin\gamma - m\dot{W}_h\cos\gamma, \end{aligned} \tag{3.1}$$

$$\begin{aligned} \dot{x} &= V\cos\gamma + W_x, \\ \dot{h} &= V\sin\gamma + W_h. \end{aligned} \tag{3.2}$$

Here, α is the attack angle; W_x and W_h are the longitudinal and vertical components of the wind velocity, respectively; g is the acceleration of gravity; m the aircraft mass; δ the thrust inclination; T, D, and L are the thrust, the drag, and the lift forces, respectively, defined as follows:

$$T = A_0 + A_1V + A_2V^2,$$

$$D = \frac{1}{2}C_D\rho SV^2, \quad C_D = B_0 + B_1\alpha + B_2\alpha^2,$$

$$L = \frac{1}{2}C_L\rho SV^2, \quad C_L = \begin{cases} C_0 + C_1\alpha, & \alpha \le \alpha_{**} \\ C_0 + C_1\alpha + C_2(\alpha-\alpha_{**})^2, & \alpha \in [\alpha_{**}, \alpha_*]. \end{cases}$$

For simplicity, the coefficients A_i, B_i, C_i, $i = 0, 1, 2$, are assumed to be constants; α_* and α_{**} are given constants; ρ is the air density; and S is the reference surface area.

The attack angle α is considered as the control parameter restricted by $0 \leq \alpha \leq \alpha_*$.

In the case where the longitudinal and vertical components, W_x and W_h, of the wind velocity field are supposed to be known, the derivatives $\dot{W}_x$ and $\dot{W}_h$ in (3.1) are computed as follows:

$$
\begin{aligned}
\dot{W}_x &= \frac{\partial W_x}{\partial x}(V \cos\gamma + W_x) + \frac{\partial W_x}{\partial h}(V \sin\gamma + W_h), \\
\dot{W}_h &= \frac{\partial W_h}{\partial x}(V \cos\gamma + W_x) + \frac{\partial W_h}{\partial h}(V \sin\gamma + W_h).
\end{aligned}
\tag{3.3}
$$

To simulate wind velocity fields, two different models of microburst are used.

3.2.1 Microburst Model 1

The model equations are taken from Chen and Pandey (1989):

$$
W_x = \begin{cases} -k, & x \leq a \\ -k + 2k(x-a)/(b-a), & a \leq x \leq b \\ k, & x \geq b, \end{cases}
$$

$$
W_h = \begin{cases} 0, & x \leq a \\ -k(h/h_*)(x-a)/(c-a), & a \leq x \leq c \\ -k(h/h_*)(b-x)/(b-c), & c \leq x \leq b \\ 0, & x \geq b, \end{cases}
$$

where a and b are the onset and termination of windshear, respectively, $c = (a+b)/2$, and h_* is a fixed constant. The parameter k defines the intensity of the microburst.

An example of wind streamlines generated by this model is shown in Fig. 3.1.

3.2.2 Microburst Model 2

The second model of microburst is the double vortex model (see Fig. 3.2) taken from Leitmann and Pandey (1990). Two cores of radius R located symmetric about the vertical line $x = 1500$ are considered. The vortex motion of air about the centers of

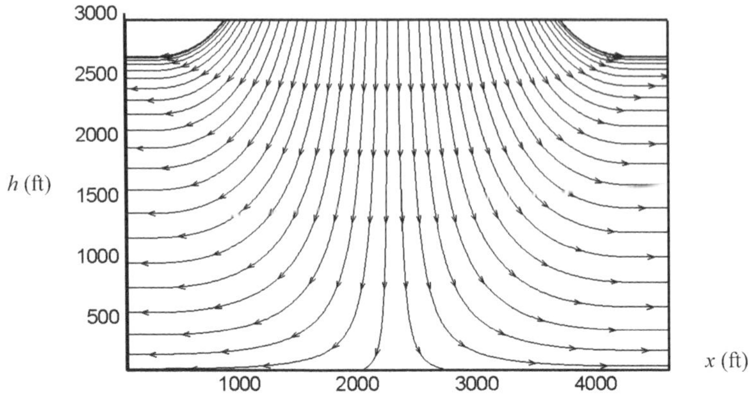

Fig. 3.1 Example of wind streamlines generated by microburst model 1

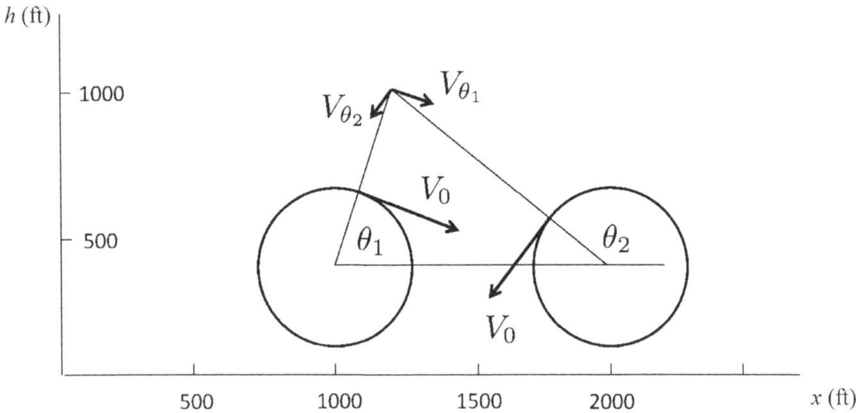

Fig. 3.2 Double vortex microburst model

the cores is assumed. Inside of each core, the tangential speed, V_θ, of wind increases linearly from zero at the center to a maximum value V_0 at $r = R$, where r is the distance from the core center. Outside of the core, V_θ decreases in inverse proportion to the distance r. In polar coordinate system located at the core center, the tangential speed of wind is

$$V_\theta = \begin{cases} V_0\, r/R, \; 0 \le r \le R, \\ V_0\, R/r, \; r > R. \end{cases}$$

At every spatial point (x, h), the horizontal (resp. vertical) component W_x (resp. W_h) of the wind velocity is computed as the projection of the vector $V_{\theta_1} + V_{\theta_2}$ (see Fig. 3.2) to the axis x (resp. h). The resulting wind velocity field is shown in Fig. 3.3.

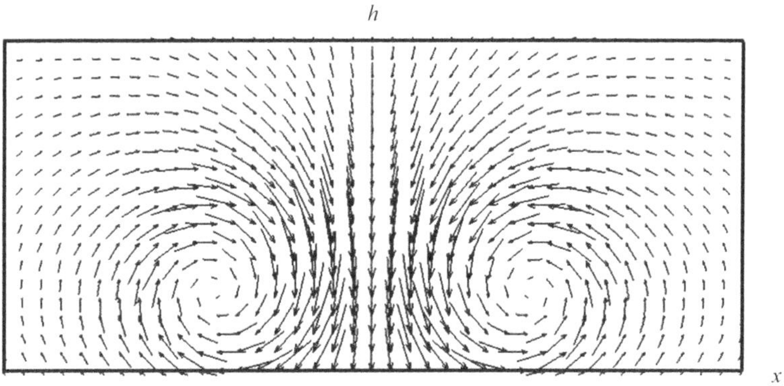

Fig. 3.3 Example of a wind velocity field generated by microburst model 2

3.3 Problem Formulations

Three variants of the problem statement are considered.

P1. The objective of the control α in system (3.1)–(3.2) is to maximize the performance index

$$J = \int_{t_0}^{t_f} \Big(V(t) \sin \gamma(t) + W_h\big(x(t), h(t)\big)\Big) dt \tag{3.4}$$

and to satisfy the state constraint

$$h(t) \geq 0, \quad t_0 \leq t \leq t_f, \tag{3.5}$$

where t_0 and t_f are the start and final times of the process, respectively. Notice that the integrand in (3.4) is the aircraft climb rate so that the functional expresses the altitude at the time t_f. The wind velocity field is supposed to be known and described by microburst model 1 or 2.

In this variant, the full four-dimensional nonlinear optimal control problem (3.1)–(3.5) is numerically solved. Notice that indeed the performance index $-J$ is to be minimized by the control α to fit the problem statement to the numerical method given in Sect. 3.4.

P2. In the second variant, a two-dimensional differential game is designed in the following manner. The functions $W_x(x, h)$ and $W_h(x, h)$ are supposed to be known and defined by microburst model 1 or 2. The horizontal distance and the altitude are determined from the assumption that the aircraft moves with a constant velocity V_r along a straight line with an inclination angle γ_v, i.e.

$$x = V_r \cos \gamma_v \cdot t, \quad h = V_r \sin \gamma_v \cdot t. \tag{3.6}$$

The expressions (3.6) are substituted into the functions $\dot{W}_x$ and $\dot{W}_h$ defined by the relations (3.3), and then the resulting expressions for $\dot{W}_x$ and $\dot{W}_h$, denoted by $\dot{W}_x|_{(3.6)}$ and $\dot{W}_h|_{(3.6)}$, are substituted into (3.1). This yields a two-dimensional differential game

$$\dot{V} = f_1(t, V, \gamma, \alpha, \gamma_v), \quad \dot{\gamma} = f_2(t, V, \gamma, \alpha, \gamma_v),$$

where

$$f_1 = m^{-1}\big(T\cos(\alpha+\delta) - D - mg\sin\gamma - m\dot{W}_x|_{(3.6)}\cos\gamma - m\dot{W}_h|_{(3.6)}\sin\gamma\big),$$
$$f_2 = (mV)^{-1}\big(T\sin(\alpha+\delta) + L - mg\cos\gamma + m\dot{W}_x|_{(3.6)}\sin\gamma - m\dot{W}_h|_{(3.6)}\cos\gamma\big).$$

The attack angle α is, as before, the control parameter of the first player, the angle γ_v is considered as a disturbance controlled by the opposite player. It is assumed that $|\gamma_v - \gamma_r| \le \gamma_*$, where γ_r is a reference value of the path inclination, and γ_* is a given bound. The objective of the first player is to minimize the functional

$$J = \int_{t_0}^{t_f} \big(V(t)\sin\gamma(t) - V_r\sin\gamma_r\big)^2 dt, \tag{3.7}$$

whereas the aim of the disturbance is the opposite. The functional (3.7) measures the deviation of the relative vertical velocity from a reference value.

P3. In the third variant of the problem statement, the wind velocity field is supposed to be unknown. We use the idea to control the climb rate, see Leitmann et al. (1993). The computation of $\ddot{h}$ yields

$$\ddot{h} = \frac{T}{m}[\cos(\alpha+\delta)\sin\gamma + \sin(\alpha+\delta)\cos\gamma] - \frac{D}{m}\sin\gamma + \frac{L}{m}\cos\gamma - g. \tag{3.8}$$

Substituting the expressions for $\sin\gamma$ and $\cos\gamma$ given by (3.2) into (3.8) and introducing new variable $z = \dot{h}$ yield the system

$$\dot{h} = z,$$
$$\dot{z} = \Big[\frac{T}{m}\cos(\alpha+\delta) - \frac{D}{m}\Big]\frac{z - W_h}{V} + \Big[\frac{T}{m}\sin(\alpha+\delta) + \frac{L}{m}\Big]\sqrt{1 - \Big(\frac{z - W_h}{V}\Big)^2} - g. \tag{3.9}$$

The control parameter of the first player is, as before, the attack angle α. The second player has the vertical wind velocity W_h at his disposal. To avoid the extension of system (3.9) by adding an equation for V, this variable is placed at the disposal of the second player too, which is in accordance with the guaranteed control concept. Moreover, such a formulation is reasonable because the relative velocity V is strongly affected by the wind velocity.

The objective of the first player is to minimize the payoff functional (3.7), the aim of the second player is the opposite. The bounds on the control parameters are the following:

$$0 \le \alpha \le \alpha_*, \quad V \in [V_r - 50, V_r], \quad W_h \in [-100, 0]. \tag{3.10}$$

Notice that only negative deviations of V and W_h from their reference values are assumed in (3.10), because negative deviations are mostly dangerous, whereas positive ones are favorable.

3.4 Numerical Method

Let us outline solution methods for problems P1–P3. The description will be given in terms of general nonlinear differential games with integral payoff functionals and state constraints.

3.4.1 Viscosity Solutions

Consider the following differential game:

$$\dot{x} = f(t, x, u, v), \tag{3.11}$$

where $x := (x_1, \ldots, x_n) \in \mathbb{R}^n$ is the state vector, u and v are control parameters of the first and second player, respectively, restricted as

$$u \in P \subset \mathbb{R}^p, \quad v \in Q \subset \mathbb{R}^q. \tag{3.12}$$

Here, P and Q are given compacts. The game starts at $t_0 \in [0, t_f]$ and finishes at t_f. The objective of the first player (control u) is to minimize the functional

$$J(x(\cdot)) = \int_{t_0}^{t_f} \sigma\big(t, x(t)\big) dt, \tag{3.13}$$

where $\sigma : [0, t_f] \times \mathbb{R}^n \to \mathbb{R}$ is a given function. The objective of the second player (control v) is the opposite. Moreover, the trajectories should remain in a state constraint set N given by the relation

$$N := \{(t, x) : t \in [0, t_f], \ \theta(t, x) \le \epsilon\}, \tag{3.14}$$

where $\theta : [0, t_f] \times \mathbb{R}^n \to \mathbb{R}$ is a given function, and ϵ is a fixed (small) value comparable with expected values of the functional (3.13).

It is reasonable to formalize the game (3.11)–(3.14) using the concept of feedback strategies (see Krasovskii and Subbotin 1988). Nevertheless, the existence of the equilibrium in pure feedback strategies requires the fulfillment of the following saddle point condition:

$$\max_{v \in Q} \min_{u \in P} \langle p, f(t,x,u,v) \rangle = \min_{u \in P} \max_{v \in Q} \langle p, f(t,x,u,v) \rangle \qquad (3.15)$$

for any $p \in \mathbb{R}^n$, $(t,x) \in [0,t_f] \times \mathbb{R}^n$. Unfortunately, this relation does not hold for problems $\mathbf{P}_2$ and $\mathbf{P}_3$ formulated in Sect. 3.3. On the other hand, it is known that the equilibrium, the existence of the value function, can be achieved if one of the players uses feedback counter strategies, whereas the other applies pure feedback strategies (see Krasovskii and Subbotin 1988). In our case, the second player (wind) will be permitted to measure the current value of the attack angle ("future" values are not available). This meets the concept of guaranteeing aircraft control.

Coming back to the formal description, the first player uses feedback pure strategies which are arbitrary functions

$$\mathscr{P} : [0,t_f] \times \mathbb{R}^n \to P,$$

whereas the second player applies feedback counter strategies which are functions of the form:

$$\mathscr{Q}^c : [0,t_f] \times \mathbb{R}^n \times P \to Q.$$

For any initial position $(t_0, x_0) \in [0,t_f] \times \mathbb{R}^n$ and any strategies $\mathscr{P}$ and $\mathscr{Q}^c$, two functional sets $X_1(t_0, x_0, \mathscr{P})$ and $X_2(t_0, x_0, \mathscr{Q}^c)$ are defined (see Krasovskii and Subbotin 1988).

The set $X_1(t_0, x_0, \mathscr{P})$ consists of all limits of Euler trajectories of (3.11) which are obtained when the first player chooses $u \equiv \mathscr{P}(t_i, x(t_i))$ on each interval $[t_i, t_{i+1})$ of partitions of $[t_0, t_f]$, and the second player uses admissible controls $v(t)$, $t \in [t_0, t_f]$. In doing that, all possible partitions whose diameter tends to zero and all admissible controls of the second player are exhausted. All Euler trajectories start at t_0 from the fixed initial state x_0.

The set $X_2(t_0, x_0, \mathscr{Q}^c)$ consists of all limits of Euler trajectories of (3.11) which are obtained when the first player uses admissible controls $u(t)$, $t \in [t_0, t_f]$, and the second player chooses $v \equiv \mathscr{Q}^c(t_i, x(t_i), u(t_i))$ on each interval $[t_i, t_{i+1})$ of partitions of $[t_0, t_f]$. All possible partitions whose diameter tends to zero and all admissible controls of the first player are exhausted. All Euler trajectories start at t_0 from the fixed initial state x_0.

We assume that the function f is bounded, uniformly continuous, and uniformly Lipschitzian in t and x on the set $[0,t_f] \times \mathbb{R}^n \times P \times Q$; the functions σ and θ are bounded and Lipschitzian in t, x.

It is proved in Krasovskii and Subbotin (1988) that the differential game (3.11)–(3.14) has a value function $\mathscr{V} : (t,x) \to \mathscr{V}(t,x)$ defined by the relation

$$\mathscr{V}(t,x) = \min_{\mathscr{P}} \max_{x(\cdot)\in X_1(t,x,\mathscr{P})} J(x(\cdot)) = \max_{\mathscr{Q}^c} \min_{x(\cdot)\in X_2(t,x,\mathscr{Q}^c)} J(x(\cdot)). \tag{3.16}$$

In other words, the upper value of the game coincides with the lower one for $(t,x) \in [0,t_f] \times \mathbb{R}^n$, if the second player uses counter strategies. The value function is bounded and Lipschitzian in t, x on $[0,t_f] \times \mathbb{R}^n$ (see Subbotin 1995; Subbotin and Chentsov 1981).

Define the Hamiltonian as follows in the case of counter strategies of the second player (see Krasovskii and Subbotin 1988):

$$H(t,x,p) = \min_{u\in P} \max_{v\in Q} \langle p, f(t,x,u,v)\rangle + \sigma(t,x) \tag{3.17}$$

and consider the Hamilton–Jacobi–Bellman–Isaacs equation

$$\mathscr{V}_t + H(t,x,\mathscr{V}_x) = 0, \quad \mathscr{V}(t_f,x) = 0. \tag{3.18}$$

It is a well-known fact in the theory of differential games that the value function is a viscosity solution of an appropriate Hamilton–Jacobi–Bellmann–Isaacs equation. In the case of state constraints, the following proposition, which is a particular case of Proposition 4.1 from Botkin et al. (2011a), is true.

Proposition 3.1. *A Lipschitz function $\mathscr{V}$ is the value function of differential game (3.11)–(3.14) if and only if:*

(i) for any $(t,x) \in [0,t_f] \times \mathbb{R}^n$, $\mathscr{V}(t_f,x) = 0$ and $\mathscr{V}(t,x) \ge \theta(t,x)$;
(ii) for any point $(s_0,y_0) \in [0,t_f] \times \mathbb{R}^n$ and any function $\varphi \in \mathbb{C}^1$ such that $\mathscr{V} - \varphi$ attains a local minimum at (s_0,y_0), the following inequality holds

$$\frac{\partial\varphi}{\partial t}(s_0,y_0) + H(s_0,y_0,\frac{\partial\varphi}{\partial y}(s_0,y_0)) \le 0; \tag{3.19}$$

(iii) for any point $(s_0,y_0) \in [0,t_f] \times \mathbb{R}^n$ such that $\mathscr{V}(s_0,y_0) > \theta(s_0,y_0)$ and any function $\varphi \in \mathbb{C}^1$ such that $\mathscr{V} - \varphi$ attains a local maximum at (s_0,y_0), the following inequality holds

$$\frac{\partial\varphi}{\partial t}(s_0,y_0) + H(s_0,y_0,\frac{\partial\varphi}{\partial y}(s_0,y_0)) \ge 0. \tag{3.20}$$

3.4.2 Finite-Difference Scheme

To compute viscosity solutions of (3.18) satisfying Proposition 3.1, the following finite difference scheme is applied.

Let $\tau, h_1, \ldots, h_n$ be time and space discretization step sizes, and F be an operator defined on continuous functions as

$$F(\mathscr{V};t,\tau)(x) = \min_{u\in P}\max_{v\in Q} \mathscr{V}\big(x+\tau f(t,x,u,v)\big) + \tau\sigma(t,x). \tag{3.21}$$

Put $\Lambda = t_f/\tau + 1$, $t_\ell = \ell\tau$, $\ell = 0,\dots,\Lambda$ and introduce the following notation:

$$\mathscr{V}^\ell(x_{i_1},\dots,x_{i_n}) = \mathscr{V}(t_\ell, i_1h_1,\dots,i_nh_n), \quad \theta^\ell(x_{i_1},\dots,x_{i_n}) = \theta(t_\ell, i_1h_1,\dots,i_nh_n).$$

Denote $h := (h_1,\dots,h_n)$ and $|h| = \max\{h_1,\dots,h_n\}$. Let $\mathscr{L}_h$ be an interpolation operator that maps grid functions to continuous functions and satisfies the estimate

$$\|\mathscr{L}_h[\tilde{\phi}] - \phi\| \le C\,|h|^2\,\|D^2\phi\| \tag{3.22}$$

for any smooth function ϕ. Here, $\tilde{\phi}$ is the restriction of ϕ to the grid, $\|\cdot\|$ the point-wise maximum norm, $D^2\phi$ the Hessian matrix of ϕ, and C is an independent constant.

Notice that estimate (3.22) is typical for interpolation operators (see, e.g., Mößner and Reif 2009). Roughly speaking, interpolation operators reconstruct the value and the gradient of interpolated functions, and therefore the expected error is of order $|h|^2$.

As an example, consider a multilinear interpolation operator constructed in the following way (see Botkin et al. 2011b).

Let $m \in \overline{1,2^n}$ be an integer, and $(j_1^m,\dots,j_n^m)$ the binary representation of m so that j_i^m is either 0 or 1. Thus, each multiindex $(j_1^m,\dots,j_n^m)$ represents a vertex of the unit cube in $\mathbb{R}^n$, and m counts the vertices. Introduce the following functions

$$\omega_m(x_1,\dots,x_n) = \prod_{i=1}^{n}(1-x_i)^{1-j_i^m}x_i^{j_i^m}, \quad m = 1,\dots,2^n. \tag{3.23}$$

Note that the i th member in the product (3.23) is either $1-x_i$ or x_i depending on the value of j_i^m. Consider a point $x = (x_1,\dots,x_n) \in \mathbb{R}^n$. Denote by $\underline{x}_i$ the lower and by $\overline{x}_i = \underline{x}_i + h_i$ the upper grid points of the i th axis such that $\underline{x}_i \le x_i \le \overline{x}_i$. Let $\tilde{\phi}_m$, $m = 1,\dots,2^n$ be the values of a grid function in the vertices of the n-brick $\prod_{i=1}^n[\underline{x}_i,\overline{x}_i]$ (the vertices are ordered in the same way as the vertices of the unit n-cube above). The multilinear interpolation of $\tilde{\phi}$ at $(x_1,\dots,x_n)$ is

$$\mathscr{L}_h[\tilde{\phi}](x) = \sum_{m=1}^{2^n}\tilde{\phi}_m\cdot\omega_m\left(\frac{x_1-\underline{x}_1}{h_1},\dots,\frac{x_n-\underline{x}_n}{h_n}\right).$$

Consider the following grid scheme:

$$\mathscr{V}^{\ell-1} = \max\left\{F\big(\mathscr{L}_h[\mathscr{V}^\ell];t_\ell,\tau\big),\theta^\ell\right\},\ \mathscr{V}^\Lambda = 0, \quad \ell = \Lambda,\Lambda-1,\dots,1. \tag{3.24}$$

Here, $F\big(\mathscr{L}_h[\mathscr{V}^\ell]; t_\ell, \tau\big)$ is a continuous function which is assumed to be restricted to the grid and then compared with the grid function θ^ℓ. Thus, the right-hand side of the first equation of (3.24) returns a grid function.

Theorem 3.1. *The time-space grid function obtained by (3.24) converges pointwise to the value function of the game (3.11)–(3.14) as τ and $|h|$ tend to zero under the condition $|h|/\tau \le C$, where C is any positive constant. The convergence rate is*

$$\max(\sqrt{\tau}, \sqrt{|h|}).$$

Proof. First, note that the problem (3.11)–(3.12) with the cost functional (3.13) and state constraint (3.14) is equivalent to the problem

$$\begin{aligned} \dot{x}_i &= f_i(t, x, u, v), \quad i \in \overline{1, n}, \\ \dot{x}_{n+1} &= \sigma(t, x) \end{aligned} \tag{3.25}$$

with the payoff functional

$$\max\Big\{x_{n+1}(t_f),\ \max_{t\in[t_0,t_f]} \theta\big(t, x(t)\big)\Big\}. \tag{3.26}$$

Really, if the value function of the game (3.25) and (3.26) at $(t_0, x_0, 0)$ is less than or equal to ϵ, then the first player guarantees the validity of the conditions

$$x_{n+1}(t_f) = \int_{t_0}^{t_f} \sigma\big(t, x(t)\big)dt \le \epsilon, \quad \theta\big(t, x(t)\big) \le \epsilon, \quad t \in [t_0, t_f],$$

independently on the behavior of the second player. Vice versa, if the guaranteed result of the first player in the game (3.11)–(3.13) is less than or equal to ϵ at (t_0, x_0) and the state constraint (3.14) holds, then the value function of the game (3.25) and (3.26) at $(t_0, x_0, 0)$ is less than or equal to ϵ. Thus, we deal with a particular case of a more general problem considered in Botkin et al. (2011a).

Second, the operator $F\big(\mathscr{L}_h[\cdot]; t_\ell, \tau\big)$ is monotone and possesses the following consistency (generator) property:

$$\left|\frac{F(\mathscr{L}_h[\tilde{\phi}], t, \tau)(x) - \phi(x)}{\tau} - H\big(t, x, D\phi(x)\big)\right| \le C_1 \cdot \|D^2\phi\|(\tau + |h|^2/\tau) \tag{3.27}$$

for every $\phi \in C_b^2(\mathbb{R}^n)$. The monotonicity of the operator $F\big(\mathscr{L}_h[\cdot]; t_\ell, \tau\big)$ holds due to the following monotonicity properties of the interpolation operator $\mathscr{L}_h$ and the minimax step operator $F(\cdot; t_\ell, \tau)$:

$$\begin{aligned} \mathscr{L}_h[\tilde{\phi}_1](x) &\ge \mathscr{L}_h[\tilde{\phi}_2](x), \quad x \in \mathbb{R}^n, \\ F(\phi_1; t_\ell, \tau)(x) &\ge F(\phi_2; t_\ell, \tau)(x), \quad x \in \mathbb{R}^n, \end{aligned}$$

where $\tilde{\phi}_1$ and $\tilde{\phi}_2$ are arbitrary grid functions such that $\tilde{\phi}_1(x^h) \geq \tilde{\phi}_2(x^h)$ for all grid nodes x^h, and ϕ_1 and ϕ_2 are arbitrary functions such that $\phi_1(x) \geq \phi_2(x)$ for all $x \in \mathbb{R}^n$.

The property (3.27) follows from inequality (3.22) and the boundedness of the function f which provide the estimates

$$|\mathscr{L}_h[\tilde{\phi}](x + f(t, x, u, v)) - \phi(x + \tau f(t, x, u, v))| \leq C\,|h|^2\,\|D^2\phi\|,$$

$$|\phi(x + \tau f(t, x, u, v)) - \phi(x) - \tau\langle D\phi(x), f(t, x, u, v)\rangle| \leq C'\tau^2\|D^2\phi\|$$

that hold for all $t \in [0, t_f]$, $x \in \mathbb{R}^n$, $u \in P$, and $v \in Q$. These estimates along with the definitions of the Hamiltonian H, see (3.17), and the operator $F(\cdot; t_\ell, \tau)$, see (3.21), prove (3.27).

Using the consistency property (3.27) and arguing as in Botkin et al. (2011a), we arrive at the estimates (5.12) and (5.18) of Botkin et al. (2011a) with $\tau + |h|^2/\tau$ instead of $|P|$. The assumption $|h|/\tau \leq C$ provides the proof of Theorem 3.1. □

Notice that, in contrast to the upwind operator presented in Botkin et al. (2011a), the grid scheme (3.24) does not require small values of τ compared to $|h|$. This advantage allows us to reduce the total number of time steps, which is very important when the time interval of the game is relatively large, and each time step is resource consuming.

3.4.3 Control Design

The control design proceeds in the following way. When computing the grid approximation of the value function, see (3.24), minimizing (maximizing) values of the control parameters of the first (second) players, see (3.21) are computed at each grid node and stored on a hard disk at each sampling time instant (in our case, only optimizing values of the attack angle are stored). In the simulation process, the control is computed as a weighted linear combination of the control values at the nodes of the grid cell in which the current four-dimensional state vector lies. The weights are determined on the base of relative coordinates of the state vector in the cell.

Thus, if the current state of the system at a time instant t_ℓ is x, the grid function u^ℓ with values

$$u^{\ell}_{i_1 i_2 \dots i_n} = \arg\min_{u \in P} \max_{v \in Q} \mathscr{V}(x_{i_1 i_2 \dots i_n} + \tau f(t_\ell, x_{i_1 i_2 \dots i_n}, u, v))$$

is extracted from the hard disc. The control $u(t_\ell, x)$ is computed as $\mathscr{L}_h[u^\ell](x)$.

3.5 Simulation Results

This section describes numerical results obtained by the application of the above-described grid method to problems **P1–P3**. Numerical values of the model coefficients correspond to Boeing-727 are taken from Chen and Pandey (1989).

The calculations are performed on a Linux SMP-computer with 8xQuad-Core AMD Opteron processors (Model 8384, 2.7 GHz) and shared 64 Gb memory. The programming language C with OpenMP (Open Multiprocessing) support is used. The efficiency of the parallelization is up to 80%.

For problem **P1** where we operate in the four-dimensional state space (x, h, V, γ), the grid size is $200 \times 200 \times 100 \times 20$. For problem **P2** (**P3**) with only two state variables V and γ (h and $\dot{h}$), the grid size is 2000×400 (1000×100). It should be stressed that the reduction of the grid size to 200×40 does not change the quality of the control. However, the runtime is less than one second in this case, which makes possible to develop diverse adaptive real-time control algorithms.

In Figs. 3.4 and 3.5, results of the simulation of system (3.1)–(3.2) with the control law obtained by solving problems **P1** and **P2** are presented. The wind velocity field is described by microburst model 1. The solid lines correspond to problem **P1**, whereas the dash and gray lines are related to problem **P2**. The horizontal axes measure time in seconds. The vertical axes in Fig. 3.4a–d measure the altitude h (feet), the aircraft relative velocity V (feet/s), the path inclination γ (deg), and the angle of attack α (deg), respectively. The value of the parameter k defining the intensity of the microburst is equal to 60. In Fig. 3.5a, b, the altitude and the angle of attack versus time are presented for a weaker windshear $(k = 50)$.

Figure 3.6 shows an example of the value function computed for problem **P3**.

Figure 3.7 shows simulation results for system (3.1)–(3.2). The control scheme is based on solving problem **P3**, and the wind velocity is generated by microburst model 2. Two different microburst intensities defined by the parameter V_0 are considered: $V_0 = 100$ and $V_0 = 140$. Figures 3.7a, b show the altitude and the aircraft velocity, respectively. The black lines correspond to $V_0 = 100$, the dash lines are related to the case $V_0 = 140$. In the case $V_0 = 140$, the realizations of the longitudinal and vertical wind velocities along the aircraft trajectory (Fig. 3.7c) and the realization of the (averaged) attack angle (Fig. 3.7d) are additionally presented. The averaging of the control is necessary because of practical infeasibility of a bang-bang control obtained by solving differential game (3.9)–(3.10).

Note that our simulation results are in a good agreement with those of Chen and Pandey (1989) where a robust take-off control based on Lyapunov's stability theory is designed for the aircraft dynamics given by (3.1)–(3.2). Besides, our results are in conformity with those of Turova (1991) where a control based on the computation of switch lines in an appropriate two-dimensional linear differential game is constructed.

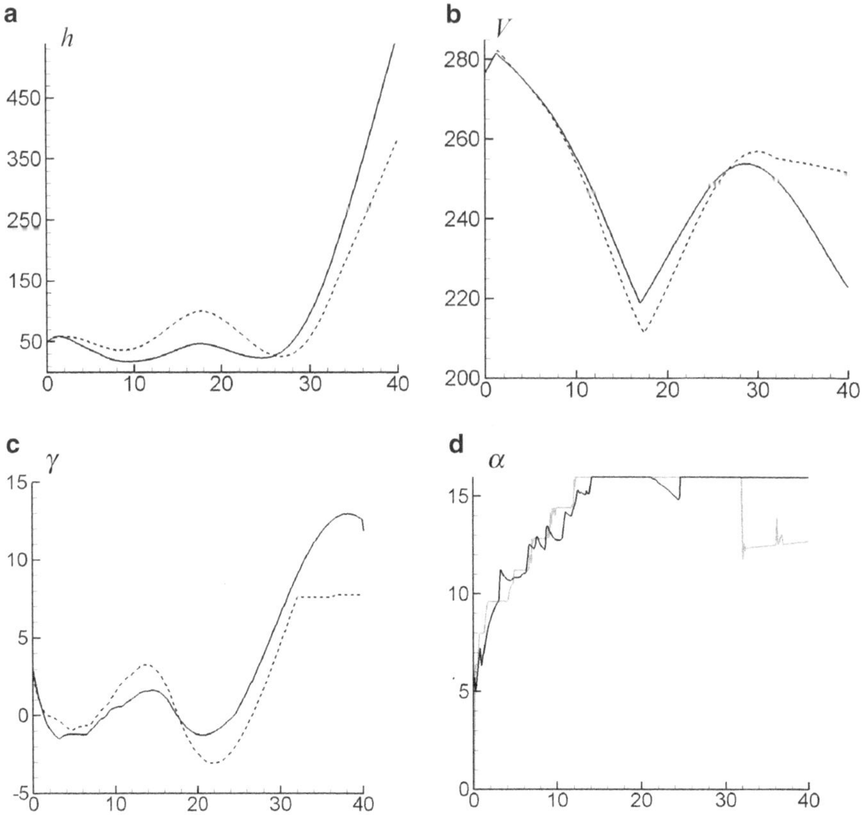

Fig. 3.4 Simulation results for problems **P1** (*solid black lines*) and **P2** (*dash or gray lines*); microburst intensity $k = 60$: (**a**) the altitude h (feet), (**b**) the aircraft relative velocity V (feet/s), (**c**) the path inclination γ (deg), (**d**) the angle of attack α (deg)

3.6 Conclusion

The current investigation shows that numerical methods can be successfully applied to nonlinear (up to four-dimensional) optimal control problems and differential games with terminal and non-terminal payoff functionals and state constraints. The paper demonstrates that the approach based on the direct solution of nonlinear problems yields control laws that are effective against severe wind disturbances during take-off. These results are competitive with known approaches based on robust control theory and numerical methods related to linearization and application of linear differential games. The future work will be concentrated on the effective numerical treatment of higher-dimensional nonlinear problems using sparse tensor representation of grid functions.

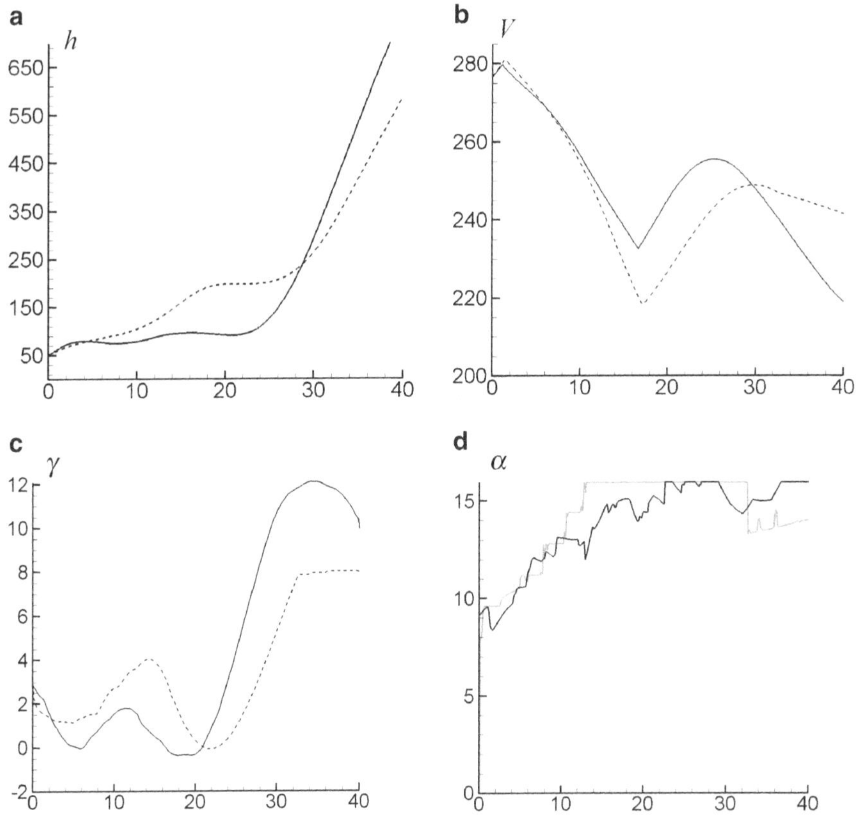

Fig. 3.5 Simulation results for problems **P1** (*solid black lines*) and **P2** (*dash or gray lines*); the microburst intensity $k = 50$: (**a**) the altitude h (feet), (**b**) the aircraft relative velocity V (feet/s), (**c**) the path inclination γ (deg), (**d**) the angle of attack α (deg)

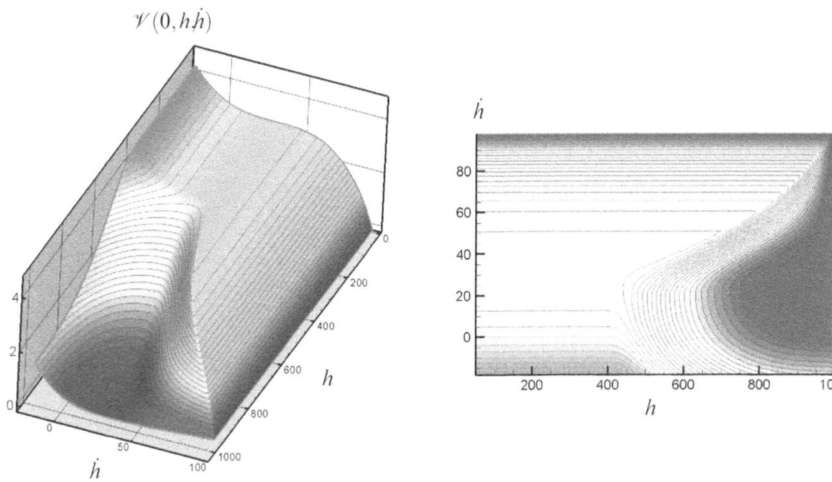

Fig. 3.6 Example of computed value function for problem **P3**

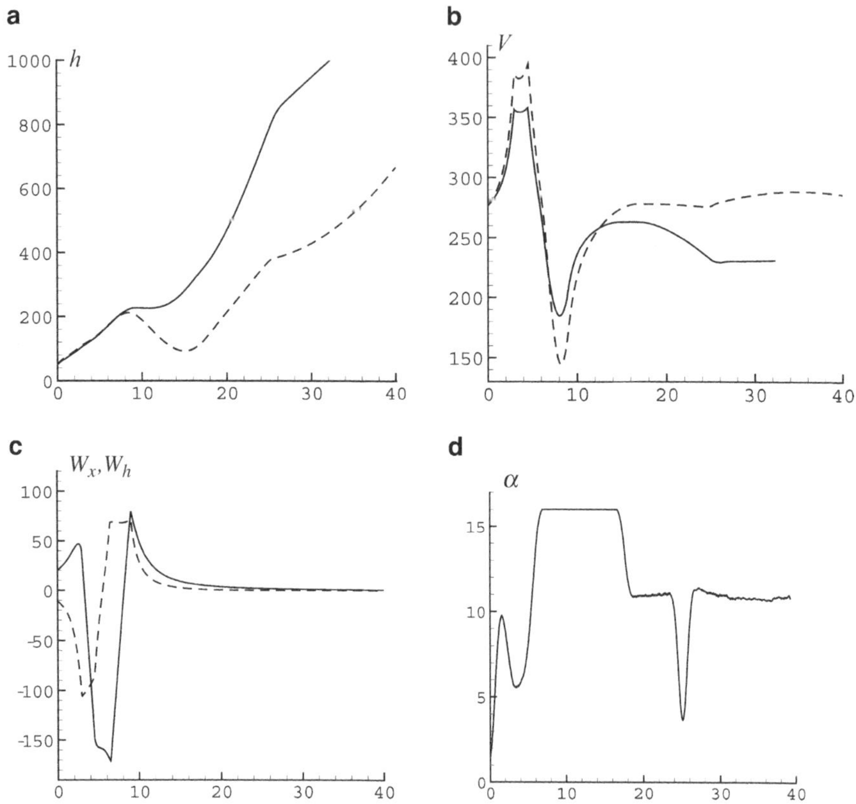

Fig. 3.7 Simulation results for problem **P3**; microburst intensities $V_0 = 100$ and $V_0 = 140$: (**a**) the altitude h (feet), (**b**) the aircraft relative velocity V (feet/s), (**c**) the wind velocities W_x and W_h (feet/s), *dashed and sold lines*, respectively (**d**) the angle of attack α (deg)

Acknowledgments The authors are grateful to anonymous reviewer for his constructive and helpful comments.

References

Botkin N, Kein V, Patsko V, Turova V, Zarkh M (1993) Differential games and aircraft control problems in the presence of wind disturbances. Izv Ross Akad Nauk, Tekhn Kibernet 1:68–76, translated as J Comput Syst Sci Int 32(3):110–117 (1994)

Botkin N, Hoffmann KH, Mayer N, Turova V (2011a) Approximation schemes for solving disturbed control problems with non-terminal time and state constraints. Analysis 31:355–379

Botkin N, Hoffmann KH, Turova V (2011b) Stable numerical schemes for solving Hamilton–Jacobi–Bellman–Isaacs equations. SIAM J Sci Comp 33(2):992–1007

Chen Y, Pandey S (1989) Robust control strategy for take-off performance in a windshear. Optim Control Appl Methods 10(1):65–79

Krasovskii N, Subbotin A (1974) Positional differential games. Nauka, Moscow, in Russian

Krasovskii N, Subbotin A (1988) Game-theoretical control problems. Springer, New York

Leitmann G, Pandey S (1990) Aircraft control under conditions of windshear. In: Leondes C (ed) Control and dynamic systems, vol 34. Academic Press, New York, pp 1–79

Leitmann G, Pandey S (1991) Aircraft control for flight in an uncertain environment: Takeoff in windshear. J Optim Theory Appl 70(1):25–55

Leitmann G, Pandey S, Ryan E (1993) Adaptive control of aircraft in windshear. Int J Robust Nonlinear Control 3:133–153

Miele A, Wang T, Melvin W (1986a) Guidance strategies for near-optimum take-off performance in a windshear. J Optim Theory Appl 50(1):1–47

Miele A, Wang T, Melvin W (1986b) Optimal take-off trajectories in the presence of windshear. J Optim Theory Appl 49(1):1–45

Mößner B, Reif U (2009) Error bounds for polynomial tensor product interpolation. Computing 86(2–3):185–197

Subbotin A (1995) Generalized solutions of first order PDEs: The dynamical optimization perspective. Birkhäuser, Boston

Subbotin A, Chentsov A (1981) Optimization of guaranteed result in control problems. Nauka, Moscow, in Russian

Turova V (1991) Take-off control in a windshear. Tech. Rep., Preprint of the Inst Math Mech, Ekaterinburg

Chapter 4
Existence of Optimal Controls for a Bi-Level Optimal Control Problem

Dean A. Carlson

Abstract In this paper we consider a class of bi-level optimal control problems. Both the upper and lower level problems are formulated as ordinary optimal control models of Lagrange type. Our goal is to formulate and prove a general existence theorem for an optimal solution based on classical compactness, convexity and seminormality conditions originating in the work of L. Tonelli for ordinary calculus of variations problems and extended to optimal control problems by L. Cesari, R.T. Rockafellar, L. Berkovitz and others. A distinguishing feature of our result is that we do not require the lower level problem to have a unique optimal solution corresponding to each admissible strategy of the upper level problem.

Keywords Optimal control • Bi-level optimization • Existence • Leader-follower • Game theory • Dynamic optimization

4.1 Introduction

The existence of optimal controls for problems involving ordinary differential equations has a rich history going back to the beginning of the twentieth century with the seminal work of L. Tonelli in the calculus of variations. This was the beginning of what became known as the direct method which, for the first time, adapted the Weierstrass theorem for continuous functions defined on compact sets to lower semicontinuous functionals defined on a function space over a compact set. Prior to that time, the theory of the calculus of variations primarily consisted of necessary conditions for optimality and relatively few sets of sufficient conditions for optimality. The existence of a solution to a specific problem relied on an

D.A. Carlson (✉)
American Mathematical Society, Mathematical Reviews, 416 Fourth Street,
Ann Arbor, MI 48103, USA
e-mail: dac@ams.org

V. Křivan and G. Zaccour (eds.), *Advances in Dynamic Games*, Annals of the International Society of Dynamic Games 13, DOI 10.1007/978-3-319-02690-9_4,

indirect method in which a candidate solution was first obtained using the necessary conditions and then its existence was established by showing it satisfied a sufficient condition. Tonelli's direct method did not rely on the necessary conditions and did not require a candidate solution. Instead the method consisted in showing that a minimizing sequence of feasible trajectories forms a compact set in an appropriate topology and that the objective functional is lower semicontinuous with respect to that same topology. This type of result depends on a delicate balance between these two goals to choose the correct topology. Namely, if the topology is too strong, the semicontinuity properties will hold, but the conditions ensuring the compactness properties will not hold. Conversely, if the topology is too weak, the opposite problem arises. Fortunately since Tonelli's time, this problem has been carefully considered and a lot is known concerning this issue. In particular, for the purposes considered here, it is important to remark that with the advent of optimal control theory in the 1950s, Tonelli's existence theory was extended to allow for a corresponding theory for problems involving control functions. This theory was developed primarily by L. Cesari and his students, but also had significant contributions given by C. Olech, R.T. Rockafellar, L. Berkovitz, E. Balder, and many others. To see these contributions as well as earlier ones in the calculus of variations the reader is directed to the book by Cesari (1983). One distinguishing feature between the existence theory of calculus of variations and the corresponding theory for optimal control is the presence of the control functions. The desired optimal control is obtained via a measurable selection theorem, once a convergent sequence of minimizing trajectories is obtained. That is, there is no explicit convergence of the optimal controls.

In this paper we adapt Cesari's techniques to establish the existence of optimal controls for a class of bi-level optimal control problems in which both the upper and lower level problems are optimal control problems of Lagrange-type involving ordinary differential equations. The lower level problem may be viewed as a parametric control problem which depends on an admissible trajectory y of the upper level problem. On the other hand, the upper level problem has a functional dependence on the lower level problem in that the control system and cost objective depend explicitly on the optimal trajectory of the lower level problem. These problems have also been investigated as Stackelberg or hierarchal differential games in which the upper level problem is considered to be "controlled" by the leader and the lower level problem is "controlled" by the follower.

4.2 The Problem and Basic Hypotheses

The problem we consider is a bi-level optimal control problem in which the upper level problem is an ordinary optimal control problem that depends on the optimal state trajectory of a lower level ordinary optimal control problem which in turn depends on the state trajectory of the upper level problem. More specifically, we describe the upper level problem (UP) as follows:

$$\text{minimize} \quad G(y,x,v) := \int_a^b g_0(t,y(t),x(t),v(t))\,dt$$

subject to

$$\begin{aligned} \dot{y}(t) &= g(t,y(t),x(t),v(t)), \quad \text{a.e. } a \le t \le b, \\ (y(a),y(b)) &\in \mathscr{B}_Y, \\ v(t) &\in V(t,y(t),x(t)), \quad \text{a.e. } a \le t \le b, \\ (t,y(t),x(t)) &\in \mathscr{A}_Y, \quad \text{for } a \le t \le b, \\ x &\in \Gamma(y), \end{aligned} \tag{4.1}$$

where $\mathscr{A}_Y = [a,b] \times A_Y \subset \mathbb{R} \times \mathbb{R}^n \times \mathbb{R}^k$ is closed, $\mathscr{B}_Y \subset \mathbb{R}^{2n}$ is closed and bounded, $V : \mathscr{A}_Y \to 2^{\mathbb{R}^m}$ is a set-valued mapping with a closed graph and (g_0, g) are given functions satisfying some appropriate hypotheses given in the next section. The notation $x \in \Gamma(y)$ indicates that, for a given continuous function $y : [a,b] \to \mathbb{R}^n$, $x : [a,b] \to \mathbb{R}^k$ is an optimal state trajectory of the following lower level problem (LP):

$$\text{minimize} \quad F(y,x,u) := \int_a^b f_0(t,y(t),x(t),u(t))\,dt$$

subject to

$$\begin{aligned} \dot{x}(t) &= f(t,x(t),u(t)), \quad \text{a.e. } a \le t \le b, \\ (x(a),x(b)) &\in \mathscr{B}_X, \\ u(t) &\in U(t,x(t)), \quad \text{a.e. } a \le t \le b, \\ (t,x(t)) &\in \mathscr{A}_X, \quad \text{for } a \le t \le b, \end{aligned} \tag{4.2}$$

where $\mathscr{A}_X = [a,b] \times A_X \subset \mathbb{R} \times \mathbb{R}^k$ is closed, $\mathscr{B}_X \subset \mathbb{R}^{2k}$ is compact, $U : \mathscr{A}_X \to 2^{\mathbb{R}^l}$ is a set-valued mapping with a closed graph and (f_0, f) are given functions satisfying some appropriate hypotheses given in the next section.

For a fixed continuous function y, the lower level problem (4.2) is a standard optimal control problem and a variety of existence theorems exist. On the other hand, for the upper level problem (4.1), there are coupled constraints that involve the optimal state trajectory x of the corresponding lower level problem (4.2), which leads to difficulties when one tries to apply the direct method.

Remark 4.1. We note that the lower problem (4.2) depends on the trajectory y of the upper problem (4.1) only through the objective functional. Further, we notice that in the upper problem we have included constraints that depend on the optimal state trajectory of the lower level problem. This seems to indicate that the Leader reacts to the Follower's optimal strategy. Formally, this constraint could also have been placed as a constraint in the lower problem, where one could view it as the

Follower reacting to the Leader. For reasons to become clear in the proof of our result, we chose the former viewpoint.

Remark 4.2. Another interpretation of the above bi-level problem as a hierarchical game is to view the Follower as being "behind the scenes" in that he/she wishes to force the perceived leader to a desired outcome. In this case the goal of the lower level problem is to use a strategy that is best for himself/herself so that the perceived leader is encouraged to select their strategy to insure that it, combined with the strategy of the lower level problem, solves the bi-level optimal control problem. A potential application of such a model is when the lower level problem represents a lobbyist and the upper level problem represents the government. The weak coupling between the lower and upper level problems reflects the fact that the lobbyist wishes to achieve its goal without having the government unduly influence its strategy. That is, the lobbyist wishes to achieve its goal without being over regulated by the government.

4.3 Admissible Pairs, Growth Conditions, and Lower Closure

To investigate the existence of an optimal solution to the bi-level optimal control problem (4.1), (4.2) we impose additional hypotheses. We begin by letting

$\mathscr{D}_Y = \{(t, y, x, v) : (t, y, x) \in \mathscr{A}_Y,\ v \in V(t, y, x)\} \subset [a, b] \times \mathbb{R}^{n+k+m+l}$,
$\mathscr{D}_X = \{(t, y, x, u) : (t, y, x) \in \mathscr{A}_Y,\ (t, x) \in \mathscr{A}_X,\ u \in U(t, x)\} \subset [a, b] \times \mathbb{R}^{n+k+l}$ and
$\tilde{\mathscr{D}}_X = \{(t, x, u) : (t, x) \in \mathscr{A}_X,\ u \in U(t, x)\} \subset [a, b] \times \mathbb{R}^{k+l}$,

and impose the following conditions:

A. The function $g_0 : \mathscr{D}_Y \to \mathbb{R}$ is a Lebesgue normal integrand. That is, $g_0(t, \cdot, \cdot, \cdot)$ is lower semicontinuous for almost all $t \in [a, b]$ and measurable with respect to the σ-algebra generated by Cartesian products of Lebesgue measurable subsets of $[a, b]$ and the Borel measurable subsets of $\mathbb{R}^{n+k+m}$.
B. The functions $f_0 : \mathscr{D}_X \to \mathbb{R}$, $g : \mathscr{D}_Y \to \mathbb{R}^n$ and $f : \tilde{\mathscr{D}}_X \to \mathbb{R}^k$ satisfy the Carathéodory conditions, namely that $t \mapsto f_0(t, y, x, v)$, $t \mapsto g(t, y, x, v)$ and $t \mapsto f(t, x, u)$ are Lebesgue measurable and that $f_0(t, \cdot, \cdot, \cdot)$, $g(t, \cdot, \cdot, \cdot)$ and $f(t, \cdot, \cdot)$ are continuous for almost all $t \in [a, b]$.

Remark 4.3. Notice that the sets $\mathscr{D}_Y$, $\mathscr{D}_X$ and $\tilde{\mathscr{D}}_X$ are closed sets.

Definition 4.1. For a fixed continuous function $y : [a, b] \to \mathbb{R}^n$, we say a pair of functions $\{x, u\} : [a, b] \to \mathbb{R}^{k+l}$ is admissible for the lower level problem (4.2) relative to y, if $x : [a, b] \to \mathbb{R}^k$ is absolutely continuous, $u : [a, b] \to \mathbb{R}^l$ is Lebesgue measurable, $(t, x(t)) \in \mathscr{A}_X$ for all $t \in [a, b]$, $(x(a), x(b)) \in \mathscr{B}_X$, $u(t) \in U(t, x(t))$ for almost every $t \in [a, b]$, $\dot{x}(t) = f(t, x(t), u(t))$ for almost every $t \in [a, b]$ and $t \mapsto f_0(t, y(t), x(t), u(t))$ is Lebesgue integrable on $[a, b]$.

Definition 4.2. For a fixed continuous function $y : [a,b] \to \mathbb{R}^n$ we say an admissible pair for (4.2) $\{x^*, u^*\} : [a,b] \to \mathbb{R}^{k+l}$ relative to y is optimal for (4.2) relative to y if

$$\int_a^b f_0(t, y(t), x^*(t), u^*(t))\, dt \leq \int_a^b f_0(t, y(t), x(t), u(t))\, dt$$

for all admissible pairs $\{x, u\}$ for (4.2) relative to y.

Definition 4.3. For a fixed continuous function $x : [a,b] \to \mathbb{R}^k$, we say a pair of functions $\{y, v\} : [a,b] \to \mathbb{R}^{n+m}$ is admissible for the upper level problem (4.1) relative to x, if $y : [a,b] \to \mathbb{R}^n$ is absolutely continuous, $v : [a,b] \to \mathbb{R}^m$ is Lebesgue measurable, $(t, y(t), x(t)) \in \mathscr{A}_Y$ for all $t \in [a,b]$, $(y(a), y(b)) \in \mathscr{B}_Y$, $(v(t), u(t)) \in V(t, y(t), x(t))$ for almost every $t \in [a,b]$, $\dot{y}(t) = g(t, y(t), x(t), v(t))$ for almost every $t \in [a,b]$ and $t \mapsto g_0(t, y(t), x(t), v(t))$ is Lebesgue integrable on $[a,b]$.

Definition 4.4. A "pair" of functions $\{(y,x),(v,u)\} : [a,b] \to \mathbb{R}^{n+k} \times \mathbb{R}^{m+l}$ is admissible for the bi-level optimal control problem if $\{y, v\}$ is an admissible pair relative to x for the upper problem (4.1) and $\{x, u\}$ is a *optimal* for (4.2) relative to y.

Definition 4.5. We say an admissible pair $\{(y^*, x^*), (v^*, u^*)\}$ is an optimal solution for the bi-level optimal control problem if

$$\int_a^b g(t, y^*(t), x^*(t), v^*(t))\, dt \leq \int_a^b g(t, y(t), x(t), v(t))\, dt$$

for all admissible pairs $\{(y,x),(v,u)\}$ for the bi-level optimal control problem.

Remark 4.4. Following standard conventions, given admissible pairs $\{x, u\}$, $\{y, v\}$ and $\{(y,x),(u,v)\}$ we will refer to x, y, and (y,x) as admissible trajectories and to u, v, and (v,u) as admissible controls (or strategies)

Our goal is to provide conditions so that the above bi-level problem has an optimal solution. This assumes that the above problem is well-posed in the sense that the set of admissible pairs $\{(y,x),(v,u)\}$ for the bi-level optimal control problem is nonempty. This is a standard assumption when studying existence of optimal solutions, and we will follow standard protocol and make this assumption here.

The approach to existence followed by Tonelli and his successors is the so-called direct method, in which one shows that the objective functional is bounded below so that the infimum is finite, that given any minimizing sequence of admissible pairs that it is relatively compact in an appropriate topology, and that the objective functional is lower semicontinuous with respect to this topology. In optimal control theory these properties are established through the use of lower closure theorems and growth conditions.

We begin with a discussion of growth conditions and relative weak compactness in the space $AC([a,b];\mathbb{R}^N)$ of absolutely continuous functions $w : [a,b] \to \mathbb{R}^N$.

4.3.1 Relative Weak Compactness

The first observation is to notice that the Fundamental Theorem of Calculus implies that the space $AC([a,b];\mathbb{R}^N)$ is isomorphic to $\mathbb{R} \times L^1([a,b];\mathbb{R}^N)$ through the formula

$$w(t) = w(a) + \int_a^t \dot{w}(s)\,ds.$$

Conversely if $(w_a, z) \in \mathbb{R}^N \times L^1([a,b];\mathbb{R}^N)$, then $\tilde{w} \in AC([a,b];\mathbb{R}^N)$, uniquely defined by the formula

$$\tilde{w}(t) = w_a + \int_a^t z(s)\,ds,$$

is such that $\tilde{w}(a) = w_a$ and $\dot{\tilde{w}}(t) = z(t)$ for almost all $t \in [a,b]$.

With this observation we have the following definition.

Definition 4.6. We say that a sequence $\{w_j\}_{j\in\mathbb{N}}$ of functions in $AC([0,T];\mathbb{R}^N)$ converges weakly to $w \in AC([a,b];\mathbb{R}^N)$ if and only if there exists a sequence of $\{t_j\}_{j\in\mathbb{N}} \subset [a,b]$ converging to $t_\infty \in [a,b]$ such that $w_j(t_j) \to w(t_\infty)$ as $j \to \infty$ and $\{\dot{w}_j\}_{n\in\mathbb{N}}$ converges weakly in $L^1([a,b];\mathbb{R}^N)$ to $\dot{w}$ as $j \to \infty$, that is for any $\psi \in L^\infty([a,b];\mathbb{N})$ one has

$$\lim_{j\to+\infty} \int_a^b \langle \dot{w}_j(s) - \dot{w}(s), \psi(s)\rangle\,ds = 0,$$

where $\langle\cdot,\cdot\rangle$ denotes the usual inner product in $\mathbb{R}^N$.

Remark 4.5. For our purposes, since the endpoints $a < b$ are fixed, it will always be the case that $t_j = a$ for all $j \in \mathbb{N}$ so that $w_j(a) \to w(a)$ as $j \to \infty$.

The conditions for weak compactness in $L^1([a,b];\mathbb{R}^n)$ are well known and in particular we have the following due to Dunford and Pettis.

Theorem 4.1. *Let Z be a family of functions in $L^1([a,b];\mathbb{R}^N)$. Then the family Z is weakly relatively compact in $L^1([a,b];\mathbb{R}^N$ if and only if the family Z is equiabsolutely integrable. That is, for every $\epsilon > 0$ there exists a $\delta > 0$ such that for every measurable set $E \subset [a,b]$ that has Lebesgue measure $meas(E) < \delta$ one has*

$$\int_E |z(s)|\,ds < \epsilon$$

for every $z \in Z$.

Proof. See Dunford and Schwartz (1958, pp. 292). □

For the optimal control problems considered here it will also be convenient to state the following theorem.

Theorem 4.2. *Consider a family Ω consisting of pairs of function $\{\eta, w\}$, with $\eta \in L^1([a,b];\mathbb{R})$ and $w \in AC([a,b];\mathbb{R}^N)$, such that for any element $\{\eta, w\} \in \Omega$ one has*

$$\int_a^b \eta(s)\, ds \leq M,$$

for some fixed constant $M > 0$. Further suppose for every $\epsilon > 0$ there exists a function $\psi_\epsilon \in L^1([a,b];[0,+\infty))$ such that

$$|\dot{w}(t)| \leq \psi_\epsilon(t) + \epsilon\, \eta(t) \quad \textit{for a.e. } t \in [a,b], \tag{4.3}$$

for every pair $\{\eta, w\} \in \Omega$. Then the class of functions $w \subset AC([a,b];\mathbb{R}^N)$ for which there exists an $\eta \in L^1([a,b];\mathbb{R})$ so that $\{\eta, w\} \in \Omega$ is such that the corresponding family of integrable functions $\{\dot{w}\}$ is equiabsolutely integrable (and hence relatively weakly compact in $L^1([a,b];\mathbb{R}^N)$). Moreover, if for each $\{\eta, w\} \in \Omega$ it is the case that $\{w(a)\} \subset \mathbb{R}^N$ is compact, then the family $\{w\}$ is relatively weakly compact in $AC([a,b];\mathbb{R}^N)$.

Proof. See Cesari (1983, Theorem 10.4ii). □

In view of the above theorem we impose the following growth conditions.

C. For every $\epsilon > 0$ there exist a function $\psi_\epsilon \in L^1([a,b];[0,+\infty)$ such that

$$|f(t,x,u)| + |g(t,y,x,v)| \leq \psi_\epsilon(t) + \epsilon \min\{f_0(t,y,x,u), g_0(t,y,x,v)\},$$

for almost all $t \in [a,b]$ and all (t,y,x,u,v) in the appropriate domain.

D. There exists a Lebesgue integrable function $M : [a,b] \to [0,+\infty)$ and a continuous function $K : \mathbb{R}^n \to [0,+\infty)$ such that

$$|f_0(t,y,x,u)| \leq M(t)K(y)$$

for almost all $t \in [a,b]$ with $(t,y,x,u) \in \mathscr{D}_X$.

Remark 4.6. Clearly, our choice of growth condition **C** is motivated by the above discussion. We remark that there are other growth conditions which could be appropriately modified to obtain the desired compactness properties. For a discussion of alternatives, we refer the reader to Cesari (1983, Sects. 10.3 and 10.4).

Remark 4.7. One consequence of the growth condition **D** is that it allows us to ensure that every pair of functions $\{x,u\} : [a,b] \to \mathbb{R}^{k+l}$, for which $x \in AC([a,b];\mathbb{R}^k)$, u is Lebesgue measurable, $\dot{x}(t) = f(t,x(t),u(t))$ a.e. on $[a,b]$,

$(t, x(t)) \in \mathscr{A}_X$ for all $t \in [a, b]$, $(x(a), x(b)) \in \mathscr{B}_X$ and $u(t) \in U(t, x(t))$ a.e. $t \in [a, b]$, is admissible relative to any continuous function $y : [a, b] \to \mathbb{R}^n$ for which $(t, y(t), x(t), u(t)) \in \mathscr{D}_X$ a.e $t \in [a, b]$ since it is clear from **D**, that $t \mapsto f_0(t, y(t), x(t), u(t))$ is Lebesgue integrable.

4.3.2 Lower Closure

The appropriate lower semicontinuity properties are obtained through the use of lower closure theorems for certain types of set-valued mappings. For the problem considered here we consider two set-valued mappings, $\tilde{Q}_U : \mathscr{A}_Y \to 2^{\mathbb{R}\times\mathbb{R}^n}$ and $\tilde{Q}_L : \widehat{\mathscr{A}_X} \to 2^{\mathbb{R}\times\mathbb{R}^k}$, where $\widehat{\mathscr{A}_X} \doteq \{(t, y, x) : (t, x) \in \mathscr{A}_X\}$, given, respectively, by the formulas

$$\tilde{Q}_U(t, y, x) = \{(z^0, z) : z^0 \geq g_0(t, y, x, v),\ z = g(t, y, x, v),\ v \in V(t, y, x)\}, \tag{4.4}$$

$$\tilde{Q}_L(t, y, x) = \{(z^0, z) : z^0 \geq f_0(t, y, x, u),\ z = f(t, x, u),\ u \in U(t, x)\}. \tag{4.5}$$

One can notice that if $\{y, v\}$ (resp., $\{x, u\}$) is admissible for the upper (lower) problem relative to a continuous function x (resp., y) and if one defines $z^0(t) = g_0(t, y(t), x(t), v(t))$ (resp., $z^0(t) = f_0(t, y(t), x(t), u(t))$), then one has $(z^0(t), \dot{y}(t)) \in \tilde{Q}_U(t, y(t), x(t))$ (resp., $(z^0(t), \dot{x}(t)) \in \tilde{Q}_L(t, y(t), x(t))$) for almost all $t \in [a, b]$.

Given this observation we now consider a general lower closure theorem for an arbitrary set-valued mapping $\mathscr{R} : \hat{\mathscr{A}} \to 2^{\mathbb{R}\times\mathbb{R}^M}$, where $\hat{\mathscr{A}} = [a, b] \times \mathscr{A} \subset [a, b] \times \mathbb{R}^N$ is closed. In particular we make the following assumption concerning $\mathscr{R}$

E. For each $(t, x) \in \hat{\mathscr{A}}$ the set $\mathscr{R}(t, x)$ is nonempty, closed and convex.
F. For each $t \in [a, b]$ the set-valued mapping $x \mapsto \mathscr{R}(t, x)$ satisfies the Kuratowski upper semicontinuity property at each point $\bar{x} \in \mathscr{A}$. That is,

$$\mathscr{R}(t, \bar{x}) = \bigcap_{\delta > 0} \mathrm{cl} \bigcup \{\mathscr{R}(t, x) : |x - \bar{x}| < \delta\}.$$

Theorem 4.3. *Let $\hat{\mathscr{A}} = [a, b] \times \mathscr{A}$ be closed and let $\mathscr{R} : \hat{\mathscr{A}} \to 2^{\mathbb{R}\times\mathbb{R}^M}$ be a set-valued mapping satisfying assumptions **E** and **F**. Further let ψ, x, η_j, ψ_j, x_j, λ and λ_j $(j = 1, 2 \ldots)$ be given measurable functions such that $\psi, \psi_j \in L^1([a, b]; \mathbb{R}^M)$, $\eta_j \in L^1([a, b]; \mathbb{R})$, $x, x_j : [a, b] \to \mathbb{R}^N$ are such that $x_j \to x$ in measure on $[a, b]$ as $j \to \infty$, $\psi_j \to \psi$ weakly in $L^1([a, b]; \mathbb{R}^M)$ as $j \to \infty$,*

$$(t, x_j(t)) \in \hat{\mathscr{A}} \text{ and } (\eta_j(t), \psi_j(t)) \in \mathscr{R}(t, x_j(t)) \quad \text{a.e. } t \in [a, b], \quad j = 1, 2, \ldots,$$

$$-\infty < i = \liminf_{j\to\infty} \int_a^b \eta_j(s)\,ds < +\infty,$$

$$\eta_j(t) \geq \lambda_j(t), \quad \lambda, \lambda_j \in L^1([a,b];\mathbb{R}), \quad \lambda_j \to \lambda \text{ weakly in } L^1([a,b];\mathbb{R}).$$

Then there is a function $\eta \in L^1([a,b];\mathbb{R})$ *such that*

$$(t, x(t)) \in \mathscr{A}, \quad (\eta(t), \psi(t)) \in \mathscr{R}(t, x(t)) \text{ for a.e. } t \in [a,b] \text{ and } \int_a^b \eta(s)\,ds \leq i.$$

Proof. See Cesari (1983, Theorem 10.7.i). □

To finish our prerequisites we need the following measurable selection theorem due to Kuratowski and Ryll-Nardzewski.

Theorem 4.4. *Let* $t \mapsto \mathscr{T}(t) \subset \mathbb{R}^N$, $t \in [a,b]$, *be a set-valued mapping such that* $T(t)$ *is closed for a.e.* $t \in [a,b]$ *and such that for every open set* $G \subset \mathbb{R}^N$ *the set* $\{t \in [a,b] : T(t) \cap G\}$ *is Lebesgue measurable (i.e.* $t \to \mathscr{T}(t)$ *is a closed valued, Lebesgue measurable set-valued mapping), then there exists a Lebesgue measurable function* $\tau : [a,b] \to \mathbb{R}^N$ *such that* $\tau(t) \in T(t)$ *for almost every* $t \in [a,b]$.

Proof. See Cesari (1983, Sect. 8.3). □

We now have the necessary prerequisites to prove our results.

4.4 The Existence Theorems

In this section we prove two existence theorems. The first is an existence theorem for the lower level problem (4.2) for a fixed continuous function $y : [a,b] \to \mathbb{R}^n$ and the second is an existence theorem for the bi-level optimal control problem (4.1), (4.2). The first of these results is to provide conditions which support the assumption that the feasible set of the bi-level problem is nonempty. The second, of course, is the theorem we set out to prove.

4.4.1 Existence of a Solution to the Lower Level Problem

In this section we fix a continuous function $y : [a,b] \to \mathbb{R}^n$ and consider the optimal control problem described by (4.2). With y fixed, the lower level problem is a standard Lagrange type optimal control problem and we will easily adapt known existence theorems to provide existence in this case.

Theorem 4.5. *Assume that the optimal control problem (4.2) has at least one admissible pair relative to the fixed continuous function* $y : [a,b] \to \mathbb{R}^n$, *and that the functions* f_0 *and* f *satisfy assumption* ***B*** *and are such that for every* $\epsilon > 0$ *there* *exists an integrable function* $\psi_\epsilon : [a,b] \to [0,\infty)$ *such that* $|f(t,x,u)| \le \psi_\epsilon(t) + \epsilon\, f_0(t,y(t),x,u)$ *for almost all* $t \in [a,b]$. *Further assume that the set-valued mapping* $\mathscr{R} : \hat{\mathscr{A}} \to 2^{\mathbb{R}\times\mathbb{R}^k}$ *defined, by using (4.5), as* $\mathscr{R}(t,x) := \tilde{Q}_L(t,y(t),x)$ *where* $\hat{\mathscr{A}} := \{(t,x) : (t,y(t),x) \in \mathscr{A}_Y,\ (t,x) \in \mathscr{A}_X\}$, *satisfies assumptions* ***E*** *and* ***F***. *Then there exists an admissible pair* $\{x^*,u^*\}$ *relative to* y *for (4.2) that is optimal for (4.2) relative to* y.

Proof. We begin by observing that for any admissible pair $\{x,u\}$ relative to y for the lower level problem (4.2) the growth condition, by taking $\epsilon = 1$, gives us

$$-\int_a^b \psi_1(t)\,dt \le \int_a^b f_0(t,y(t),x(t),u(t))\,dt,$$

which implies that (4.2) has a finite infimum. This mean there exists a sequence $\{x_j,u_j\}$ of admissible pairs (relative to y) such that

$$\lim_{j\to\infty}\int_a^b f_0(t,y(t),x_j(t),u_j(t))\,dt = \inf_{\mathrm{LP}}(y),$$

where $\inf_{\mathrm{LP}}(y)$ denotes the infimum of the lower level problem with y fixed. For each $j \in \mathbb{N}$ define $\eta_j : [a,b] \to \mathbb{R}$ by the formula $\eta_j(t) = f_0(t,y(t),x_j(t),u_j(t))$ for $t \in [a,b]$. We observe that, as a consequence of our assumptions, the family $\{\eta_j,x_j\}$ satisfies the growth condition (4.3). This fact, combined with the compactness of $\mathscr{B}_X$, allows us to assume without loss of generality that there exists an absolutely continuous function $x^* : [a,b] \to \mathbb{R}^k$ such that $\{x_j\}$ converges weakly in $AC([a,b];\mathbb{R}^k)$ to x^*. This means we can write, for almost all $t \in [a,b]$ and all $j = 1,2,\ldots,$

$$(t,x_j(t)) \in \hat{\mathscr{A}} \quad\text{and}\quad (\eta_j(t),\dot{x}_j(t)) \in \mathscr{R}(t,x_j(t)) = \tilde{Q}_L(t,y(t),x_j(t))$$

and

$$-\infty < \inf_{LP}(y) = \liminf_{j\to\infty}\int_a^b \eta_j(s)\,ds < +\infty.$$

Further we observe that for all $j \in \mathbb{N}$ we have $\eta_j(t) \ge -\psi_1(t)$ for almost all $t \in [a,b]$. We now can apply the lower closure theorem, Theorem 4.3, by taking $\lambda(t) \equiv \lambda_j(t) \equiv -\psi_1(t)$, to conclude that there exists an integrable function $\eta^* : [a,b] \to \mathbb{R}$ so that

$$(t,x^*(t)) \in \hat{\mathscr{A}} \text{ and } (\eta^*(t),\dot{x}^*(t)) \in \mathscr{R}(t,x^*(t)) = \tilde{Q}_L(t,y(t),x^*(t)) \text{ a.e. } t \in [a,b]$$

and

$$\int_a^b \eta^*(s)\,ds \le \inf_{LP}(y) = \liminf_{j\to\infty}\int_a^b \eta_j(s)\,ds.$$

We now show that we can associate a control $u^* : [a,b] \to \mathbb{R}^l$ with the absolutely continuous function x^* so that $\{x^*, u^*\}$ is admissible relative to y and, moreover, that it is optimal for the lower level problem (4.2) with y fixed. To this end, for each $t \in [a,b]$ define the set

$$\begin{aligned}\mathscr{T}(t) &= \big\{u \in U(t, x^*(t)) : \eta^*(t) \geq f_0(t, y(t), x^*(t), u),\ \dot{x}^*(t) \\ &= f(t, x^*(t), u)\big\} \subset \mathbb{R}^l.\end{aligned}$$

It is an easy matter to show that $t \mapsto \mathscr{T}(t)$ is a closed valued, Lebesgue measurable, set-valued mapping. Therefore by an application of the measurable selection theorem, Theorem 4.4, there exists a Lebesgue measurable function $u^* : [a,b] \to \mathbb{R}^l$ such that $u^*(t) \in \mathscr{T}(t)$ for almost every $t \in [a,b]$. With this measurable selection we have that the pair $\{x^*, u^*\}$ satisfies

$$\begin{aligned}\dot{x}^*(t) &= f(t, x^*(t), u^*(t)), \quad \text{a.e. } t \in [a,b], \\ (x^*(a), x^*(b)) &\in \mathscr{B}_X, \\ u^*(t) &\in U(t, x^*(t)), \quad \text{a.e. } t \in [a,b], \\ (t, x^*(t)) &\in \mathscr{A}_X, \quad \text{for all } t \in [a,b],\end{aligned}$$

and since $t \mapsto f_0(t, y(t), x^*(t), u^*(t))$ satisfies

$$-\psi_1(t) \leq f_0(t, y(t), x^*(t), u^*(t)) \leq \eta^*(t), \quad \text{a.e. } t \in [a,b]$$

we see that it is Lebesgue integrable which implies that $\{x^*, u^*\}$ is an admissible pair for the lower level problem (4.2) relative to y. Moreover, we also have that

$$-\int_a^b \psi_1(s)\,ds \leq \int_a^b f_0(s, y(s), x^*(s), u^*(x))\,ds \leq \int_a^b \eta^*(s)\,ds \leq \inf_{\mathrm{LP}}(y),$$

implying that this pair is optimal for (4.2) relative to y. □

Remark 4.8. In the above theorem the assumption that the function f_0 satisfies the Carathéodory conditions may be weakened to assuming only that f_0 is a Lebesgue normal integrand (see assumption **A**).

4.4.2 The Existence of an Optimal Solution for the Bi-Level Problem

Theorem 4.6. *Assume that the functions f_0, g_0, g, and f satisfy the assumptions **A**, **B** and the growth conditions **C** and **D**. Further assume that the set-valued maps $\tilde{Q}_U$ and $\tilde{Q}_L$ given by (4.4) and (4.5), respectively, satisfy the assumptions **E** and **F**. Then if the bi-level optimal control problem (4.1),(4.2) has at least one admissible* *pair* $\{(y, x), (v, u)\}$, *there exists an optimal admissible pair* $\{(y^*, x^*), (v^*, u^*)\}$ *for the bi-level optimal control problem.*

Proof. As in the proof of the existence of the lower level problem, the growth condition **C** allows us to conclude that the infimum for the bi-level optimal control problem is finite. Thus, we know there exists a sequence of admissible pairs $\{(y_j, x_j), (v_j, u_j)\}$ for the bi-level problem such that

$$\lim_{j\to\infty} \int_a^b g_0(t, y_j(t), x_j(t), v_j(t))\, dt = \inf_{\text{UP-LP}},$$

where $i \doteq \inf_{\text{UP-LP}}$ denotes the infimum for the bi-level optimal control problem. Moreover, as a consequence of the growth condition **C** and the compactness of $\mathscr{B}_Y$ and $\mathscr{B}_X$, an application of Theorem 4.2 with $\eta_j(t) = g_0(t, y_j(t), x_j(t), v_j(t))$ and $w_j(t) = (y_j(t), x_j(t))$ for all $j = 1, 2, \ldots$, allows us to assume that there exists absolutely continuous functions $y^*\colon [a, b] \to \mathbb{R}^n$ and $x^*\colon [a, b] \to \mathbb{R}^k$ such that $y_j \to y^*$ and $x_j \to x^*$ weakly as $j \to \infty$ in $AC([a, b]; .\mathbb{R}^n)$ and $AC([a, b]; \mathbb{R}^k)$, respectively. Now observe that for almost all $t \in [a, b]$ and all $j = 1, 2, \ldots$, we have

$$(\eta_j(t), \dot{y}_j(t)) \in \tilde{Q}_U(t, y_j(t), x_j(t)) \quad \text{and} \quad (\xi_j(t), \dot{x}_j(t)) \in \tilde{Q}_L(t, y_j(t), x_j(t)),$$

where $\xi_j(t) = f_0(t, y_j(t), x_j(t), u_j(t))$. Appealing to the Lower Closure Theorem 4.3, applied with $\mathscr{R} = \tilde{Q}_U$ and $\mathscr{R} = \tilde{Q}_L$, and $\lambda(t) = \lambda_j(t) = -\psi_1(t)$ we can conclude that there exists integrable functions $\eta^* : [a, b] \to \mathbb{R}$ and $\xi^* : [a, b] \to \mathbb{R}$ such that

$$(\eta^*(t), \dot{y}^*(t)) \in \tilde{Q}_U(t, y^*(t), x^*(t)) \quad \text{and} \quad (\xi^*(t), \dot{x}^*(t)) \in \tilde{Q}_L(t, y^*(t), x^*(t)),$$

for almost all $t \in [a, b]$, and that

$$\int_a^b \eta^*(t)\, dt \le i \quad \text{and} \int_a^b \xi(t)\, dt \le \liminf_{j\to\infty} \int_a^b f_0(t, y_j(t), x_j(t), u_j(t))\, dt.$$

Now define the set-valued maps $T_Y : [a, b] \to 2^{\mathbb{R}^m}$ and $T_X : [a, b] \to 2^{\mathbb{R}^l}$ by the formulas

$$\begin{aligned} T_Y(t) &= \big\{v : \eta^*(t) \ge g_0(t, y^*(t), x^*(t), v), \dot{y}^*(t) = g(t, y^*(t), x^*(t), v), \\ &\qquad v \in V(t, y^*(t), x^*(t))\big\} \\ T_X(t) &= \big\{u : \xi^*(t) \ge f_0(t, y^*(t), x^*(t), u), \dot{x}^*(t) = f(t, y^*(t), x^*(t), u), \\ &\qquad u \in U(t, x^*(t))\big\}. \end{aligned}$$

It is easy to see that both of these set-valued mappings are closed valued, Lebesgue measurable, and nonempty. Thus by an application of the Measurable Selection

Theorem 4.4 there exists Lebesgue measurable functions $v^* : [a,b] \to \mathbb{R}^m$ and $u^* : [a,b] \to \mathbb{R}^l$ such that we have the following

$$\begin{aligned} \dot{y}^*(t) &= g(t, y^*(t), x^*(t), v^*(t)), \quad \text{a.e. } a \le t \le b, \\ (y^*(a), y^*(b)) &\in \mathscr{B}_Y, \\ v^*(t) &\in V(t, y^*(t), x^*(t)), \quad \text{a.e. } a \le t \le b, \\ (t, y^*(t), x^*(t)) &\in \mathscr{A}_Y, \quad \text{for } a \le t \le b, \end{aligned}$$

and

$$\begin{aligned} \dot{x}^*(t) &= f(t, x^*(t), u^*(t)), \quad \text{a.e. } a \le t \le b, \\ (x^*(a), x^*(b)) &\in \mathscr{B}_X, \\ u^*(t) &\in U(t, x^*(t)), \quad \text{a.e. } a \le t \le b, \\ (t, x^*(t)) &\in \mathscr{A}_X, \quad \text{for } a \le t \le b. \end{aligned}$$

In addition we have

$$\int_a^b g_0(t, y^*(t), x^*(t), v^*(t)) \le \int_a^b \eta^*(t)\, dt \le \inf_{\text{UP-LP}}$$

and

$$\begin{aligned} \int_a^b f_0(t, y^*(t), x^*(t), u^*(t))\, dt &\le \int_a^b \xi^*(t)\, dt \\ &\le \liminf_{j\to\infty} \int_a^b f_0(t, y_j(t), x_j(t), u_j(t))\, dt. \end{aligned}$$

From these facts it is clear that if we can show that the pair $\{x^*, u^*\}$ is an optimal solution of the lower level problem (4.2) relative to y^* we can conclude that the pair $\{(y^*, x^*), (v^*, u^*)\}$ is an optimal solution to the bi-level optimal control problem. To see that this is the case we observe that by the optimality of $\{x_j, u_j\}$ relative to y_j we have that

$$\int_a^b f_0(t, y_j(t), x_j(t), u_j(t))\, dt \le \int_a^b f_0(t, y_j(t), x(t), u(t))\, dt,$$

holds for any admissible pair $\{x, u\}$ relative to y_j. This means that

$$\int_a^b f_0(t, y^*(t), x^*(t), u^*(t))\, dt \le \liminf_{j\to\infty} \int_a^b f_0(t, y_j(t), x(t), u(t))\, dt.$$

By our remarks concerning the growth condition **D** we know that any pair $\{x, u\}$ that is admissible for (4.2) relative to y^* is admissible relative to any y.

This means for any such pair $\{x, u\}$ we have, by the Carathéodory conditions, $f_0(t, y_j(t), x(t), u(t)) \to f_0(t, y^*(t), x(t), u(t))$ for almost every $t \in [a, b]$ as $j \to +\infty$ so that by an application of Lebesgue's dominated convergence theorem we have

$$\int_a^b f_0(t, y^*(t), x^*(t), u^*(t))\, dt \le \int_a^b f_0(t, y^*(t), x(t), u(t))\, dt$$

as desired. □

Remark 4.9. The assumptions and hypotheses used to establish the above two existence results are the classical convexity and seminormality conditions found in the existence theory of the calculus of variations originating in the works of L. Tonelli and E.J. McShane and further continued in the realm of optimal control theory by L. Cesari, L.D. Berkovitz, R.T. Rockafellar, and others. The first result (for the lower level problem) could have been proved by appealing to one of a number of standard existence results found in Cesari (1983) since for a fixed continuous function y this problem is an ordinary optimal control problem of Lagrange type. We chose to give a more direct proof to give a flavor of the approach as well as to ease the reader into the somewhat more complicated proof for the full bi-level problem.

For the second existence result, we had to strengthen the hypotheses slightly to affect the proof. In particular we refer to the strengthened regularity and growth hypotheses imposed on the lower level objective integrand f_0.

4.5 Conclusions

In this paper we considered a class of bi-level optimal control problems in which the lower and upper level problems were described by ordinary optimal control problems of Lagrange type. In particular we focused on sufficient conditions for the existence of an optimal solution based on classical convexity, seminormality conditions and compactness conditions (i.e., growth conditions) found in the existence theory of optimal control and the calculus of variations. In particular, our results do not require the lower level problem to have a unique optimal solution relative to each admissible trajectory of the upper level problem which is apparently a common assumption in earlier existence results for such problems.

References

Cesari L (1983) Optimization-theory and applications: problems with ordinary differential equations. In: Applications of applied mathematics, vol 17. Springer, New York

Dunford N, Schwartz JT (1958) Linear operators. I. General theory. In: Pure and applied mathematics, vol 7. Interscience Publishers, New York, with the assistance of W.G. Bade and R.G. Bartle

Chapter 5
Static Linear-Quadratic Gaussian Games

Meir Pachter

Abstract In this paper a simple static two-player linear-quadratic game where the players have private information is addressed. The players have private information, however each players is able to formulate an expression for his expected payoff, without the need, a la Harsanyi, to provide a prior probability distribution function of the game's parameter, and without recourse to the player Nature. Hence, the closed-form solution of the game is obtained. It is shown that in this special case of a one-stage linear-quadratic game where the players have private information, the solution is similar in structure to the solution of the game with complete information, namely, the deterministic linear-quadratic game, and the linear-quadratic game with partial information, where the information about the game's parameter is shared by the players. It is shown that the principle of certainty equivalence holds.

Keywords Linear-Quadratic Gaussian Games • Private information • Imperfect information • Perfect information • Certainty equivalence • Static games

5.1 Introduction

This paper is a first step in an attempt at bringing closer together the dynamic games paradigm and the theory of games, which historically have developed along separate lines. Dynamic game theorists have traditionally emphasized control theoretic aspects and the backward induction/dynamic programming solution method, whereas game theorists have focused on information economics, that is, the role of information in games.

M. Pachter
Department of Electrical and Computer Engineering, Air Force Institute of Technology, Wright-Patterson AFB, OH 45433, USA
e-mail: meir.pachter@afit.edu

V. Křivan and G. Zaccour (eds.), *Advances in Dynamic Games*, Annals of the International Society of Dynamic Games 13, DOI 10.1007/978-3-319-02690-9_5,

Linear-Quadratic Dynamic Games (LQDG) with perfect information have received a great deal of attention (Başar and Bernhard 2008; Başar and Olsder 1995; Engwerda 2005). In these works, the concepts of state, and state feedback, are emphasized and the solution method entails backward induction, a.k.a., dynamic programming. In previous work (Pachter and Pham 2013) a static LQG team problem was addressed. In this paper a static LQDG, where each player has private information, is considered. Specifically, the simplest linear-quadratic game with incomplete/partial information is addressed: a one-stage, two-player, "zero-sum," Linear-Quadratic Gaussian Game (LQGG) is solved.

In this paper a simple static linear-quadratic game where the players have private information, however each players is able to formulate an expression for his expected payoff, without the need to provide a prior probability distribution function of the game's parameter and without recourse to the player Nature, is analyzed. Thus, in Sect. 5.2 the static linear-quadratic Gaussian game, where the players have private information, is introduced. The solution of the baseline game with perfect information is given in Sect. 5.3 and the solution of the game with imperfect information is given in Sect. 5.4. The scenario where the players have private information is analyzed in Sect. 5.5, and the complete solution of the game is given in Sect. 5.6. Concluding remarks are made in Sect. 5.7.

5.2 LQGG Problem Statement

The following linear-quadratic game, a static, two-player, "zero-sum" game, is considered. The players are P and E and their respective control variables are u and v. It is a one-stage game with linear "dynamics"

$$x_1 = Ax_0 + Bu_0 + Cv_0 \,, \quad x_0 \equiv x_0 \,, \tag{5.1}$$

where the state $x_0, x_1 \in R^n$. The P and E players' controls are $u \in R^{m_u}$ and $v \in R^{m_v}$. The payoff function is quadratic:

$$J = x_1^T Q_F x_1 + u_0^T R_u u_0 - v_0^T R_v v_0 \tag{5.2}$$

where the Q_F, R_u, and R_v weighing matrices are real, symmetric, and positive definite. Both players are cognizant of the A, B, C, Q_F, R_u, and R_v data.

Player P strives to minimize the payoff/cost function (5.2) and player E strives to maximize the payoff (5.2).

The initial state information available to player P is

$$x_0 \sim \mathcal{N}(\overline{x}_0^{(P)}, P_0^{(P)}) \,, \tag{5.3}$$

where the vector $\overline{x}_0^{(P)} \in R^n$ and the $n \times n$ covariance matrix $P_0^{(P)}$ is real, symmetric, and positive definite. The initial state information available to player E is

$$x_0 \sim \mathcal{N}(\overline{x}_0^{(E)}, P_0^{(E)}) \,, \tag{5.4}$$

where the vector $\overline{x}_0^{(E)} \in R^n$ and the $n \times n$ covariance matrix $P_0^{(E)}$ is real, symmetric, and positive definite. The $P_0^{(P)}$ and $P_0^{(E)}$ data is public knowledge—only the $\overline{x}_0^{(P)}$ and $\overline{x}_0^{(E)}$ information is proprietary to the respective P and E players. This is tantamount to saying that players P and E took separate measurements of the initial state x_0, yet the accuracy of the instruments they used is known; however, the actual measurements $\overline{x}_0^{(P)}$ and $\overline{x}_0^{(E)}$ are the respective P and E players' private information.

Since the pertinent random variables are Gaussian, we shall refer to the game (5.1)–(5.4) as a Linear-Quadratic Gaussian Game (LQGG).

5.3 Linear-Quadratic Game with Perfect Information

It is instructive to first analyze the perfect information version of the linear-quadratic game (5.1) and (5.2).

If the initial state x_0 is known to both players, we have a game with perfect information.

The closed-form solution of Linear-Quadratic Dynamic Games with perfect information, a.k.a., deterministic Linear-Quadratic Dynamic Games (LQDGs), is derived in Pachter and Pham (2010, Theorem 2.1). The Schur complement concept (Zhang 2005) was used in (Pachter and Pham 2010) to invert a blocked $(m_u+m_v)\times(m_u+m_v)$ matrix and derive *explicit* formulae for the P and E players' optimal strategies. The said matrix contains four blocks and its diagonal blocks are $m_u \times m_u$ and $m_u \times m_u$ matrices. One can improve on the results of Pachter and Pham (2010) by noting that a matrix with four blocks has *two* Schur complements, say S_B and S_C.

Concerning the linear-quadratic game (5.1) and (5.2), where the initial state/game parameter x_0 is known to both players and thus the game is a game with perfect information, the following holds.

Theorem 5.1. *A necessary and sufficient condition for the existence of a solution to the zero-sum game* (5.1) *and* (5.2) *with perfect information is*

$$R_v > C^T Q_F C \tag{5.5}$$

A Nash equilibrium/saddle point exists and the players' optimal strategies are the linear state feedback control laws

$$u_0^*(x_0) = -S_B^{-1}(Q_F)B^T[I + Q_F C(R_v - C^T Q_F C)^{-1}C^T]Q_F A \cdot x_0\,, \tag{5.6}$$

$$v_0^*(x_0) = (R_v - C^T Q_F C)^{-1}C^T\{I - Q_F B S_B^{-1}(Q_F)B^T$$
$$[I + Q_F C(R_v - C^T Q_F C)^{-1}C^T]\}Q_F A \cdot x_0$$

An alternative formula for the optimal strategy of player E is

$$v^*(x_0) = -S_C^{-1}(Q_F)C^T[I - Q_F B(R_u + B^T Q_F B)^{-1} B^T]Q_F A\, x_0 \quad (5.7)$$

The value of the game

$$V_0(x_0) = x_0^T P_1 x_0\,, \quad (5.8)$$

where the matrix

$$\begin{aligned} P_1 = A^T\{Q_F - Q_F[BS_B^{-1}(Q_F)B^T + BS_B^{-1}(Q_F)B^T Q_F C(R_v - C^T Q_F C)^{-1}C^T \\ + C(R_v - C^T Q_F C)^{-1}C^T Q_F B S_B^{-1}(Q_F)B^T \\ + C(R_v - C^T Q_F C)^{-1}C^T Q_F B S_B^{-1}(Q_F)B^T Q_F C(R_v - C^T Q_F C)^{-1}C^T \\ + C(C^T Q_F C - R_v)^{-1}C^T]Q_F\}A \end{aligned} \quad (5.9)$$

In (5.6) *and* (5.9),

$$S_B(Q_F) \equiv B^T Q_F B + R_u + B^T Q_F C(R_v - C^T Q_F C)^{-1} C^T Q_F B \quad (5.10)$$

is the first Schur complement of the blocked matrix and

$$S_C(Q_F) \equiv -[R_v - C^T Q_F C + C^T Q_F B(R_u + B^T Q_F B)^{-1} B^T Q_F C]$$

is the second Schur complement of the blocked matrix.

Remark 5.1. Using both Schur complements of the blocked matrix renders the respective P and E players' strategies, (5.6) and (5.7), "symmetric."

5.4 Linear-Quadratic Gaussian Game with Imperfect Information

If in (5.3) and (5.4) $P_0^{(P)} = P_0^{(E)} = P_0$ and the P and E players' information $\overline{x}_0^{(P)} = \overline{x}_0^{(E)} = \overline{x}_0$ is public knowledge, we have on hand a linear-quadratic game with imperfect information; this is tantamount to saying that both players, together, took the measurement of the initial state and the outcome was

$$x_0 \sim \mathcal{N}(\overline{x}_0, P_0) \quad (5.11)$$

This is a stochastic game.

The closed-form solution of Linear-Quadratic Dynamic Games with imperfect information proceeds as follows.

Using (5.1) and (5.2), we calculate the payoff function

$$J(u_0, v_0; x_0) = x_0^T A^T Q_F A x_0 + u_0^T (R_u + B^T Q_F B) u_0 - v_0^T (R_v - C^T Q_F C) v_0 \\ + 2u_0^T B^T Q_F A x_0 + 2v_0^T C^T Q_F A x_0 + 2u_0^T B^T Q_F C v_0 \qquad (5.12)$$

The random variable at work is the initial state x_0. The players calculate the expected payoff function

$$\begin{aligned} \overline{J}(u_0, v_0; \overline{x}_0) &\equiv E_{x_0} \left(J(u_0, v_0; x_0) \mid \overline{x}_0 \right) \\ &= \overline{x}_0^T A^T Q_F A \overline{x}_0 + Trace(A^T Q_F A P_0) + u_0^T (R_u + B^T Q_F B) u_0 \\ &- v_0^T (R_v - C^T Q_F C) v_0 + 2u_0^T B^T Q_F A \overline{x}_0 + 2v_0^T C^T Q_F A \overline{x}_0 \\ &+ 2u_0^T B^T Q_F C v_0 \end{aligned} \qquad (5.13)$$

The expected payoff function $\overline{J}(u_0, v_0; \overline{x}_0)$ is convex in u_0 and concave in v_0. Differentiation in u_0 and v_0 yields a coupled linear system in the decision variables u_0 and v_0. Its solution is obtained using the Schur complement concept and it yields the optimal P and E strategies. The following holds.

Theorem 5.2. *A necessary and sufficient condition for the existence of a solution to the zero-sum game* (5.1) *and* (5.2) *with imperfect information, that is, a game where the initial state information* (5.11) *is available to both P and E, is that condition* (5.5) *holds. The respective optimal P and E strategies are given by* (5.6) *and* (5.7), *where x_0 is replaced by $\overline{x}_0$. The value of the game is*

$$V_0(\overline{x}_0) = \overline{x}_0^T P_1 \overline{x}_0 + Trace(A^T Q_F A P_0) \,, \qquad (5.14)$$

where, as before, the real symmetric matrix P_1 is given by (5.9).

Similar to LQG optimal control, in the game with imperfect information the separation principle/certainty equivalence holds.

5.5 Linear-Quadratic Gaussian Game with Private Information

The initial state x_0 features in the payoff function (5.12). The players' information on the initial state x_0 is now private information: Player P believes the initial state to be

$$x_0 \sim \mathcal{N}(\overline{x}_0^{(P)}, P_0^{(P)}) \qquad (5.15)$$

whereas player E believes the initial state to be

$$x_0 \sim \mathcal{N}(\overline{x}_0^{(E)}, P_0^{(E)}) \qquad (5.16)$$

This is tantamount to stipulating that players P and E took separate measurements of the initial state x_0. Assuming that the quality of the players' instruments used to take the measurements is public knowledge—we refer to the measurement error covariances $P_0^{(E)}$ and $P_0^{(E)}$—the private information of the players P and E are their respective measurements, $\overline{x}_0^{(P)}$and $\overline{x}_0^{(E)}$. The measurement recorded by player E, $\overline{x}_0^{(E)}$, is his private information and is not shared with player P. Hence, as far as player P is concerned, an E player with the private information $\overline{x}_0^{(E)} = x$ is an E player of *type* x. Thus, the P player's information on the game is incomplete. Similarly, the measurement recorded by player P, $\overline{x}_0^{(P)}$ is his private information and is not shared with the E player. Therefore, as far as the E player is concerned, a player P with the private information $\overline{x}_0^{(P)} = y$ is a P player of *type* y; also the E player's information on the game is incomplete.

We are analyzing what appears to be a game with incomplete information. In the process of planning his strategy, the player's opposition *type* is not known to him. However, although the information is incomplete, a Bayesian player can nevertheless assess, based on the private information available to him, the probability that the opposition he is facing is of a certain *type*. Consequently, the player can calculate the expectation of the payoff functional, conditioned on his private information.

The strategies available to player P are mappings $f : R^n \rightarrow R^{m_u}$ from his information set into his actions set; thus, the action of player P is

$$u_0 = f(\overline{x}_0^{(P)}) \tag{5.17}$$

Similarly, the strategies available to the E player are mappings $g : R^n \rightarrow R^{m_v}$ from his information set into his actions set; thus, the action of player E is

$$v_0 = g(\overline{x}_0^{(E)}) \tag{5.18}$$

From player P's vantage point, the action v_0 of player E is a random variable because from player P's vantage point, the measurement $\overline{x}_0^{(E)}$ used by player E to form his control v_0, is a random variable. Similarly, from player E's vantage point, the action u_0 of player P is a random variable.

Consider the decision process of player P whose private information is $\overline{x}_0^{(P)}$.

From player P's perspective, the random variables at work are x_0 and $\overline{x}_0^{(E)}$. Player P is confronted with a *stochastic* optimization problem and he calculates the expectation of the payoff function (5.12), conditional on his private information $\overline{x}_0^{(P)}$,

$$\overline{J}^{(P)}(u_0, g(\cdot); \overline{x}_0^{(P)}) = E_{x_0, \overline{x}_0^{(E)}} (J(u_0, g(\overline{x}_0^{(E)}); x_0) \mid \overline{x}_0^{(P)}) \tag{5.19}$$

It is important to realize that by using in the calculation of his expected cost in (5.19) player's E *strategy* $g(\overline{x}_0^{(E)})$, rather than player E's *control* v_0, player P has eliminated the possibility of an infinite regress in reciprocal reasoning. Thus, player P calculates

$$\begin{aligned}
\overline{J}^{(P)}(u_0, g(\cdot); \overline{x}_0^{(P)}) = & (\overline{x}_0^{(P)})^T A^T Q_F A \overline{x}_0^{(P)} + Trace(A^T Q_F A P_0^{(P)}) \\
& + u_0^T (R_u + B^T Q_F B) u_0 \\
& + 2u_0^T B^T Q_F A \overline{x}_0^{(P)} + 2E_{x_0, \overline{x}_0^{(E)}} (g^T(\overline{x}_0^{(E)}) C^T Q_F A x_0 \mid \overline{x}_0^{(P)}) \\
& - E_{\overline{x}_0^{(E)}} (g^T(\overline{x}_0^{(E)})(R_v - C^T Q_F C) g(\overline{x}_0^{(E)}) \mid \overline{x}_0^{(P)}) \\
& + 2u_0^T B^T Q_F C E_{\overline{x}_0^{(E)}} (g(\overline{x}_0^{(E)}) \mid \overline{x}_0^{(P)})
\end{aligned} \tag{5.20}$$

Player P calculates the expectations with respect to the random variable $\overline{x}_0^{(E)}$, which feature in (5.20). To this end, player P models his measurement $\overline{x}_0^{(P)}$ of the initial state x_0, and player E's measurement $\overline{x}_0^{(E)}$ of the initial state x_0, as follows.

$$\overline{x}_0^{(P)} = x_0 + w_P , \tag{5.21}$$

where x_0 is the true initial state and w_P is player P's measurement error, whose statistics are

$$w_P \sim \mathcal{N}(0, P_0^{(P)})$$

Similarly, player E's measurement

$$\overline{x}_0^{(E)} = x_0 + w_E , \tag{5.22}$$

where x_0 is the true initial state and w_E is player E's measurement error, whose statistics are

$$w_E \sim \mathcal{N}(0, P_0^{(E)})$$

Furthermore, the Gaussian random variables w_P and w_E are independent.

From player P's point of view, $\overline{x}_0^{(E)}$ is a random variable, but $\overline{x}_0^{(P)}$ is not. Subtracting (5.21) from (5.22), player P concludes that as far as he is concerned, player E's measurement upon which he will decide on his control v_0 is the random variable

$$\overline{x}_0^{(E)} = \overline{x}_0^{(P)} + \tilde{w} , \tag{5.23}$$

where the random variable

$$\tilde{w} \equiv w_E - w_P ; \tag{5.24}$$

in other words

$$\overline{x}_0^{(E)} \sim \mathcal{N}(\overline{x}_0^{(P)}, P_0^{(P)} + P_0^{(E)}) \tag{5.25}$$

Consider now the calculation of the expectations which feature in (5.20).

$$E_{\overline{x}_0^{(E)}}\left(g(\overline{x}_0^{(E)}) \mid \overline{x}_0^{(P)}\right) = E_{\tilde{w}}\left(g(\overline{x}_0^{(P)} + \tilde{w})\right) \tag{5.26}$$

where the random variable

$$\tilde{w} \sim \mathcal{N}(0, P_0^{(P)} + P_0^{(E)}) \tag{5.27}$$

Similarly, the expectation

$$E_{\overline{x}_0^{(E)}}\left(g^T(\overline{x}_0^{(E)})(R_v - C^T Q_F C)g(\overline{x}_0^{(E)}) \mid \overline{x}_0^{(P)}\right) = E_{\tilde{w}}\left(g^T(\overline{x}_0^{(P)} + \tilde{w})(R_v - C^T Q_F C)g(\overline{x}_0^{(P)} + \tilde{w})\right) \tag{5.28}$$

In addition, since

$$x_0 = \overline{x}_0^{(P)} - w_P\,, \tag{5.29}$$

the expectation

$$\begin{aligned} E_{x_0,\overline{x}_0^{(E)}}\left(g^T(\overline{x}_0^{(E)})C^T Q_F A x_0 \mid \overline{x}_0^{(P)}\right) &= E_{w_E,w_P}\left(g^T(\overline{x}_0^{(P)}+w_E-w_P)C^T Q_F A(\overline{x}_0^{(P)}-w_P)\right) \\ &= E_{\tilde{w}}\left(g^T(\overline{x}_0^{(P)} + \tilde{w})\right)C^T Q_F A\overline{x}_0^{(P)} \\ &\quad -E_{w_E,w_P}\left(g^T(\overline{x}_0^{(P)} + w_E - w_P)C^T Q_F A w_P\right) \end{aligned} \tag{5.30}$$

Inserting (5.26), (5.28), and (5.30) into (5.20) yields the expression for player P's expected cost in response to player E's strategy $g(\cdot)$, as a function of his decision variable u_0,

$$\begin{aligned} \overline{J}^{(P)}(u_0, g(\cdot); \overline{x}_0^{(P)}) &= (\overline{x}_0^{(P)})^T A^T Q_F A\overline{x}_0^{(P)} + Trace(A^T Q_F A P_0^{(P)}) \\ &\quad + u_0^T(R_u + B^T Q_F B)u_0 \\ &\quad + 2u_0^T B^T Q_F A\overline{x}_0^{(P)} + 2E_{\tilde{w}}\left(g^T(\overline{x}_0^{(P)} + \tilde{w})\right)C^T Q_F A\overline{x}_0^{(P)} \\ &\quad - 2E_{w_E,w_P}\left(g^T(\overline{x}_0^{(P)} + w_E - w_P)C^T Q_F A w_P\right) \\ &\quad - E_{\tilde{w}}\left(g^T(\overline{x}_0^{(P)} + \tilde{w})(R_v - C^T Q_F C)g(\overline{x}_0^{(P)} + \tilde{w})\right) \\ &\quad + 2u_0^T B^T Q_F C E_{\tilde{w}}\left(g(\overline{x}_0^{(P)} + \tilde{w})\right) \end{aligned} \tag{5.31}$$

Consider now the decision process of player E whose private information is $\overline{x}_0^{(E)}$.

From player E's perspective, the random variables at work are x_0 and $\overline{x}_0^{(P)}$. Player E is confronted with a *stochastic* optimization problem and he calculates the expectation of the payoff function (5.12), conditioned on his private information $\overline{x}_0^{(E)}$,

$$\overline{J}^{(E)}(f(\cdot), v_0; \overline{x}_0^{(E)}) \equiv E_{x_0,\overline{x}_0^{(P)}}\left(J(f(\overline{x}_0^{(P)}), v_0; x_0) \mid \overline{x}_0^{(E)}\right) \tag{5.32}$$

As before, it is important to realize that by using in the calculation of his expected cost in (5.32) player P's strategy $f(\overline{x}_0^{(P)})$, rather than player P's decision variable u_0, player E has eliminated the possibility of an infinite regress in reciprocal reasoning. Thus, player E calculates

$$\begin{aligned}\overline{J}^{(E)}(f(\cdot), v_0; \overline{x}_0^{(E)}) &= (\overline{x}_0^{(E)})^T A^T Q_F A \overline{x}_0^{(E)} + Trace(A^T Q_F A P_0^{(E)}) \\ &- v_0^T (R_v - C^T Q_F C) v_0 \\ &+ 2 v_0^T C^T Q_F A \overline{x}_0^{(E)} + E_{\overline{x}_0^{(P)}} (f^T(\overline{x}_0^{(P)})(R_u + B^T Q_F B) f(\overline{x}_0^{(P)}) \mid \overline{x}_0^{(E)}) \\ &+ 2E_{x_0, \overline{x}_0^{(P)}} (f^T(\overline{x}_0^{(P)}) B^T Q_F A x_0 \mid \overline{x}_0^{(E)}) \\ &+ 2 v_0^T C^T Q_F B E_{\overline{x}_0^{(P)}} (f(\overline{x}_0^{(P)}) \mid \overline{x}_0^{(E)}) \end{aligned} \tag{5.33}$$

Player E calculates the expectations with respect to the random variable $\overline{x}_0^{(P)}$, which feature in (5.33). To this end, player E models his measurement $\overline{x}_0^{(E)}$ of the initial state x_0 using (5.22), and he models player P's measurement $\overline{x}_0^{(P)}$ of the initial state x_0 using (5.21).

From player E's point of view, $\overline{x}_0^{(P)}$ is a random variable, but $\overline{x}_0^{(E)}$ is not. Subtracting (5.22) from (5.21), player E concludes that as far as he is concerned, player P's measurement upon which he will decide on his control u_0 is the random variable

$$\overline{x}_0^{(P)} = \overline{x}_0^{(E)} - \tilde{w} \tag{5.34}$$

In other words

$$\overline{x}_0^{(P)} \sim \mathcal{N}(\overline{x}_0^{(E)}, P_0^{(P)} + P_0^{(E)}) \tag{5.35}$$

Consider now the calculation of the expectations which feature in (5.33).

$$E_{\overline{x}_0^{(P)}} (f(\overline{x}_0^{(P)}) \mid \overline{x}_0^{(E)}) = E_{\tilde{w}} (f(\overline{x}_0^{(E)} - \tilde{w})) \tag{5.36}$$

Similarly, the expectation

$$\begin{aligned}E_{\overline{x}_0^{(P)}} (f^T(\overline{x}_0^{(P)})(R_u + B^T Q_F B) f(\overline{x}_0^{(P)}) \mid \overline{x}_0^{(E)}) = E_{\tilde{w}} (f^T(\overline{x}_0^{(E)} - \tilde{w})(R_u \\ + B^T Q_F B) f(\overline{x}_0^{(E)} - \tilde{w}))\end{aligned} \tag{5.37}$$

In addition, since

$$x_0 = \overline{x}_0^{(E)} - w_E , \tag{5.38}$$

the expectation

$$E_{x_0,\overline{x}_0^{(P)}}\,(f^T(\overline{x}_0^{(P)})B^T Q_F A x_0 \mid \overline{x}_0^{(E)}) = E_{w_E,w_P}\,(f^T(\overline{x}_0^{(E)}+w_P-w_E)B^T$$
$$Q_F A(\overline{x}_0^{(E)}-w_E))$$
$$= E_{\tilde{w}}\,(f^T(\overline{x}_0^{(E)}-\tilde{w}))B^T Q_F A\overline{x}_0^{(E)}$$
$$- E_{w_E,w_P}\,(f^T(\overline{x}_0^{(E)}+w_P - w_E)B^T Q_F A w_E) \tag{5.39}$$

Inserting (5.36), (5.37), and (5.39) into (5.33) yields the expression for player E's expected payoff in response to player P's strategy $f(\cdot)$, as a function of his decision variable v_0,

$$\overline{J}^{(E)}(f(\cdot), v_0; \overline{x}_0^{(E)}) = (\overline{x}_0^{(E)})^T A^T Q_F A\overline{x}_0^{(E)} + Trace(A^T Q_F A P_0^{(E)})$$
$$- v_0^T(R_v - C^T Q_F C)v_0 + 2v_0^T C^T Q_F A\overline{x}_0^{(E)}$$
$$+ E_{\tilde{w}}\,(f^T(\overline{x}_0^{(E)} - \tilde{w})(R_u + B^T Q_F B) f(\overline{x}_0^{(E)} - \tilde{w}))$$
$$+ 2E_{\tilde{w}}\,(f^T(\overline{x}_0^{(E)} - \tilde{w}))B^T Q_F A\overline{x}_0^{(E)}$$
$$- 2E_{w_E,w_P}\,(f^T(\overline{x}_0^{(E)} + w_P - w_E)B^T Q_F A w_E)$$
$$+ 2v_0^T C^T Q_F B E_{\tilde{w}}\,(f(\overline{x}_0^{(E)} - \tilde{w})) \tag{5.40}$$

The cost of player P is now given by (5.31) and the payoff of Player E is given by (5.40). Imperfect information leads to a nonzero-sum game formulation. Consequently, one id interested in a Nash equilibrium, a.k.a., Person By Person Satisfactory (PBPS) strategies.

Next, player P calculates his response to player E's strategy $g(\overline{x}_0^{(E)})$. Thus, given the information $\overline{x}_0^{(P)}$, player P minimizes his expected cost (5.31); the minimization is performed in the *decision variable* u_0. The cost function is quadratic in the decision variable. Thus, the optimal decision variable u_0^* must satisfy the equation

$$u_0^* = -(R_u + B^T Q_F B)^{-1} B^T Q_F (A\overline{x}_0^{(P)} + C E_{\tilde{w}}\,(g(\overline{x}_0^{(P)} + \tilde{w})\,))$$

In other words, the optimal response of player P to player E's strategy $g(\cdot)$ is

$$f^*(\overline{x}_0^{(P)}) = -(R_u + B^T Q_F B)^{-1} B^T Q_F (A\overline{x}_0^{(P)} + C E_{\tilde{w}}\,(g(\overline{x}_0^{(P)} + \tilde{w})\,))\ \ \forall\, \overline{x}_0^{(P)} \in R^n$$

Similarly, player E calculates his optimal response to player P's strategy $f(\overline{x}_0^{(P)})$. Thus, given the information $\overline{x}_0^{(E)}$, player E maximizes his expected payoff (5.40); the maximization is performed in the *decision variable* v_0. The cost function is quadratic in the decision variable. Thus, the optimal decision variable v_0^* must satisfy the equation

$$v_0^* = (R_v - C^T Q_F C)^{-1} C^T Q_F (A\overline{x}_0^{(E)} + B E_{\tilde{w}} \, (f(\overline{x}_0^{(E)} - \tilde{w}) \,))$$

In other words, the optimal response of player E to player P's strategy $f(\cdot)$ is

$$g^*(\overline{x}_0^{(E)}) = (R_v - C^T Q_F C)^{-1} C^T Q_F (A\overline{x}_0^{(E)} + B E_{\tilde{w}} \, (f(\overline{x}_0^{(E)} - \tilde{w}) \,)) \;\; \forall \, \overline{x}_0^{(E)} \in R^n$$

Hence, the respective optimal strategies $f^*(\cdot)$ and $g^*(\cdot)$ of players P and E satisfy the set of two coupled equations (5.41) and (5.42),

$$f^*(\overline{x}_0^{(P)}) = -(R_u + B^T Q_F B)^{-1} B^T Q_F (A\overline{x}_0^{(P)} + C E_{\tilde{w}} \, (g^*(\overline{x}_0^{(P)} + \tilde{w}) \,)) \quad \forall \, \overline{x}_0^{(P)} \in R^n \tag{5.41}$$

$$g^*(\overline{x}_0^{(E)}) = (R_v - C^T Q_F C)^{-1} C^T Q_F (A\overline{x}_0^{(E)} + B E_{\tilde{w}} \, (f^*(\overline{x}_0^{(E)} - \tilde{w}) \,)) \quad \forall \, \overline{x}_0^{(E)} \in R^n \tag{5.42}$$

The expectation

$$E_{\tilde{w}} \, (f(\overline{x}_0^{(E)} - \tilde{w}) \,) = \frac{1}{(2\pi)^{\frac{n}{2}} \sqrt{\det(P_0^{(P)} + P_0^{(E)})}} \int \cdots \int_{R^n} f(\overline{x}_0^{(E)} - \tilde{w}) e^{-\frac{1}{2}\tilde{w}^T (P_0^{(P)} + P_0^{(E)})^{-1} \tilde{w}} d\tilde{w}$$

It is convenient to use the notation for the multivariate Gaussian distribution with covariance P (> 0),

$$G(x; P) \equiv \frac{1}{(2\pi)^{\frac{n}{2}} \sqrt{\det(P)}} e^{-\frac{1}{2}x^T P^{-1} x}$$

whereupon

$$E_{\tilde{w}} \, (f(\overline{x}_0^{(E)} - \tilde{w}) \,) = [f * G(P_0^{(P)} + P_0^{(E)})](\overline{x}_0^{(E)})$$

Similarly, the expectation

$$E_{\tilde{w}} \, (g(\overline{x}_0^{(P)} + \tilde{w}) \,) = [g * G(P_0^{(P)} + P_0^{(E)})](\overline{x}_0^{(P)})$$

Using the convolution notation in (5.41) and (5.42), one obtains

$$f^*(\overline{x}_0^{(P)}) = -(R_u + B^T Q_F B)^{-1} B^T Q_F (A\overline{x}_0^{(P)} + Cg^* * G(P_0^{(P)} + P_0^{(E)})) \ \forall\, \overline{x}_0^{(P)} \in R^n$$

$$g^*(\overline{x}_0^{(E)}) = (R_v - C^T Q_F C)^{-1} C^T Q_F (A\overline{x}_0^{(E)} + Bf^* * G(P_0^{(P)} + P_0^{(E)})) \ \forall\, \overline{x}_0^{(E)} \in R^n$$

Thus, the functions

$$f^*(x) = -(R_u + B^T Q_F B)^{-1} B^T Q_F (Ax + C[g^* * G(P_0^{(P)} + P_0^{(E)})](x)), \tag{5.43}$$

and

$$g^*(x) = (R_v - C^T Q_F C)^{-1} C^T Q_F (Ax + B[f^* * G(P_0^{(P)} + P_0^{(E)})](x)) \quad \forall\, x \in R^n \tag{5.44}$$

Inserting (5.44) into (5.43) and suppressing the dependence of the Gaussian p.d.f. on the covariance matrix yields

$$\begin{aligned} f^*(x) &= -(R_u + B^T Q_F B)^{-1} B^T Q_F A\, x \\ &\quad - (R_u + B^T Q_F B)^{-1} B^T Q_F C (R_v - C^T Q_F C)^{-1} C^T Q_F A\, x * G \\ &\quad - (R_u + B^T Q_F B)^{-1} B^T Q_F C (R_v - C^T Q_F C)^{-1} C^T Q_F B\ f^* * G * G \\ &\quad \forall\, x \in R^n \end{aligned} \tag{5.45}$$

Similarly, inserting (5.43) into (5.44) yields

$$\begin{aligned} g^*(x) &= (R_v - C^T Q_F C)^{-1} C^T Q_F A\, x \\ &\quad - (R_v - C^T Q_F C)^{-1} C^T Q_F B (R_u + B^T Q_F B)^{-1} B^T Q_F A\, x * G \\ &\quad - (R_v - C^T Q_F C)^{-1} C^T Q_F B (R_u + B^T Q_F B)^{-1} B^T Q_F C\ g^* * G * G \\ &\quad \forall\, x \in R^n \end{aligned} \tag{5.46}$$

The convolution operation is associative. We shall require the following

Lemma 5.1. *The Gaussian kernel is self-similar, namely,*

$$G(P) * G(P) = G(2P) \tag{5.47}$$

Proof.

$$G(x;P) * G(x;P) = \int \cdots \int_{R^n} \frac{1}{(2\pi)^{\frac{n}{2}}\sqrt{\det(P)}} e^{-\frac{1}{2}(x-y)^T P^{-1}(x-y)}$$
$$\frac{1}{(2\pi)^{\frac{n}{2}}\sqrt{\det(P)}} e^{-\frac{1}{2}y^T P^{-1} y} dy$$
$$= \int \cdots \int_{R^n} \left(\frac{1}{(2\pi)^{\frac{n}{2}}\sqrt{\det(P)}}\right)^2 e^{-\frac{1}{2}[(x-y)^T P^{-1}(x-y) + y^T P^{-1} y]} dy$$

We calculate

$$\begin{aligned}(x-y)^T P^{-1}(x-y) + y^T P^{-1} y &= 2y^T P^{-1} y - 2x^T P^{-1} y + x^T P^{-1} x \\ &= y^T (\frac{1}{2}P)^{-1} y - 2(\frac{1}{2}x)^T (\frac{1}{2}P)^{-1} y \\ &+ (\frac{1}{2}x)^T (\frac{1}{2}P)^{-1} (\frac{1}{2}x) + \frac{1}{2}x^T P^{-1} x \\ &= (y - \frac{1}{2}x)^T (\frac{1}{2}P)^{-1} (y - \frac{1}{2}x) + x^T (2P)^{-1} x\end{aligned}$$

Hence

$$\int \cdots \int_{R^n} \left(\frac{1}{(2\pi)^{\frac{n}{2}}\sqrt{\det(P)}}\right)^2 e^{-\frac{1}{2}[(x-y)^T P^{-1}(x-y) + y^T P^{-1} y]} dy_1 \ldots dy_n$$
$$= \int \cdots \int_{R^n} \frac{1}{(2\pi)^{\frac{n}{2}}\sqrt{\det(\frac{1}{2}P)}} e^{-\frac{1}{2}(y-\frac{1}{2}x)^T (\frac{1}{2}P)^{-1}(y-\frac{1}{2}x)}$$
$$dy_1 \ldots dy_n \cdot (\frac{1}{2})^{\frac{n}{2}} \frac{1}{(2\pi)^{\frac{n}{2}}\sqrt{\det(P)}} e^{-\frac{1}{2}x^T (2P)^{-1} x}$$
$$= 1 \cdot (\frac{1}{2})^{\frac{n}{2}} \frac{1}{(2\pi)^{\frac{n}{2}}\sqrt{\det(P)}} e^{-\frac{1}{2}x^T (2P)^{-1} x}$$
$$= \frac{1}{(2\pi)^{\frac{n}{2}}\sqrt{\det(2P)}} e^{-\frac{1}{2}x^T (2P)^{-1} x}$$
$$= G(2P)$$

□

We also calculate

$$\begin{aligned} x * G(P) &= \int_{R^n} (x-y) G(y;P) dy \\ &= x \qquad \forall P > 0 \end{aligned} \tag{5.48}$$

Inserting (5.47) and (5.48) into (5.45) yields a Fredholm integral equation of the second kind for the optimal strategy of player P,

$$f^*(\overline{x}_0^{(P)}) = -(R_u + B^T Q_F B)^{-1} B^T Q_F [I + C(R_v - C^T Q_F C)^{-1} C^T Q_F] A\, \overline{x}_0^{(P)} \\ - (R_u + B^T Q_F B)^{-1} B^T Q_F C(R_v - C^T Q_F C)^{-1} C^T Q_F B\, G(2P) * f^* \\ \forall\, \overline{x}_0^{(P)} \in R^n \tag{5.49}$$

Similarly, inserting (5.47) and (5.48) into (5.46) yields a Fredholm integral equation of the second kind for the optimal strategy of player E,

$$g^*(\overline{x}_0^{(E)}) = (R_v - C^T Q_F C)^{-1} C^T Q_F [I - B(R_u + B^T Q_F B)^{-1} B^T Q_F] A\, \overline{x}_0^{(E)} \\ - (R_v - C^T Q_F C)^{-1} C^T Q_F B(R_u + B^T Q_F B)^{-1} B^T Q_F C\, G(2P) * g^* \\ \forall\, x_0^{(E)} \in R^n \tag{5.50}$$

The Fredholm equations of the second kind (5.49) and (5.50) are of the convolution type and the kernel is a Gaussian function.

If the state's measurement error covariances are "small," namely, $P_0^{(P)} < 1$ and $P_0^{(E)} < 1$ and therefore the Gaussian distribution approaches a delta function, from (5.49) and (5.50) we conclude that the P and E strategies satisfy the equations

$$f^*(\overline{x}_0^{(P)}) = -(R_u + B^T Q_F B)^{-1} B^T Q_F [I + C(R_v - C^T Q_F C)^{-1} C^T Q_F] A \overline{x}_0^{(P)} \\ - (R_u + B^T Q_F B)^{-1} B^T Q_F C(R_v - C^T Q_F C)^{-1} C^T Q_F B f^*(\overline{x}_0^{(P)}) \\ \forall\, \overline{x}_0^{(P)} \in R^n \tag{5.51}$$

and

$$g^*(\overline{x}_0^{(E)}) = (R_v - C^T Q_F C)^{-1} C^T Q_F [I - B(R_u + B^T Q_F B)^{-1} B^T Q_F] A\, \overline{x}_0^{(E)} \\ - (R_v - C^T Q_F C)^{-1} C^T Q_F B(R_u + B^T Q_F B)^{-1} B^T Q_F C\, g^*(\overline{x}_0^{(E)}) \\ \forall\, \overline{x}_0^{(E)} \in R^n \tag{5.52}$$

From (5.51) and (5.52) we therefore obtain players' P and E optimal strategies, which are explicitly given by

$$f^*(\overline{x}_0^{(P)}) = -[I + (R_u + B^T Q_F B)^{-1} B^T Q_F C(R_v - C^T Q_F C)^{-1} C^T Q_F B]^{-1} (R_u \\ - B^T Q_F B)^{-1} B^T Q_F [I + C(R_v - C^T Q_F C)^{-1} C^T Q_F] A\, \overline{x}_0^{(P)} \\ = -[R_u + B^T Q_F B + B^T Q_F C(R_v - C^T Q_F C)^{-1} C^T Q_F B]^{-1} B^T [I \\ + Q_F C(R_v - C^T Q_F C)^{-1} C^T Q_F] A\, \overline{x}_0^{(P)} \forall\, \overline{x}_0^{(P)} \in R^n$$

that is, the optimal strategy of player P is

$$f^*(\overline{x}_0^{(P)}) = -S_B^{-1}(Q_F)B^T[I + Q_F C(R_v - C^T Q_F C)^{-1} C^T Q_F] A\,\overline{x}_0^{(P)}$$
$$\forall\, \overline{x}_0^{(P)} \in R^n \tag{5.53}$$

Similarly,

$$\begin{aligned} g^*(\overline{x}_0^{(E)}) &= [I + (R_v - C^T Q_F C)^{-1} C^T Q_F B(R_u + B^T Q_F B)^{-1} B^T Q_F C]^{-1} (R_v \\ &\quad - C^T Q_F C)^{-1} C^T Q_F [I - B(R_u + B^T Q_F B)^{-1} B^T Q_F] A\,\overline{x}_0^{(E)} \\ &= [R_v - C^T Q_F C + C^T Q_F B(R_u + B^T Q_F B)^{-1} B^T Q_F C]^{-1} C^T [I \\ &\quad - Q_F B(R_u + B^T Q_F B)^{-1} B^T] Q_F A\,\overline{x}_0^{(E)} \quad \forall\, \overline{x}_0^{(E)} \in R^n \end{aligned}$$

that is, the optimal strategy of player E is

$$g^*(\overline{x}_0^{(E)}) = -S_C^{-1}(Q_F)C^T[I - Q_F B(R_u + B^T Q_F B)^{-1} B^T] Q_F A\,\overline{x}_0^{(E)}$$
$$\forall\, \overline{x}_0^{(E)} \in R^n \tag{5.54}$$

where the Schur complement

$$S_C(Q_F) \equiv -[R_v - C^T Q_F C + C^T Q_F B(R_u + B^T Q_F B)^{-1} B^T Q_F C] \tag{5.55}$$

Indeed, having calculated the functions $f^*(x)$ and $g^*(x)$, we obtained the optimal strategies of players P and E by setting $x := \overline{x}_0^{(P)}$ in $f^*(x)$ and $x := \overline{x}_0^{(E)}$ in $g^*(x)$. In the limiting case of Gaussian distributions with small covariance matrices, the players' optimal strategies (5.53) and (5.54) are linear in the players' respective measurements.

Equations (5.53) and (5.54) are identical to the respective (5.6) and (5.7) in Theorem 5.1—we have recovered the perfect information result of Theorem 5.1. This makes sense—the initial state's measurements of both players are very accurate and thus the game is almost deterministic. Thus, one could have argued that when the covariances are "small," namely, $P_0^{(P)} << 1$ and $P_0^{(E)} << 1$, that is, $\overline{x}_0^{(P)} \approx \overline{x}_0^{(E)} \approx x_0$, one can re-use the deterministic state feedback strategies (5.6) and (5.7) of players P and E given by Theorem 5.1—simply set $x_0 := \overline{x}_0^{(P)}$ in (5.6) and $x_0 := \overline{x}_0^{(E)}$ in (5.7).

5.6 Linear Strategies

The Fredholm integral equations of the second kind, (5.49) and (5.50), are linear integral equations. Furthermore, the "forcing terms"/inputs on the R.H.S. of (5.49) and (5.50) are linear in $\overline{x}_0^{(P)}$ and $\overline{x}_0^{(E)}$, respectively. Consequently, the solution $f^*(\cdot)$ of (5.49) is linear in $\overline{x}_0^{(P)}$ and the solution $g^*(\cdot)$ of (5.50) is linear in $\overline{x}_0^{(E)}$—think of linear integral operators as infinite dimensional matrices. Hence, postulate that the players' optimal strategies are linear—in other words,

$$f(\overline{x}_0^P) = F_u \overline{x}_0^P \tag{5.56}$$

and

$$g(\overline{x}_0^E) = F_v \overline{x}_0^E \,, \tag{5.57}$$

where the yet to be determined constant gains F_u and F_v are $m_u \times n$ and $m_v \times n$ matrices, respectively. Constant gain strategies (5.56) and (5.57) which satisfy the respective second kind Fredholm integral equations of the convolution type with a Gaussian kernel, (5.49) and (5.50), can be found. This is due to the fact that, according to (5.48), the convolution of the state vector x with a Gaussian function returns the state vector x. In the process of deriving the equations which yield the gains F_u and F_v, the necessary and sufficient conditions for the existence of a solution are obtained.

The optimal gains F_u^* and F_v^* are obtained as follows. Insert (5.56) into (5.49) and insert (5.57) into (5.50):

$$\begin{aligned}
F_u^* \,\overline{x}_0^{(P)} &= -(R_u + B^T Q_F B)^{-1} B^T Q_F [I + C(R_v - C^T Q_F C)^{-1} C^T Q_F] A \,\overline{x}_0^{(P)} \\
&\quad - (R_u + B^T Q_F B)^{-1} B^T Q_F C (R_v - C^T Q_F C)^{-1} C^T Q_F B \; F_u^* x * G(2P) \\
&= -(R_u + B^T Q_F B)^{-1} B^T Q_F [I + C(R_v - C^T Q_F C)^{-1} C^T Q_F] A \,\overline{x}_0^{(P)} \\
&\quad - (R_u + B^T Q_F B)^{-1} B^T Q_F C (R_v - C^T Q_F C)^{-1} C^T Q_F B \; F_u^* \,\overline{x}_0^{(P)} \\
&\quad \forall \, \overline{x}_0^{(P)} \in R^n
\end{aligned}$$

Therefore

$$\begin{aligned}
F_u^* &= -[I + (R_u + B^T Q_F B)^{-1} B^T Q_F C (R_v - C^T Q_F C)^{-1} C^T Q_F B]^{-1} (R_u \\
&\quad + B^T Q_F B)^{-1} B^T Q_F [I + C(R_v - C^T Q_F C)^{-1} C^T Q_F] A \\
&= -S_B^{-1}(Q_F) B^T Q_F [I + C(R_v - C^T Q_F C)^{-1} C^T Q_F] A
\end{aligned} \tag{5.58}$$

Similarly

$$
\begin{aligned}
F_v^*\,\overline{x}_0^{(E)} &= (R_v - C^T Q_F C)^{-1} C^T Q_F [I - B(R_u + B^T Q_F B)^{-1} B^T Q_F] A\,\overline{x}_0^{(E)} \\
&\quad - (R_v - C^T Q_F C)^{-1} C^T Q_F B (R_u + B^T Q_F B)^{-1} B^T Q_F C\ F_v^* x * G(2P) \\
&= (R_v - C^T Q_F C)^{-1} C^T Q_F [I - B(R_u + B^T Q_F B)^{-1} B^T Q_F] A\,\overline{x}_0^{(E)} \\
&\quad - (R_v - C^T Q_F C)^{-1} C^T Q_F B (R_u + B^T Q_F B)^{-1} B^T Q_F C\ F_v^* \overline{x}_0^{(E)} \\
&\quad \forall\ x_0^{(E)} \in R^n
\end{aligned}
$$

and

$$
\begin{aligned}
F_v^* &= [I + (R_v - C^T Q_F C)^{-1} C^T Q_F B (R_u + B^T Q_F B)^{-1} B^T Q_F C]^{-1} (R_v \\
&\quad - C^T Q_F C)^{-1} C^T Q_F [I - B(R_u + B^T Q_F B)^{-1} B^T Q_F] A \\
&= -S_C^{-1}(Q_F) C^T Q_F [I - B(R_u + B^T Q_F B)^{-1} B^T Q_F] A
\end{aligned} \tag{5.59}
$$

We have found constant gain strategies F_u^* and F_v^* which satisfy the respective second kind Fredholm integral equations of the convolution type with a Gaussian kernel, (5.49) and (5.50). This is due to the fact that, according to (5.48), the convolution of the state vector x with a Gaussian function returns the state vector x. Furthermore, (5.48) holds, irrespective of the covariance P. Hence, the constant gains are *not* dependent on the cumulative covariance P of the measurement errors and also apply in the limiting case of a deterministic scenario—in other words, the optimal constant gains F_u and F_v are exactly as in (5.6) and (5.7), and certainty equivalence holds. Having obtained the optimal strategies, one can now calculate the respective value functions of players P and E by evaluating the expectations in (5.31) and (5.40):

Consider (5.31), the expected cost $\overline{J}^{(P)}(u_0, g(\cdot); \overline{x}_0^{(P)})$ of player P first. The expectations

$$
E_{\tilde{w}}\ (g(\overline{x}_0^{(P)} + \tilde{w})) = F_v^* \overline{x}_0^{(P)}\ , \tag{5.60}
$$

$$
E_{w_E, w_P}\ (g^T(\overline{x}_0^{(P)} + w_E - w_P) C^T Q_F A w_P) = -Trace(P_0^{(P)} A^T Q_F C F_v^*)\ , \tag{5.61}
$$

and

$$
\begin{aligned}
&E_{\tilde{w}}\ (g^T(\overline{x}_0^{(P)} + \tilde{w})(R_v - C^T Q_F C) g(\overline{x}_0^{(P)} + \tilde{w})) \\
&\quad = (\overline{x}_0^{(P)})^T (F_v^*)^T (R_v - C^T Q_F C) F_v^* \overline{x}_0^{(P)} \\
&\quad + Trace((F_v^*)^T (R_v - C^T Q_F C) F_v^* (P_0^{(P)} + P_0^{(E)}))
\end{aligned} \tag{5.62}
$$

Inserting (5.60)–(5.62) into (5.31) yields, with some abuse of notation, the value function of player P,

$$\begin{aligned}
V_0^{(P)} &= (\overline{x}_0^{(P)})^T A^T Q_F A \overline{x}_0^{(P)} + Trace(A^T Q_F A P_0^{(P)}) + u_0^T (R_u + B^T Q_F B) u_0 \\
&+ 2u_0^T B^T Q_F A \overline{x}_0^{(P)} + 2(\overline{x}_0^{(P)})^T (F_v^*)^T C^T Q_F A \overline{x}_0^{(P)} \\
&+ 2Trace(P_0^{(P)} A^T Q_F C F_v^*) \\
&- (\overline{x}_0^{(P)})^T (F_v^*)^T (R_v - C^T Q_F C) F_v^* \overline{x}_0^{(P)} \\
&- Trace((F_v^*)^T (R_v - C^T Q_F C) F_v^* (P_0^{(P)} + P_0^{(E)})) \\
&+ 2u_0^T B^T Q_F C F_v^* \, \overline{x}_0^{(P)}
\end{aligned} \tag{5.63}$$

Also, in (5.63)

$$\begin{aligned}
u_0^* &= -(R_u + B^T Q_F B)^{-1} B^T Q_F (A \overline{x}_0^{(P)} + C E_{\tilde{w}} \, (g(\overline{x}_0^{(P)} + \tilde{w}) \,)) \\
&= -(R_u + B^T Q_F B)^{-1} B^T Q_F (A \overline{x}_0^{(P)} + C \; F_v^* \overline{x}_0^{(P)}) \\
&= -(R_u + B^T Q_F B)^{-1} B^T Q_F (A + C \; F_v^*) \overline{x}_0^{(P)}
\end{aligned} \tag{5.64}$$

Inserting (5.59) and (5.64) into (5.63) yields the value function of player P. The value function $V_0^{(P)}(\overline{x}_0^{(P)})$ of player P is quadratic in $\overline{x}_0^{(P)}$. It is of the form

$$V_0^{(P)}(\overline{x}_0^{(P)}) = (\overline{x}_0^{(P)})^T M \overline{x}_0^{(P)} + c^{(P)}$$

where M is an $n \times n$ real, symmetric matrix and $c^{(P)}$ is a constant. While the matrix M is complex in appearance, note that it is not dependent on the covariances $P_0^{(P)}$ and $P_0^{(E)}$ of the players' state measurement errors. Hence, we conclude that the matrix

$$M = P_1 \, ,$$

where the matrix P_1 is given by (5.11). The constant

$$\begin{aligned}
c^{(P)} = Trace(\; & A^T Q_F A P_0^{(P)} + 2P_0^{(P)} A^T Q_F C F_v^* \\
& - (F_v^*)^T (R_v - C^T Q_F C) F_v^* (P_0^{(P)} + P_0^{(E)}) \,) \, ,
\end{aligned} \tag{5.65}$$

where the gain F_v^* is given by (5.59).

Next, consider the expected cost $\overline{J}^{(E)}(f(\cdot), v_0; \overline{x}_0^{(E)})$ of player E, (5.40). The expectations

$$
\begin{aligned}
E_{\tilde{w}}\left(f^T(\overline{x}_0^{(E)} - \tilde{w})(R_u + B^T Q_F B) f(\overline{x}_0^{(E)} - \tilde{w})\right) \\
&= (\overline{x}_0^{(E)} - \tilde{w})^T (F_u^*)^T (R_u + B^T Q_F B) F_u^* \overline{x}_0^{(E)} \\
&+ Trace((F_u^*)^T (R_u + B^T Q_F B) F_u^* (P_0^{(P)} \\
&+ P_0^{(E)}))
\end{aligned} \tag{5.66}
$$

$$
E_{\tilde{w}}\left(f(\overline{x}_0^{(E)} - \tilde{w})\right) = F_u^* \overline{x}_0^{(E)} \tag{5.67}
$$

and

$$
E_{w_E, w_P}\left(f^T(\overline{x}_0^{(E)} + w_P - w_E) B^T Q_F A w_E\right) = -Trace((F_u^*)^T B^T Q_F P_0^{(E)}) \tag{5.68}
$$

Inserting (5.66)–(5.68) into (5.40) yields the value function of player E

$$
\begin{aligned}
V_0^{(E)} &= (\overline{x}_0^{(E)})^T A^T Q_F A \overline{x}_0^{(E)} + Trace(A^T Q_F A P_0^{(E)}) - v_0^T (R_v - C^T Q_F C) v_0 \\
&+ 2 v_0^T C^T Q_F A \overline{x}_0^{(E)} + (\overline{x}_0^{(E)})^T (F_u^*)^T (R_u + B^T Q_F B) F_u^* \overline{x}_0^{(E)} \\
&+ Trace((F_u^*)^T (R_u + B^T Q_F B) F_u^* (P_0^{(P)} + P_0^{(E)})) \\
&+ 2 (\overline{x}_0^{(E)})^T (F_u^*)^T B^T Q_F A \overline{x}_0^{(E)} \\
&+ 2 Trace((F_u^*)^T B^T Q_F P_0^{(E)}) \\
&+ 2 v_0^T C^T Q_F B F_u^* \overline{x}_0^{(E)}
\end{aligned} \tag{5.69}
$$

Also, in (5.69),

$$
\begin{aligned}
v_0^* &= (R_v - C^T Q_F C)^{-1} C^T Q_F (A \overline{x}_0^{(E)} + B E_{\tilde{w}}\left(f(\overline{x}_0^{(E)} - \tilde{w})\right)) \\
&= (R_v - C^T Q_F C)^{-1} C^T Q_F (A + B F_u^*) \overline{x}_0^{(E)}
\end{aligned} \tag{5.70}
$$

Inserting (5.58) and (5.70) into (5.69) yields the value function of player E. The value function $V_0^{(E)}(\overline{x}_0^{(E)})$ of player E is quadratic in $\overline{x}_0^{(E)}$. Similar to the value function of player P, it is of the form

$$
V_0^{(E)}(\overline{x}_0^{(E)}) = (\overline{x}_0^{(P)})^T P_1 \overline{x}_0^{(P)} + c^{(E)}
$$

The constant

$$
\begin{aligned}
c^{(E)} = Trace(\, A^T Q_F A P_0^{(E)} + (F_u^*)^T (R_u + B^T Q_F B) F_u^* (P_0^{(P)} + P_0^{(E)}) \\
+ 2 (F_u^*)^T B^T Q_F P_0^{(E)}\,)
\end{aligned} \tag{5.71}
$$

where the gain F_u^* is given by (5.58).

These results are summarized in

Theorem 5.3. *A necessary and sufficient condition for the existence of a solution to the game* (5.1) *and* (5.2) *where the players have private information, that is, a game where the initial state information of player P is specified in* (5.15) *and the initial state information of player E is specified in* (5.16) *is that condition* (5.5) *holds. Certainty equivalence holds and the optimal P strategy is given by* (5.6) *where x_0 is replaced by $\overline{x}_0^{(P)}$ and the optimal E strategy is given by* (5.7) *where x_0 is replaced by $\overline{x}_0^{(E)}$. The value function of player P is*

$$V_0^{(P)}(\overline{x}_0^{(P)}) = (\overline{x}_0^{(P)})^T P_1 \overline{x}_0^{(P)} + c^{(P)} \tag{5.72}$$

and the value function of player E

$$V_0^{(E)}(\overline{x}_0^{(E)}) = (\overline{x}_0^{(P)})^T P_1 \overline{x}_0^{(P)} + c^{(E)} \tag{5.73}$$

The matrix P_1 in (5.72) *and* (5.73) *is specified by* (5.9) *and the constant terms in* (5.72) *and* (5.73) , *$c^{(P)}$ and $c^{(E)}$, are specified in* (5.65) *and* (5.71), *respectively.*

5.7 Conclusion

A static two-player linear-quadratic game where the players have private information on the game's parameter, is addressed. The players have private information, however each player is able to formulate an expression for his expected payoff, without the need, a la Harsanyi, to provide a prior probability distribution function of the game's parameter, and without recourse to the player Nature. Hence, the closed-form solution of the game is possible. It is shown that in this special case of a one-stage linear-quadratic game where the players have private information, the solution is similar in structure to the solution of the game with complete information, namely, the deterministic linear-quadratic game, and the solution of the linear-quadratic game with partial information, where the information about the game's parameter is shared by the players. The principle of certainty equivalence holds. The analysis in this paper shows the way to possible extensions of the theory to multi-stage linear-quadratic dynamic games.

References

Başar T, Bernhard P (2008) H^{∞}—Optimal control and related minimax design problems: A dynamic game approach. Birkhäuser, Boston

Başar T, Olsder GJ (1995) Dynamic non-cooperative game theory. Academic Press, San Diego, CA

Engwerda J (2005) LQ Dynamic optimization and differential games. Wiley, Chichester

Pachter M, Pham K (2010) Discrete-time linear-quadratic dynamic games. J Optim Theory Appl 146(1):151–179

Pachter M, Pham K (2013) Static teams and stochastic games. In: Sorokin A, Pardalos PM (eds) Dynamics of information systems: Algorithmic approaches. Springer Proceedings in Mathematics & Statistics, vol 51. Springer, New York

Zhang F (2005) The Schur complement and its applications. Springer, New York

Chapter 6
Interior Convergence Under Payoff Monotone Selections and Proper Equilibrium: Application to Equilibrium Selection

Dai Zusai

Abstract An interior convergent path in an evolutionary dynamic can be seen as a sequence of perturbed quasi-equilibria to refine an equilibrium. In this note, we investigate a variety of conditions to establish the connection between interior convergence in regular payoff monotone selections and versions of proper equilibrium and use the connection for equilibrium selection.

Keywords Evolutionary dynamics • Payoff monotone selections • Proper equilibrium

6.1 Introduction

In equilibrium refinement, we consider a sequence of completely mixed strategy profiles that satisfy a certain quasi-equilibrium condition and then define a refined equilibrium as a limit of such a sequence. For example, a trembling-hand perfect equilibrium is a limit of perturbed equilibria; a proper equilibrium is a limit of strategy profiles in which a player is more likely to choose a better strategy than a worse one. In evolutionary game theory, we could interpret such a sequence as an interior convergent path under a certain kind of dynamics. In this paper, we consider the relationship between proper equilibrium and interior convergence under monotone selections such as the replicator dynamic.

Hofbauer (1995) verifies that, in a linear population game, a trembling-hand perfect equilibrium is a limit of an interior convergent path under the best response

D. Zusai (✉)
Department of Economics, Temple University, Philadelphia, PA 19122, USA
e-mail: zusai@temple.edu

V. Křivan and G. Zaccour (eds.), *Advances in Dynamic Games*, Annals of the International Society of Dynamic Games 13, DOI 10.1007/978-3-319-02690-9_6,

dynamic (BRD) and vice versa.[1] Under the BRD, only the optimal strategy increases its share of the players in the population. So each social state on the path can be seen as a perturbed equilibrium, where most of players take the optimal strategy. Thus it is natural to have the link between interior convergence under the BRD and a trembling-hand perfect equilibrium.

Under the BRD, any suboptimal strategy decreases its share of players at a constant decaying rate however close its payoff is to the optimal payoff. On the other hand, a monotone selection requires the share of a worse strategy to decay faster than that of a better strategy. Thus we would expect the limit of an interior convergent path under a monotone selection to be a proper equilibrium. However, in a simple two-stage chain-store game (Cressman 2003, p. 291), there is an interior convergent path to a non-perfect/proper Nash equilibrium. So we cannot obtain a general connection.

In this paper, we present a sufficient condition that guarantees the connection between interior convergence in regular monotone selections and proper equilibrium. First, we prove that the limit of an interior convergent path under a monotone selection is a weaker version of proper equilibrium, which is still stronger than Nash equilibrium. Under a certain condition on the path and its limit, our theorem is extended to verify that the limit is a proper equilibrium.

Here we employ a slightly weaker definition of payoff monotonicity than the conventional definition so as to include the tempered best response dynamic (tBRD), proposed by Zusai (2012). In the tBRD, an agent is more likely to revise his strategy, as his current strategy becomes more disadvantageous compared to the optimal strategy; when he revises it, he switches to the current optimal strategy like in the standard BRD. The tBRD satisfies our definition of payoff monotonicity, while the standard BRD does not. Using the results in this paper, Zusai (2012) contrasts these two dynamics and argues that the introduction of payoff-dependent revision rates makes both medium and long-run outcomes more consistent with equilibrium refinement.

This paper proceeds as follows. The next section is devoted to set up the games and the dynamics. In Sect. 6.3, we present several theorems to connect monotone selections and proper equilibrium. We apply them to equilibrium selection in a couple of examples in Sect. 6.4. The last section concludes the paper.

6.2 The Games and Dynamics

6.2.1 Population Games

We first introduce a general framework of a large population game and then interpret finite-player normal-form games as population games. A population game is played in a society with large populations of infinitely many anonymous agents. Here,

[1]Hendon et al. (1996) prove that the limit state under a fictitious play is a sequential equilibrium in an extensive form game. They consider two kinds of fictitious play in a sequential-move game and the theorem applies to both. *Local* fictitious play is the one played by "agents" in an agent-normal form, while *sequential* fictitious play is the one played in a normal form of the sequential-move game.

anonymity means that the aggregate distribution of strategies determines the payoffs of each strategy.

A society is composed of P populations $\mathscr{P} := \{1, \dots, P\}$. Each population is a unit mass of infinitely many agents with the same strategy set and the same payoff function.[2] Each agent in population $p \in \mathscr{P}$ chooses a strategy s from $\mathscr{S}^p := \{1, \dots, S^p\}$. Let $S := \sum_{p \in \mathscr{P}} S^p$ be the total number of strategies in all populations.

Denote by $x_s^p \in [0, 1]$ the mass/share of strategy-s players in population p. The state of population p is represented by a column vector $\mathbf{x}^p := (x_1^p, \dots, x_{S^p}^p)$ in $\mathscr{X}^p := \Delta \mathscr{S}^p = \{\mathbf{x}^p \in [0, 1]^{S^p} | \sum_{s \in \mathscr{S}^p} x_s^p = 1\}$.[3] The social state is represented by a column vector $\mathbf{x} := (\mathbf{x}^1, \dots, \mathbf{x}^P)$ in $\mathscr{X} := \prod_{p \in \mathscr{P}} \mathscr{X}^p$.[4] We omit superscripts for p, when it is clear from the context or the society consists of only one population ($P = 1$).

The payoff of each strategy is a function of the social state. Given the state $\mathbf{x} \in \mathscr{X}$, $F_s^p(\mathbf{x})$ is the payoff for a player of strategy $s \in \mathscr{S}^p$ in population p. Define payoff functions $\mathbf{F}^p : \mathscr{X} \to \mathbb{R}^{S^p}$ and $\mathbf{F} : \mathscr{X} \to \mathbb{R}^S$ by column vectors $\mathbf{F}^p(\mathbf{x}) := (F_1^p(\mathbf{x}), \dots, F_{S^p}^p(\mathbf{x}))$ and $\mathbf{F}(\mathbf{x}) := (\mathbf{F}^1(\mathbf{x}), \dots, \mathbf{F}^P(\mathbf{x}))$ for each $p \in \mathscr{P}$ and $\mathbf{x} \in \mathscr{X}$. In summary, a population game is mathematically defined by $\mathbf{F} : \mathscr{X} \to \mathbb{R}^S$. Assume the continuous differentiability of $\mathbf{F}$.

As usual, a Nash equilibrium is a state where (almost) every agent takes an optimal strategy. Formally, a social state $\mathbf{x} \in \mathscr{X}$ is a *Nash equilibrium*, if for each $p \in \mathscr{P}$ and $s \in \mathscr{S}^p$

$$F_s^p(\mathbf{x}) < F_*^p(\mathbf{x}) \Longrightarrow x_s^p = 0.$$

Here $F_*^p(\mathbf{x}) := \max_{s' \in \mathscr{S}^p} F_{s'}^p(\mathbf{x})$ is p's payoff from an optimal strategy in state $\mathbf{x}$. Let $b^p(\mathbf{x}) := \arg\max_{s' \in \mathscr{S}^p} F_{s'}^p(\mathbf{x})$ be the set of p's optimal strategies (the pure best responses) in $\mathbf{x}$. For a strategy profile $\mathbf{s} = (s^1, \dots, s^P) \in \mathscr{S}$, $b^{-1}(\mathbf{s}) := \{\mathbf{x} \in \mathscr{X} | s^p \in b^p(\mathbf{x})\ \forall p\}$ is the set of the social states to which $\mathbf{s}$ is the best response, i.e. the best response region of $\mathbf{s}$.

An interior social state $\mathbf{x}^\varepsilon \in \mathring{\mathscr{X}} := \mathscr{X} \cap (0, 1)^S$ is an ε-perfect equilibrium with $\varepsilon > 0$, if for each $p \in \mathscr{P}$ and $s \in \mathscr{S}^p$

$$F_s^p(\mathbf{x}^\varepsilon) < F_*^p(\mathbf{x}^\varepsilon) \Longrightarrow x_s^{\varepsilon,p} < \varepsilon.$$

A *(trembling-hand) perfect equilibrium* is the limit of a sequence $\{\mathbf{x}^n\}_{n \in \mathbb{N}}$ of ε^n-perfect equilibria with $\varepsilon^n \to 0$. Furthermore, $\mathbf{x}^\varepsilon \in \mathring{\mathscr{X}}$ is an ε-proper equilibrium with $\varepsilon > 0$, if for each $p \in \mathscr{P}$ and $s, \hat{s} \in \mathscr{S}^p$

---

[2]The assumption of unit mass is made just for notational simplicity. We could easily extend the model and the results to general cases where different populations have different masses.

[3]For a finite set $\mathscr{Z} = \{1, \dots, Z\}$, we define $\Delta\mathscr{Z}$ as $\Delta\mathscr{Z} := \{(\rho_1, \dots, \rho_Z) \in [0, 1]^Z | \sum_{z \in \mathscr{Z}} \rho_z = 1\}$, i.e. the set of all probability distributions on $\mathscr{Z}$.

[4]A bold letter represents a column vector. Precisely $\mathbf{x}$ is a column vector $(x_1^1, \dots, x_{S^1}^1, x_1^2, \dots, x_{S^2}^2, \dots, x_1^P, \dots, x_{S^P}^P)$.

$$F_s^p(\mathbf{x}^\varepsilon) < F_{\hat{s}}^p(\mathbf{x}^\varepsilon) \implies x_s^{\varepsilon,p} < \varepsilon x_{\hat{s}}^{\varepsilon,p}.$$

A *proper equilibrium* is the limit of a sequence $\{\mathbf{x}^n\}_{n\in\mathbb{N}}$ of ε^n-proper equilibria with $\varepsilon^n \to 0$.

Example 6.1 (Single-population random matching in a symmetric normal-form game). Consider a symmetric two-player game in a normal form: each player chooses a pure strategy from $\mathscr{S} = \{1, \ldots, S\}$ and gets $\Pi_{s\hat{s}} \in \mathbb{R}$ when he takes strategy $s \in \mathscr{S}$ and the opponent takes $\hat{s} \in \mathscr{S}$. Let $\boldsymbol{\Pi}$ be the payoff matrix in which $\Pi_{s\hat{s}}$ is in the s-th row and the $\hat{s}$-th column.

To use $\boldsymbol{\Pi}$ to define a population game, we imagine a single population of agents who are randomly matched: $P = 1$. An agent chooses a strategy from $\mathscr{S}$; then he meets another agent from the same population and they play the normal-form game $\boldsymbol{\Pi}$. When he decides on a strategy, the agent does not know which opponent he will face; so to evaluate the payoff from each strategy, he uses the expected payoff. The payoff function $\mathbf{F} : \mathscr{X} \to \mathbb{R}^S$ in this population game is thus defined as

$$F_s(\mathbf{x}) := \sum_{\hat{s}\in\mathscr{S}} \Pi_{s\hat{s}} x_{\hat{s}} \quad \text{for each } s \in \mathscr{S}, \mathbf{x} \in \mathscr{X}.$$

Notice that the best response to social state $\mathbf{x}$ in this population game $\mathbf{F}$ coincides with the best response in the normal-form game $\boldsymbol{\Pi}$ when the opponent takes mixed strategy $\mathbf{x}$; hence, Nash equilibria of $\mathbf{F}$ coincide with symmetric Nash equilibria of $\boldsymbol{\Pi}$. Likewise the equilibrium concepts in $\mathbf{F}$ coincide with the symmetric ones in $\boldsymbol{\Pi}$.

Example 6.2 (Multi-population random matching in a normal-form game). Let us consider a general multi-player game in a normal form. Now we explicitly distinguish players and allow them to have different strategy sets and different payoff matrices. Each player $p \in \mathscr{P} := \{1, \ldots, P\}$ chooses a pure strategy from $\mathscr{S}^p = \{1, \ldots, S^p\}$ and gets payoff $U_{\mathbf{s}}^p \in \mathbb{R}$ from strategy profile $\mathbf{s} = (s^1, \ldots, s^P) \in \mathscr{S}$, i.e. when each player $q \in \mathscr{P}$ takes strategy $s^q \in \mathscr{S}^q$. Denote by $\mathbf{U}^p$ the payoff matrix or alignment of player p, i.e. $\mathbf{U}^p := \{U_{\mathbf{s}}^p | \mathbf{s} \in \mathscr{S}\}$ and $\mathbf{U} := (\mathbf{U}^p)_{p\in\mathscr{P}}$.[5]

To interpret the normal-form game $\mathbf{U}$ as a population game, we imagine a society with P populations $\mathscr{P}$. One agent is randomly chosen from each population $p \in \mathscr{P}$ and they are matched and then play the normal-form game $\mathbf{U}$. As before, the payoff in this random-matching game is defined from the expected payoff in the normal-form game $\mathbf{U}$:

$$F_{s^p}^p(\mathbf{x}) := \sum_{\mathbf{s}^{-p}=(s^q)_{q\neq p}\in\mathscr{S}^{-p}} U_{s^p,\mathbf{s}^{-p}}^p \prod_{q\neq p} x_{s^q}^q$$

for each of p's strategy $s^p \in \mathscr{S}^p$ and each of the social state $\mathbf{x} \in \mathscr{X}$.

[5]Notice that a symmetric game $\boldsymbol{\Pi} \in \mathbb{R}^{S\times S}$ with the pure strategy set $\mathscr{S}$ is expressed as a general two-player game as $U_{s\hat{s}}^1 := \Pi_{s\hat{s}}$ and $U_{s\hat{s}}^2 = \Pi_{\hat{s}s}$ with $\mathscr{S}^1 = \mathscr{S}^2 := \mathscr{S}$. That is, it is a bimatrix game with $(\mathbf{U}^1, \mathbf{U}^2) = (\boldsymbol{\Pi}, \boldsymbol{\Pi}')$.

Population p's best response to social state $\mathbf{x}$ in this population game $\mathbf{F}$ coincides with player p's best response in the normal-form game $\mathbf{U}$ when the opponent players take mixed strategy profile $\mathbf{x}^{-p} := (\mathbf{x}^q)_{q \neq p}$; hence, the set of Nash equilibria of this random-matching game $\mathbf{F}$ coincides with the set of *all* (both symmetric and asymmetric) Nash equilibria of the normal-form $\mathbf{U}$. The other equilibrium concepts in $\mathbf{F}$ also match with the ones in $\mathbf{U}$.

6.2.2 Monotone Selections

We consider continuous-time evolutionary dynamics on $\mathscr{X}$. In an evolutionary dynamic, each agent recurrently revises his strategy according to a certain revision protocol. In large populations of infinitely many agents, their aggregate behavior is described by a deterministic dynamic on the space of the social state $\mathscr{X}$.

An evolutionary dynamic is said to satisfy payoff monotonicity and called a *(payoff) monotone selection*, if any interior path $\{\mathbf{x}(t)\}_{t \in \mathbb{R}_+} \subset \mathring{\mathscr{X}}$ satisfies both of the following two conditions for almost all time $t \in \mathbb{R}_+ := [0, \infty)$, any $p \in \mathscr{P}$ and any $s, \hat{s} \in \mathscr{S}^p$:

$$F_s^p(\mathbf{x}(t)) > F_{\hat{s}}^p(\mathbf{x}(t)) \Rightarrow \frac{\dot{x}_s^p(t)}{x_s^p(t)} > \frac{\dot{x}_{\hat{s}}^p(t)}{x_{\hat{s}}^p(t)}, \tag{PM1}$$

$$s \in b^p(\mathbf{x}(t)) \Rightarrow \dot{x}_s^p(t) \geq 0. \tag{PM2}$$

Furthermore, we call it *regular* if

$$\liminf_{t\to\infty}\{F_s^p(\mathbf{x}(t)) - F_{\hat{s}}^p(\mathbf{x}(t))\} > 0 \Rightarrow \liminf_{t\to\infty}\left\{\frac{\dot{x}_s^p(t)}{x_s^p(t)} - \frac{\dot{x}_{\hat{s}}^p(t)}{x_{\hat{s}}^p(t)}\right\} > 0. \tag{PM3}$$

In the preceding literature, payoff monotonicity requires "two-sided" monotonicity:

$$F_s^p(\mathbf{x}(t)) > F_{\hat{s}}^p(\mathbf{x}(t)) \Leftrightarrow \frac{\dot{x}_s^p(t)}{x_s^p(t)} > \frac{\dot{x}_{\hat{s}}^p(t)}{x_{\hat{s}}^p(t)}. \tag{PM0}$$

For example, see Weibull (1995, Definition 4.2) and Hofbauer and Sigmund (1998, p. 88). Samuelson and Zhang (1992) and Cressman (2003, Definition 2.3.2) call it monotonicity, and Sandholm (2010, p. 163) monotone percentage growth rates.

With the invariance of $\mathscr{X}$, i.e. $\sum_{s \in \mathscr{S}^p} \dot{x}_s^p = 0$, (PM0) implies our definition of monotonicity, both of (PM1) and (PM2), but not vice versa. Hence our monotonicity is weaker than theirs.[6]

---

[6]Cressman (2003, Definition 2.3.2) defines uniform monotonicity by imposing $\exists K \geq 1$ such that $K|F_s^p(\mathbf{x}(t)) - F_{\hat{s}}^p(\mathbf{x}(t))| \geq |\dot{x}_s^p(t)/x_s^p(t) - \dot{x}_{\hat{s}}^p(t)/x_{\hat{s}}^p(t)| \geq K^{-1}|F_s^p(\mathbf{x}(t)) - F_{\hat{s}}^p(\mathbf{x}(t))|$. This implies our regularity (PM3).

Example 6.3 (Replicator dynamic). The *replicator dynamic* defined as below (Taylor and Jonker 1978) is a regular monotone selection:

$$\dot{x}_s^p = x_s^p(F_s^p(\mathbf{x}) - \bar{F}^p(\mathbf{x})) \qquad \text{for each } p \in \mathscr{P}, s \in \mathscr{S}^p, \mathbf{x} \in \mathscr{X}.$$

Here $\bar{F}^p(\mathbf{x}) := \sum_{s' \in \mathscr{S}^p} x_{s'}^p F_{s'}^p(\mathbf{x})$ is the average payoff in population p. Schlag (1998) proposes a protocol to derive the replicator dynamic from imitation. In general, imitative dynamics are payoff monotone. See Sandholm (2010, Observation 5.4.8.).

Example 6.4 (tBRD). Zusai (2012) defines the *tempered best response dynamic* as

$$\dot{\mathbf{x}}^p \in \sum_{s \in \mathscr{S}^p} x_s^p Q(\breve{F}_s^p(\mathbf{x}))(B^p(\mathbf{x}) - \mathbf{e}_s^p) \qquad \text{for each } p \in \mathscr{P}, \mathbf{x} \in \mathscr{X}.$$

Here $\breve{F}_s^p(\mathbf{x}) := F_*^p(\mathbf{x}) - F_s^p(\mathbf{x})$ is the current payoff deficit of strategy $s \in \mathscr{S}^p$, $Q : \mathbb{R}_+ \to [0,1]$ is an increasing and continuously differentiable function with $Q(0) = 0$ and $Q(\breve{\pi}) > 0$ for any $\breve{\pi} > 0$, and $B^p(\mathbf{x}) = \arg\max_{\mathbf{y} \in \mathscr{X}^p} \mathbf{y} \cdot \mathbf{F}^p(\mathbf{x})$ is the set of p's mixed best responses to the current state $\mathbf{x}$.[7] In the tBRD, an agent myopically switches to the current optimal strategy if the payoff deficit is larger than a stochastic status-quo bias, whose distribution function is Q; otherwise, he continues to play the current strategy.

While the tBRD can admit multiple transition vectors like the standard BRD when there are multiple optimal strategies, it satisfies our definition of regular payoff monotonicity (PM1)–(PM3) unlike the BRD. Note that the tBRD does not satisfy the two-sided monotonicity (PM0), because multiple optimal strategies can have different nonnegative growth rates.[8]

6.3 Monotone Selections and Properness

Despite apparently natural analogy between payoff monotonicity and properness, the limit of an interior convergent path under a payoff monotone selection may not be proper, or even not perfect. In the definition of ε-properness, the proportion of a worse strategy should be smaller than that of better one. Although payoff monotonicity makes the worse strategy *decay* faster than the better, the decaying rate can vanish when the payoff difference diminishes. If the proportion of the worse strategy is sufficiently large in the population, the vanishing decay rate may eventually keep it survive in the limit.

[7] $\mathbf{e}_a = (e_{ab})_{b=1}^n \in \mathbb{R}^n$ is a basis vector in $\mathbb{R}^n$, with $e_{aa} = 1$ and $e_{ab=0}$ for any $b \neq a$.

[8] For the same reason, the tBRD does not satisfy Cressman's uniformity.

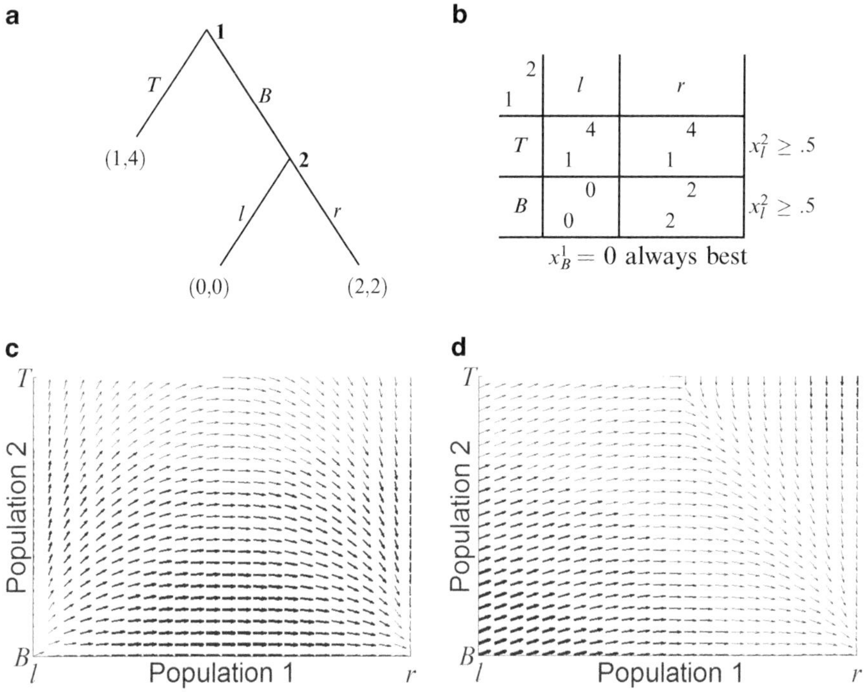

Fig. 6.1 Example 6.5. The inequality out of each column/row of the table is the condition on the social state **x** for the strategy in this column/row to be the best response. In Figs. (**c**) and (**d**), the width of an arrow shows the norm of a transition vector, while the length of an arrow is normalized. (**a**) Extensive form, (**b**) Normal form, (**c**) $\dot{\mathbf{x}}$ in replicator, (**d**) $\dot{\mathbf{x}}$ in tBRD

Example 6.5 (Cressman 2003). Consider a two-player sequential-move game in Fig. 6.1a and a two-population random matching in its normal form. While (B, r) is a subgame perfect equilibrium, there is a connected component of Nash equilibria $N_0 = \{\mathbf{x} | x_T^1 = 1, x_l^2 \geq .5\}$ that are not subgame perfect.

According to Fig. 6.1c, d, there are interior paths converging to N_0 under both the replicator dynamic and the tBRD. In the interior space $\mathring{\mathscr{X}}$ around N_0, both strategy B in population 1 and strategy l in population 2 are suboptimal and thus decrease their shares of players. But the payoff difference between l anr r vanishes as the state approaches N_0. The decaying speed of strategy l hence diminishes to zero under these dynamics, while strategy B keeps a large payoff deficit and a great decaying speed. As a result, strategy l survives while strategy B does not, when the state reaches a limit state in N_0. See also Cressman (2003, Example 9.1.2, p. 291).

Diminishing decaying rates due to vanishing payoff differences interrupt the connection between proper equilibrium and monotone selections. To retain this connection, we first neglect such vanishing payoff differences and define a weaker version of proper equilibrium, a pseudo-proper equilibrium. In a sequence converging to a pseudo-proper equilibrium, the shares of players are ranked according to the payoff ordering *at the limit*, but they can be inconsistent with the payoff ordering at each state in the sequence. This is the same idea as a weakly proper equilibrium (van Damme 1991), while we drop perfectness from its definition.

Definition 6.1 (Pseudo-proper equilibrium). An interior social state $\mathbf{x}^\varepsilon \in \mathring{\mathscr{X}}$ is an *ε-pseudo-proper equilibrium for* $\mathbf{x}^* \in \mathscr{X}$ with $\varepsilon > 0$, if for each $p \in \mathscr{P}$ and $s, \hat{s} \in \mathscr{S}^p$

$$F_s^p(\mathbf{x}^*) < F_{\hat{s}}^p(\mathbf{x}^*) \Rightarrow x_s^{\varepsilon,p} < \varepsilon x_{\hat{s}}^{\varepsilon,p}.$$

A social state $\mathbf{x}^* \in \mathscr{X}$ is a *pseudo-proper equilibrium* if there is a sequence $\{(\mathbf{x}^n, \varepsilon^n)\}_{n\in\mathbb{N}} \subset \mathring{\mathscr{X}} \times (0,\infty)$ such that each $\mathbf{x}^n$ is an ε^n-pseudo-proper equilibrium for $\mathbf{x}^*$ and $(\mathbf{x}^n, \varepsilon^n) \to (\mathbf{x}^*, 0)$ as $n \to \infty$. Further, $\mathbf{x}^*$ is a *weakly proper equilibrium*, if each $\mathbf{x}^n$ in this sequence is also an ε^n-perfect equilibrium.

Pseudo-properness is weaker than properness but stronger than Nash equilibrium. Since a proper equilibrium always exists, so does a pseudo-proper equilibrium.

Theorem 6.1. *Consider a social state $\mathbf{x}^* \in \mathscr{X}$ in a population game $\mathbf{F}$. (i) If $\mathbf{x}^*$ is a pseudo-proper equilibrium, it is a Nash equilibrium. (ii) If $\mathbf{x}^*$ is a proper equilibrium, it is pseudo-proper.*

Proof. (i) A pseudo-proper equilibrium $\mathbf{x}^*$ has a sequence of ε^n-pseudo-proper equilibria $\{\mathbf{x}^n\}$ with $\varepsilon^n \to 0$ and $\mathbf{x}^n \to \mathbf{x}^*$ as $n \to \infty$. Consider a suboptimal strategy $s \in \mathscr{S}^p \setminus b^p(\mathbf{x}^*)$ at $\mathbf{x}^*$. Then, $x_s^n < \varepsilon^n$ by ε^n-pseudo properness and $x_\cdot \leq 1$. As $n \to \infty$, we have $x_s^* = \lim x_s^n \leq 0 = \lim \varepsilon^n$. By $x_s^* \geq 0$, we have $x_s^* = 0$. Because this holds for any suboptimal strategy at $\mathbf{x}^*$, $\mathbf{x}^*$ is a Nash equilibtium.

(ii) A proper equilibrium $\mathbf{x}^*$ has a sequence of ε^n-proper equilibria $\{\mathbf{x}^n\}$. Fix $\varepsilon > 0$ arbitrarily. The continuity of $\mathbf{F}$ guarantees the existence of $N \in \mathbb{N}$ s.t. $F_s^p(\mathbf{x}^*) < F_{\hat{s}}^p(\mathbf{x}^*)$ implies $F_s^p(\mathbf{x}^n) < F_{\hat{s}}^p(\mathbf{x}^n)$ for all $n \geq N$. As $\mathbf{x}^n$ is an ε^n-proper equilibrium, this further implies $x_s^n < \varepsilon^n x_{\hat{s}}^n$; thus, at each $n \geq N$, $\mathbf{x}^n$ is an ε^n-pseudo-proper equilibrium for $\mathbf{x}^*$. Hence $\mathbf{x}^*$ is pseudo-proper. □

We can readily establish the connection between a pseudo-proper equilibrium and monotone selections: the limit of an interior convergent path in any regular monotone selection is pseudo-proper. The converse is not true, as we see in Example 6.6.

Theorem 6.2. *Consider a population game $\mathbf{F}$. Suppose that there exists an interior path $\{\mathbf{x}(t)\}_{t\in\mathbb{R}_+} \subset \mathring{\mathscr{X}}$ converging to a state $\mathbf{x}^* \in \mathscr{X}$ under a regular payoff monotone selection. Then this limit $\mathbf{x}^*$ is a pseudo-proper equilibrium.*

Proof. Consider any two strategies $s, \hat{s} \in \mathscr{S}^p$ such that $F_s^p(\mathbf{x}^*) < F_{\hat{s}}^p(\mathbf{x}^*)$. Then, by continuity of $\mathbf{F}$ and (PM3), the difference in their growth rates is bounded above by some negative constant $-\Delta g_{s\hat{s}} < 0$, after a sufficiently long time $T_{s\hat{s}}$ has passed:

$$\frac{d}{dt}\left(\log\frac{x_s(t)}{x_{\hat{s}}(t)}\right) < -\Delta g_{s\hat{s}} < 0 \quad \text{for any } t > T_{s\hat{s}}.$$

Consequently we obtain

$$\frac{x_s(t)}{x_{\hat{s}}(t)} < \frac{x_s(T_{s\hat{s}})}{x_{\hat{s}}(T_{s\hat{s}})} \exp\left(-\Delta g_{s\hat{s}}(t - T_{s\hat{s}})\right) \text{ for any } t > T_{s\hat{s}}.$$

Since the RHS converges to zero as $t \to \infty$, for any $\varepsilon > 0$ we can find another threshold moment of time $T'_{s\hat{s}}(\varepsilon) \geq T_{s\hat{s}}$ such that

$$\frac{x_s(t)}{x_{\hat{s}}(t)} < \varepsilon \quad \text{for any } t > T'_{s\hat{s}}(\varepsilon). \tag{6.1}$$

As we allow only finitely many strategies and hence finitely many pairs of strategies, the maximum of these threshold moments exists: $\bar{T}(\varepsilon) := \max\{T'_{s\hat{s}}(\varepsilon) | p \in \mathscr{P}, s, \hat{s} \in \mathscr{S}^p, F^p_s(\mathbf{x}^*) < F^p_{\hat{s}}(\mathbf{x}^*)\} < \infty$. Construct an increasing and unbounded sequence of moments of time $\{t^n\}_{n\in\mathbb{N}}$ such as $t^1 = \bar{T}(1)$ and $t^n = \max\{\bar{T}(1/n), t^{n-1}\} + 1$ for $n \geq 2$, and then a sequence $\{\mathbf{x}^n\}_{n\in\mathbb{N}}$ by choosing $\mathbf{x}(t^n)$ for $\mathbf{x}^n$. Consequently, this sequence converges to $\mathbf{x}^*$, since $t^n \to \infty$ as $n \to \infty$ and $\mathbf{x}(t) \to \mathbf{x}^*$ as $t \to \infty$. Besides, at each $n \in \mathbb{N}$, (6.1) and $t^n \geq \bar{T}(1/n)$ imply that $\mathbf{x}^n$ is an ε^n-pseudo-proper equilibrium for $\mathbf{x}^*$ with $\varepsilon^n = 1/n$. Therefore, $\mathbf{x}^*$ is a pseudo-proper equilibrium. □

Under the same idea, Samuelson and Zhang (1992) prove that, if a pure strategy is iteratively strictly dominated by pure strategies, it does not survive in the long run under any regular payoff dynamic. As such a strategy is eliminated in a Nash equilibrium, the above theorem seems a stronger statement than theirs. But they have the convergence of the share of such a strategy to zero *as a result* and do *not assume* the convergence of the social state $\mathbf{x}$, i.e. the convergence of every strategy's share.

It is well known as the "folk theorem in evolutionary game theory" that the limit of any interior convergent path in any "reasonable" evolutionary dynamic is a Nash equilibrium.[9] As a pseudo-proper equilibrium is a Nash equilibrum, the above theorem, combined with Theorem 6.1, provides an easy proof of the folk theorem.

Pseudo-proper equilibrium may not be perfect. But, if the optimal strategy does not change on the path and the limit is a pure strategy profile, the limit is weakly proper, not only pseudo-proper.

Theorem 6.3. *Consider a population game* $\mathbf{F}$. *Suppose that there exists an interior path* $\{\mathbf{x}(t)\}_{t\in\mathbb{R}_+} \subset \mathring{\mathscr{X}}$ *converging to a state* $\mathbf{x}^* \in \mathscr{X}$ *under a regular payoff monotone selection.*

[9] For example, see Cressman (2003, p. 11).

Furthermore, suppose that $b^p(\mathbf{x}(t)) = b^p(\mathbf{x}(0))$ *for any* $t \in \mathbb{R}_+$ *and* $p \in \mathscr{P}$ *and that the limit* $\mathbf{x}^*$ *is a pure strategy profile. Then this limit* $\mathbf{x}^*$ *is a weakly proper equilibrium.*

Proof. Let $s \in b^p(\mathbf{x}(t))$ be the optimal strategy for population $p \in \mathscr{P}$ at some moment of time $t \in \mathbb{R}_+$; then, it is optimal at any moment $t' \in \mathbb{R}_+$. By (PM2), an optimal strategy cannot decrease its share of players. Thus $x_s^* \geq x_s(0) > 0$. Since the limit $\mathbf{x}^*$ is a pure strategy profile, s is the only strategy that is taken by agents in population p at the limit, i.e. $x_s^* = 1$. Then, for an arbitrary $\varepsilon > 0$, there exists $T(\varepsilon) \in \mathbb{R}_+$ such that, at any $t > T(\varepsilon)$, $x_s(t) > 1 - \varepsilon$ and thus $x_{\hat{s}}(t) < \varepsilon$ for any other strategy $\hat{s} \in \mathscr{S}^p$. Hence $\mathbf{x}(t)$ is an ε-perfect equilibrium. With Theorem 6.2, this suggests that the limit $\mathbf{x}^*$ is weakly proper. □

With further additional assumptions about the limit and the path, the limit becomes proper. Condition (i) below strengthens the assumption of the invariant optimal strategies in the above theorem to the invariant payoff ordering. Condition (ii) means that, if a strategy is suboptimal in the states on the convergent path, it should be extinguished in the limit, whether or not it becomes optimal in the limit.[10] Condition (iii) prohibits the payoff difference between strategies from vanishing at the limit, unless one strategy is the optimal and the other is the second best on the path.

Theorem 6.4. *Consider a population game* $\mathbf{F}$. *Suppose that there exists an interior path* $\{\mathbf{x}(t)\}_{t\in\mathbb{R}_+} \subset \mathring{\mathscr{X}}$ *converging to a state* $\mathbf{x}^* \in \mathscr{X}$ *under a regular payoff monotone selection.*

Furthermore, suppose that for any $p \in \mathscr{P}$ *and* $s, \hat{s} \in \mathscr{S}^p$

(i) $sgn(F_s^p(\mathbf{x}(t)) - F_{\hat{s}}^p(\mathbf{x}(t))) = sgn(F_s^p(\mathbf{x}(0)) - F_{\hat{s}}^p(\mathbf{x}(0)))$ *for all* $t \in \mathbb{R}_+$,
(ii) $x_s^{*p} = 0$ *if* $F_s^p(\mathbf{x}(t)) < F_*^p(\mathbf{x}(t))$ *at any* $t \in \mathbb{R}_+$, *and*
(iii) $F_s^p(\mathbf{x}^*) = F_{\hat{s}}^p(\mathbf{x}^*)$ *implies* $F_s^p(\mathbf{x}(t)) = F_{\hat{s}}^p(\mathbf{x}(t))$ *for all* $t \in \mathbb{R}_+$, *unless either* s *or* $\hat{s}$ *is optimal at some moment of time in* $\mathbb{R}_+$.

Then, $\mathbf{x}^*$ *is a proper equilibrium.*

Proof. Consider any two strategies $s, \hat{s} \in \mathscr{S}^p$ such that $F_s^p(\mathbf{x}(t)) < F_{\hat{s}}^p(\mathbf{x}(t))$ at some moment of time $t \in \mathbb{R}_+$. Then, by (i), this holds at every moment $t \in \mathbb{R}_+$. If this payoff ordering does not change at the limit, i.e., $F_s(\mathbf{x}^*) < F_{\hat{s}}(\mathbf{x}^*)$, then we can repeat the proof of Theorem 6.2 and obtain (6.1) with some $T'_{s\hat{s}}(\varepsilon) \in \mathbb{R}_+$.

Consider a case of $F_s(\mathbf{x}^*) = F_{\hat{s}}(\mathbf{x}^*)$. By (iii), $\hat{s}$ should be optimal on the path $\{\mathbf{x}(t)\}_{t\in\mathbb{R}_+}$. By (PM2), an optimal strategy cannot decrease its share of players. So $x_{\hat{s}}^* \geq x_{\hat{s}}(0) > 0$. On the other hand, s is suboptimal on the path and thus $x_s^* = 0$ by (ii). Hence we have $x_s(t)/x_{\hat{s}}(t) \to x_s^*/x_{\hat{s}}^* = 0$. Thus, for any $\varepsilon > 0$, we can find some $T'_{s\hat{s}}(\varepsilon) \in \mathbb{R}_+$ that satisfies (6.1)

[10] In Example 6.5, interior Nash equlibria in N_0 do not satisfy condition (ii).

For each $\varepsilon > 0$, we have $\bar{T}(\epsilon) := \max\{T'_{s\hat{s}}(\varepsilon) | p \in \mathscr{P}, s, \hat{s} \in \mathscr{S}^p, F^p_s(\mathbf{x}(t)) < F^p_{\hat{s}}(\mathbf{x}(t))$ at every $t \in \mathbb{R}_+\} < \infty$. Construct an increasing and unbounded sequence $\{t^n\}_{n \in \mathbb{N}} \subset \mathbb{R}_+$ such as $t^1 = \bar{T}(1)$ and $t^n = \max\{\bar{T}(1/n), t^{n-1}\} + 1$ for $n \geq 2$; then, define a sequence $\{\mathbf{x}^n\}_{n \in \mathbb{N}} \subset \mathring{\mathscr{X}}$ as $\mathbf{x}^n := \mathbf{x}(t^n)$. At each $n \in \mathbb{N}$, $\mathbf{x}^n$ is an ε^n-proper equilibrium with $\varepsilon^n = 1/n$. (Notice that the payoff ordering is invariant on the path.) Besides, since $t^n \to \infty$ and $\varepsilon^n \to 0$ as $n \to \infty$ and $\mathbf{x}(t) \to \mathbf{x}^*$ as $t \to \infty$, $\mathbf{x}^n = \mathbf{x}(t^n)$ converges to $\mathbf{x}^*$. Therefore, the limit $\mathbf{x}^*$ is a proper equilibrium. □

Condition (iii) seems to be quite restrictive, especially when we analyze a sequential-move game in a normal form. Yet, we can apply it to an interesting class of games. For example, consider a situation where one of the players decides on whether or not to interact with the others. If he decides to interact, the other players and he choose simultaneously and independently whether or not to cooperate with each other; otherwise, each receives a status-quo payoff. Such a situation is thought of, for example, when we consider how anticipation of uncommitted cooperation/collusion or its absence affects potential entry. This is described as a two-strategy simultaneous-move game with a single player having an outside option. Theorem 6.4 is applicable to pure-strategy equilibria in such a game in a reduced normal form, as we do in Example 6.7 in the next section.

6.4 Application to Equilibrium Selection

In equilibrium refinement, properness is justified as rational players' careful consideration of payoff rankings on the assumption of similar rationality on others.[11] In evolutionary game theory, we wonder if boundedly rational agents learn to play the same outcome as fully rational players would play. Here, we see a couple of examples where proper equilibrium results in the most plausible outcome while weaker equilibria such as perfect or sequential seem implausible. By Theorem 6.4, we find that boundedly rational agents eventually reach the plausible outcome as long as they follow a monotone selection.

Example 6.6. Consider a single-population random matching game with $\mathscr{S} = \{1, 2, 3\}$ and the payoff function $\mathbf{F}$ given below (Myerson 1978).

$$\mathbf{F}(\mathbf{x}) = \begin{pmatrix} 1 & 0 & -9 \\ 0 & 0 & -7 \\ -9 & -7 & -7 \end{pmatrix} \mathbf{x} = \begin{pmatrix} x_1 - 9x_3 \\ -7x_3 \\ -9x_1 - 7x_2 - 7x_3 \end{pmatrix}.$$

All vertexes $\mathbf{e}_1, \mathbf{e}_2$, and $\mathbf{e}_3$ are Nash equilibria. But strategy 3 is weakly dominated; so $\mathbf{e}_3$ is not perfect. $\mathbf{e}_2$ is only weakly proper, and $\mathbf{e}_1$ is proper. Note that in the replicator dynamic, $(1/6, 0, 5/6)$ is also a rest point; but it is not stable and not a Nash equilibrium.

[11] For epistemological foundation of properness, see Blume et al. (1991).

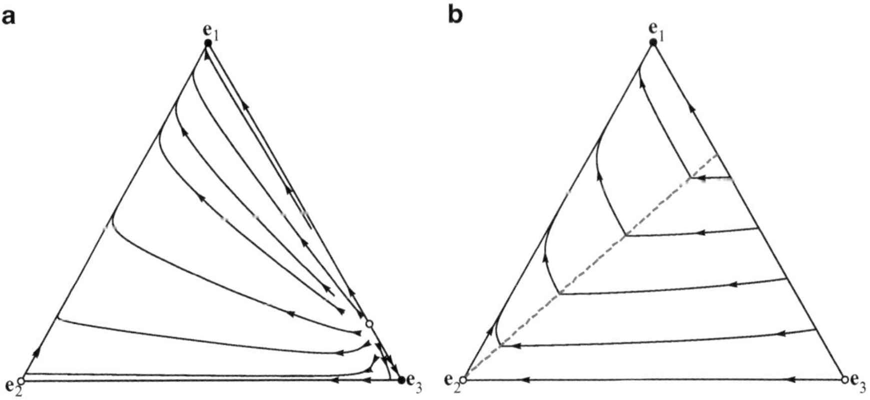

Fig. 6.2 Example 6.6. The solution paths are drawn with Dynamo. $\mathbf{e}_1$ is a proper equilibrium, $\mathbf{e}_2$ is only a weakly proper equilibrium, and $\mathbf{e}_3$ is a non-perfect Nash equilibrium. In Fig. (**b**), the optimal strategy is 1 in the region above the *gray dash line*, and 2 in the region *below*. (**a**) Paths in replicator, (**b**) Paths in tBRD

Figure 6.2 shows that all the interior paths converge to the proper equilibrium $\mathbf{e}_1$ both in the replicator dynamic and in the tBRD. We can negate interior convergence to $\mathbf{e}_2$. Suppose that there was an interior path converging to $\mathbf{e}_2$. By (PM2), strategy 2 should be optimal before the path reaches $\mathbf{e}_2$. It also implies that, if strategy 1 is optimal as well, x_1 should not decrease. Around $\mathbf{e}_2$, strategy 3 is the worse and hence x_3 should decrease in a monotone selection. These changes in x_1 and x_3 improve strategy 1's payoff relative to strategy 2's. So, if both strategies 1 and 2 are optimal at a moment of time, strategy 1 becomes uniquely optimal after this moment. Then, strategy 1 increases its share of players, remaining optimal and absorbing other players.

Therefore, if there was an interior convergent path to $\mathbf{e}_2$, strategy 2 should be the unique optimal strategy on the path. Such a path and the limit $\mathbf{e}_2$ would satisfy the assumptions in Theorem 6.4. But, the limit $\mathbf{e}_2$ is not a proper equilibrium, which contradicts with the conclusion of Theorem 6.4. Consequently, there cannot be any interior convergent path to $\mathbf{e}_2$.

This argument suggests that, regardless of the initial state, strategy 1 eventually becomes uniquely optimal and then every agent switches to it. Thus the social state converges to the proper equilibrium $\mathbf{e}_1$.

A proper equilibrium in a normal form implies a sequential equilibrium and thus a subgame perfect equilibrium in any corresponding extensive form. Theorem 6.4 helps us to select the most plausible outcome in a sequential-move game by convergence of a monotone selection, as long as the game is as simple enough as the example below.

Example 6.7. Consider a two-player sequential-move game in Fig. 6.3 (van Damme 1991, Fig. 6.5.1.). In this example, player 1 has an outside option that allows not to play a two-strategy simultaneous-move game with player 2. A typical interpretation

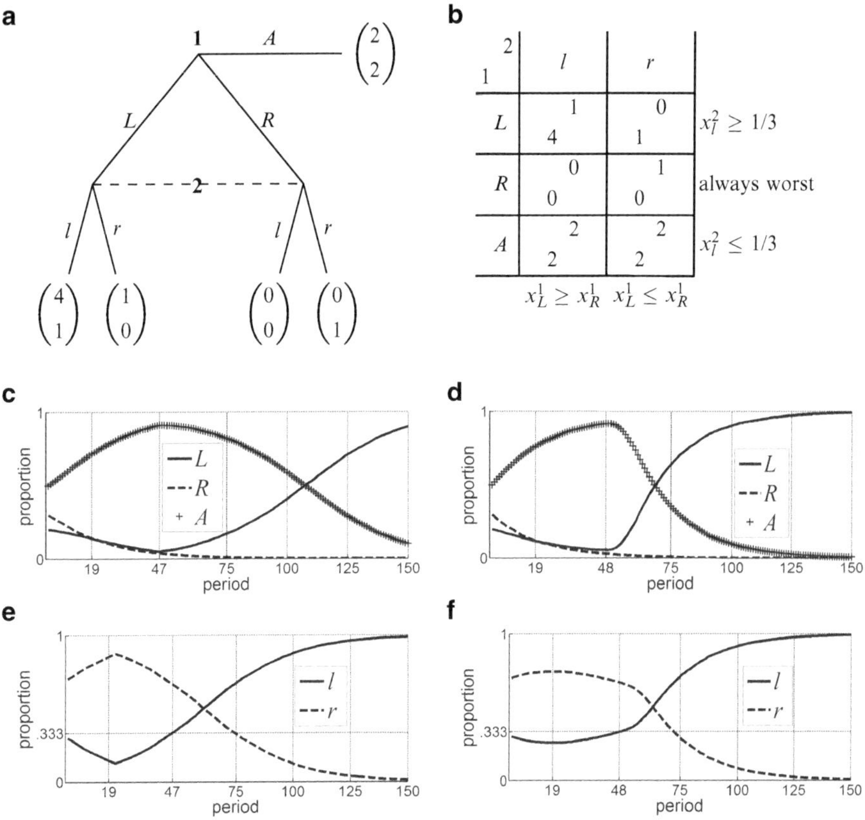

Fig. 6.3 Example 6.7. The inequality out of each column/row of the table is the condition for the strategy in this column/row to be the best response. Figs. (**c**)–(**f**) are obtained from discrete-time finite-population simulations. (**a**) Extensive form, (**b**) Normal form, (**c**) x^1 in replicator, (**d**) x^1 in tBRD, (**e**) x^2 in replicator, (**f**) x^2 in tBRD

is that player 1 is the entrant into a market and, if he decides to enter, the incumbent (player 2) and the entrant simultaneously choose whether to fight or to cooperate.

Both (A, r) and (L, l) are trembling-hand perfect in the normal form and sequential equilibria in the extensive form. (L, l) is the only proper equilibrium, while (A, r) is not. Actually (A, r) seems unreasonable if the two players are rational and the rationality is common knowledge. To choose r, player 2 should believe that player 1 plays R with higher probability than L. But R is always worse than L for player 1; hence, this belief seems against 1's rationality. We wonder if such an unreasonable outcome could be sustained in the long run of an evolutionary dynamic of boundedly rational agents.

Let us consider a two-population random matching in the normal form of this game and a regular monotone selection. We show that, even if a path starts from the interior of $b^{-1}(A, r)$, i.e. $\{\mathbf{x} \in \mathring{\mathscr{X}} | x_L^1 < x_R^1, x_l^2 < 1/3\}$, it converges to the proper equilibrium (L, l). Theorem 6.4 guarantees that the state does not reach the non-proper equilibrium $(\mathbf{e}_A^1, \mathbf{e}_r^2)$ without either inequality $x_L^1 < x_R^1$ or $x_l^2 < 1/3$ reversed. With payoff monotonicity, the former inequality, i.e. the unique optimality of r, implies that l decreases and the latter inequality keeps holding. Besides, as R is always the worst, x_R^1 should always decrease. Hence, for the path to escape from the interior of $b^{-1}(A, r)$, the inequality $x_L^1 < x_R^1$ should be reversed; the state enters the interior of $b^{-1}(A, l)$ and l becomes uniquely optimal. By payoff monotonicity, x_l^2 increases and eventually exceeds $1/3$; the social state goes into $b^{-1}(L, l)$. Then, L becomes optimal for population 1, as well as l for population 2. So the social state converges to $(\mathbf{e}_L^1, \mathbf{e}_l^2)$, i.e. the proper equilibrium (L, l).

Zusai (2012) uses this example to contrast the tBRD with the standard BRD, where an interior path converges from $b^{-1}(A, r)$ to a normal-form trembling-hand perfect equilibrium (A, r). As any regular monotone selection results in interior convergence to the proper equilibrium in this example, it shows a clear implication of payoff monotonicity, distinct from consistency with optimality alone.

6.5 Concluding Remarks

We argue the connection between interior convergence in payoff monotone selections and proper equilibrium. The connection cannot be generalized when payoff differences vanish at the limit, because of diminishing decaying rates. However, we present several versions of additional conditions to establish a natural link between them. In an example of a two-player three-strategy symmetric simultaneous-move game, the theorem helps to confirm global interior convergence to a proper equilibrium. A similar result is obtained in an example of a two-player two-strategy simultaneous-move game with one of the players having an outside option. These examples suggest that, despite very sophisticated rational reasoning implicitly imposed on its definition, a proper equilibrium can be supported as the long-run outcome from dynamic interactions of bounded rational (possibly non-optimizing) agents in large populations.

Acknowledgments I would like to thank Ross Cressman, Dimitrios Diamantaras, Josef Hofbauer, Bill Sandholm, and the participants at ISDG meeting, as well as Vlastimil Křivan and an anonymous referee, for their useful comments.

References

Blume L, Brandenburger A, Dekel E (1991) Lexicographic probabilities and equilibrium refinements. Econometrica 59(1):81–98

Cressman R (2003) Evolutionary dynamics and extensive form games. MIT Press, Cambridge, MA van Damme E (1991) Stability and perfection of Nash equilibria, 2nd edn. Springer, Berlin, Heidelberg

Hendon E, Jacobsen HJ, Sloth B (1996) Fictitious play in extensive form games. Games Econ Behav 15(2):177–202

Hofbauer J (1995) Stability for the best response dynamics, mimeo, University of Vienna

Hofbauer J, Sigmund K (1998) Evolutionary games and population dynamics. Cambridge University Press, Cambridge

Myerson RB (1978) Refinements of the Nash equilibrium concept. Int J Game Theory 7(2):73–80

Samuelson L, Zhang J (1992) Evolutionary stability in asymmetric games. J Econ Theory 57: 363–391

Sandholm WH (2010) Population games and evolutionary dynamics. MIT Press, Cambridge, MA

Sandholm WH, Dokumaci E, Franchetti F (2012) Dynamo: Diagrams for evolutionary game dynamics. http://www.ssc.wisc.edu/~whs/dynamo. Accessed 1 Jul 2012

Schlag KH (1998) Why imitate, and if so, how? a boundedly rational approach to multi-armed bandits. J Econ Theory 78:130–156

Taylor P, Jonker L (1978) Evolutionarily stable strategies and game dynamics. Math Biosci 40:145–156

Weibull JW (1995) Evolutionary game theory. MIT Press, Cambridge, MA

Zusai D (2012) Tempered best response dynamics, mimeo, Temple University

Part II
Dynamic Games: Applications

Chapter 7
Should a Retailer Support a Quality Improvements Strategy?

Pietro De Giovanni

Abstract In a one-manufacturer-one-retailer supply chain, players establish both operations and marketing strategies and coordinate the chain through the implementation of a support program. A retailer, who sets both the pricing and the advertising strategies, acts as chain leader and decides whether to support a manufacturer's operational strategy, such as quality improvements. The players share the overall chain revenues based on an exogenous, fixed sharing agreement. We compared coordinated and non-coordinated solutions in which coordination is carried out via a support program for quality improvements. While according to the literature a retailer–leader always has an economic preference for operation-based coordination, our findings reveal that: (a) low operational efficiency and effectiveness discourage the retailer's interest in coordination and (b) good sharing parameter values overcome concerns regarding operational inefficiency but not those of operational ineffectiveness.

Keywords Supply chain management • Coordination • Differential game • Advertising • Quality improvements • Support program • Feedback equilibrium

7.1 Introduction

Cachon (2003) defined supply chain coordination as the adoption of a specific contract that leads two or more players to a win–win situation. More recently, coordination has been investigated by means of several alternative mechanisms, such as a support program (Jørgensen et al. 2001), an incentive scheme (Jørgensen et al. 2006), and a combination of mechanisms (De Giovanni and Zaccour 2013).

P. De Giovanni (✉)
Department of Information, Logistics and Innovation, VU Amsterdam University, de Boelelaan 1105, 1081 HV Amsterdam, The Netherlands
e-mail: pietro.degiovanni@vu.nl

V. Křivan and G. Zaccour (eds.), *Advances in Dynamic Games*, Annals of the International Society of Dynamic Games 13, DOI 10.1007/978-3-319-02690-9_7,

Despite the devices employed, coordination has a unique objective: leading players to a payoff-Pareto-improving situation (De Giovanni and Roselli 2012).

Researchers have demonstrated that coordination is complex in the context of supply chains (von Lanzenauer and Pilz-Glombik 2002), as it is characterized by multiple interfaces among business functions (Erickson 2011). Each function has specific objectives, resources, and constraints; thus, determining optimal strategies implies considering numerous facets (Erickson 2011). This evidence has emerged in one of the most elaborate definitions of supply chain management (SCM), which was proposed by Mentzer et al. (2001): "SCM is the systematic, strategic coordination of the traditional business functions within a particular company and across businesses within the supply chain." Therefore, the interfaces among business functions in SCM have become a strategic issue, whose understanding requires a deep analysis of relationships, benefits, limitations, and countermeasures.

Research in differential games has addressed the existing interfaces between business functions and their strategies, including quality and advertising (Nair and Narasimhan 2006); quality, pricing, and advertising (De Giovanni 2011b); management accounting, advertising, and pricing (De Giovanni and Roselli 2012); quality and operational knowledge (Vörös 2006); pricing and inventory (Jørgensen 1986); quality, inventory, pricing, and advertising (El Ouardighi et al. 2008); pricing and quality (El Ouardighi and Kim 2010; Martín-Herrán et al. 2012); advertising and promotion (Jørgensen et al. 2003); contracting and operational efficiency (Kim 2003); operations and marketing (Erickson 2011); and innovation and pricing (Kim 2003). Managing interactions between business functions is critical and complex. We contribute to this stream of literature by introducing a dynamic equation of goodwill à la Nair and Narasimhan (2006) with the addition that quality influences goodwill based on its entire history. This dynamic equation seems to be very appealing nowadays, as it demonstrates that the impact of quality drastically affects the value of the brand. For instance, the recent scandals concerning non-quality food at Nestle' and Ikea have substantially damaged their respective images. Because these companies enjoy worldwide recognition, such scandals involve a loss of image that depends on the size of the brand value itself. That is, the higher the value of the brand, the higher the impact of a quality strategy.

Driven by these circumstances, we present a one-manufacturer-one-retailer Stackelberg differential game model that introduces quality, advertising and pricing strategies. The retailer—who is the leader of the chain—controls the marketing strategies, specifically: advertising efforts and pricing. The manufacturer—who is the follower—controls an operational strategy, namely, quality improvements efforts. The demand depends on both price and goodwill dynamics, and advertising and quality improvements contribute to the accumulation of goodwill over time. The players divide the total revenues based on an exogenous fixed sharing mechanism (e.g., Chintagunta and Jain 1992; Jørgensen and Zaccour 2003; Jørgensen et al. 2006). As chain leader, the retailer decides whether the supply chain should be coordinated. The game consists of the following moves: (a) the retailer announces whether the chain will be coordinated, (b) the manufacturer sets the quality improvements strategy based on the retailer's announcement, and (c) the retailer accounts for the manufacturer's strategies and determines the pricing and advertising strategies.

We investigate supply chain coordination through the implementation of a support program. In particular, we are interested in whether the retailer decides to achieve supply chain coordination by paying (supporting) a part of the manufacturer's quality improvements expenses. This type of mechanism has been extensively reported in the marketing and supply chain literature. For instance, Jørgensen et al. (2003) identified the conditions under which a manufacturer is willing to support a retailer's advertising expenditures. De Giovanni (2011b) demonstrated that the economic benefits of supporting an advertising strategy depends on advertising effectiveness and the choice of media. De Giovanni and Roselli (2012) illustrated the usefulness of a support program to overcome the limitations implied by the adoption of a revenue-sharing contract in a dynamic supply chain. Similarly, Karray and Zaccour (2006) and Jørgensen et al. (2000, 2001, 2003) developed numerous marketing support programs to evaluate the benefits obtainable through coordination. Although it is generally accepted that a support program may be payoff-Pareto-improving, He et al. (2009) demonstrated that this is not always the case. For instance, a manufacturer may offer a support program only when the retailer's margin is lower than, or close to, the manufacturer's margin.

Supporting a quality improvements strategy entails an difficult challenge for a supply chain. Quality improvements exerts both a positive role in demand—increasing sales through goodwill—and a negative role in the manufacturer's unit profit margin—boosting production costs. Such trade-offs impose strict barriers to coordination and influence several aspects of a business. For instance, Jørgensen et al. (2003) characterized a trade-off of sales promotion that has a positive impact on demand but a negative influence on goodwill. El Ouardighi et al. (2008) showed that a coordinated supply chain should balance investments in quality improvements and advertising. Jørgensen and Zaccour (2003) introduced a channel with multiple-retailer promotions that positively affect sales and negatively impact brand image. De Giovanni (2011b) highlighted the role of quality improvements in increasing the stock of goodwill and reducing the manufacturer's profits when advertising effectiveness is low. Kim (1998) illustrated the role of technology in increasing the value of dynamics production technology development while reducing sales, where the internal learning rate determines the external technology a firm needs to acquire. Resolving these trade-offs provides advantages to the coordination process in supply chains.

To evaluate the benefits of coordination, we have characterized and compared equilibria of two scenarios. In the first scenario, two firms choose their strategies non-coordinatedly and non-sequentially and the game is modeled *à la Stackelberg*. The second scenario, which is also modeled *à la Stackelberg*, characterizes coordination through a support program wherein the retailer announces a positive support program. We compared the strategies and the outcomes in the two scenarios, taking the non-coordinated scenario as a benchmark and highlighting the conditions under which supply chain coordination is worthwhile.

To address the research questions, the current paper is organized as follows. The next section describes the differential game model while the third section characterizes the equilibria in coordinated and non-coordinated scenarios. The fourth section offers a comparison of strategies and their outcomes, and the final section provides concluding remarks.

7.2 The Model and Scenarios

A conventional supply chain is composed of one manufacturer, player M, and one retailer, player R. In this text, we have referred to the manufacturer as *he* and to the retailer as *she*. We assumed that the manufacturer controls the quality improvements rate, $Q(t)$, while the retailer controls both the price, $p(t)$, and the advertising rate, $A(t)$. In this sense, we have referred to quality improvements as an operational strategy and to pricing and advertising as marketing strategies. In a non-coordinated scenario, the players establish their strategies to maximize their own payoff function. As we demonstrate, coordination in a supply chain is characterized by the integration of marketing and operational strategies through a support mechanism.

We assumed that players split the total chain revenues based on a fixed sharing mechanism (see, for instance, Chintagunta and Jain 1992; Jørgensen and Zaccour 2003; Jørgensen et al. 2006), which is characterized by a constant parameter $\phi \in (0, 1)$. The parameter value is common knowledge to both players. Leaving the fixed share an exogenous element helps in identifying the ranges of values that lead to economically successful supply chain coordination. The introduction of a sharing mechanism to split the profits among firms is common in supply chain practice, where companies belonging to the same supply chain eliminate all internal inefficiency (e.g., a double marginalization effect due to a traditional wholesale price contract) while identifying how the overall profits have to be shared (Mentzer et al. 2001).

Both the manufacturer and the retailer contribute to goodwill dynamics through their quality improvements and advertising strategies, respectively. In a dynamic framework, goodwill is investigated by means of the following dynamic equation:

$$\dot{G}(t) = aA(t) + bQ(t)\sqrt{G(t)} - \delta G(t), \qquad G(0) = G_0 \geq 0 \tag{7.1}$$

where $\delta > 0$ is the decay rate or forgetting effect of the state while G_0 is the initial stock of goodwill. $a > 0$ and $b > 0$ correspond to the marginal contributions of advertising and quality to goodwill, which are generally called *advertising* and *quality effectiveness*, respectively (De Giovanni 2011b). This way of modeling the relationships between quality improvements and goodwill is one of the proposed novelties. In contrast to Nair and Narasimhan (2006), De Giovanni (2011b), Martín-Herrán et al. (2012), and Jørgensen and Zaccour (2003), who proposed a relationship between strategies and state that no longer depends on the history of the state, we have proposed a formulation wherein quality improvements' contribution to the stock of goodwill also depends on the stock itself. An example for this formulation is reported by Doganoglu and Klapper (2006), who modeled a goodwill dynamic where an advertising strategy contributes to goodwill by considering its entire history. To check the validity of such dynamic equation, one can think about the brand damage caused by sources of non-quality. When there is a non-quality event, the negative effects on the value of the brand are as large as the prominence of the brand itself. For instance, the recent scandals that involved Nestlé and Ikea (www.nytime.com, Feb. 2013) over beef products adulterated with horse meat represent two suitable examples. The effects of (no) food safety—quality—affect (negatively) positively the image brand according to the value of the brand itself. In fact, the negative effects of these scandals on Nestlé's and Ikea's brands have spread worldwide due to their importance and recognition, with the result that quality affects goodwill according to its entire history

Customer demand, which depends on price and goodwill, is determined by:

$$D(p(t), G(t)) = \theta\sqrt{G(t)} - \beta p(t) \tag{7.2}$$

where $\beta > 0$ and $\theta > 0$ represent the effects on current sales of pricing and goodwill, respectively. According to (7.2), the retailer controls demand through the price while both players' influence the goodwill dynamics via investing in quality improvements and advertising. Marketing and operational strategies should be defined to ensure positive sales, thus setting the price such that $p(t) \leq \frac{\theta}{\beta}\sqrt{G(t)}$.

Although a quality improvements strategy contributes to goodwill dynamics, it also implies a marginal production cost, $c > 0$, for each dollar invested in quality. Thus, the manufacturer suffers when an investment in quality improvements is not sufficiently efficient. The production cost function takes the following form:

$$C_p(Q(t)) = cQ(t) \tag{7.3}$$

Production cost is an increasing function of quality improvements, so any increase in quality implies a higher production cost. The function in (7.3) has been used by Vörös (2006), De Giovanni (2011b), Fine (1986, 1988), Tapiero (1987), and Chand et al. (1996), who modeled a production cost that increases with quality improvements. Consequently, the manufacturer's unit profit margin, $\pi_M(p(t), Q(t)) = p(t)\phi - cQ(t)$, decreases in quality improvements through production costs and increases in pricing. Thus, we have referred to the effect that the cost parameter c exerts as *operational efficiency*. When the manufacturer does not invest in quality improvements, the unit profit margin coincides with the share of unit revenue.

The role of quality improvements in (7.1) and (7.3) enhances an interesting operational trade-off. On the one hand, production cost directly depends on quality improvements. However, investing in quality involves a set of operational challenges—new controls and standards, training, setup, and trials—all of which increase production costs (Roselli and De Giovanni 2012). On the other hand, quality improvements makes a positive contribution to the accumulation of goodwill (Nair and Narasimhan 2006).

Advertising and quality improvements efforts take a convex and quadratic forms:

$$C_A\left(A(t)\right) = \frac{A(t)^2}{2}; \qquad C_Q\left(Q(t)\right) = \frac{Q(t)^2}{2}, \tag{7.4}$$

For the purpose of this research, coordinated and non-coordinated scenarios have been proposed. In the non-coordinated scenario, the manufacturer is concerned with operational issue strategies, whereas the retailer only controls marketing efforts. In the coordinated scenario, the retailer supports the manufacturer's quality improvements efforts and makes use of both marketing and operational instruments simultaneously. $B(t)$ denotes the retailer's support rate, which represents the percentage of quality improvements investments that the retailer pays to the manufacturer, which takes values within the interval [0,1]. When $B(t) = 0$, players no longer coordinate the chain; when $B(t) = 1$, the retailer pays all the quality improvements expenses. When $B(t) \in (0, 1)$ the players seek to reach coordination. Assuming an infinite time horizon and a positive discount rate ρ, the manufacturer's objective functional under the coordinated scenario is:

$$J_M = \int_0^{\infty} e^{-\rho t} \left\{ (\theta\sqrt{G(t)} - \beta p(t))(p(t)\phi - cQ(t)) - \frac{1-B(t)}{2}Q(t)^2 \right\} dt \tag{7.5}$$

and the retailer's objective functional is:

$$J_R = \int_0^{\infty} e^{-\rho t} \left\{ (\theta\sqrt{G(t)} - \beta p(t))p(t)\left(1-\phi\right) - \frac{B(t)}{2}Q(t)^2 - \frac{1}{2}A(t)^2 \right\} dt \tag{7.6}$$

Using (7.1), (7.5), and (7.6), we define a two-player differential game with four controls, $A(t) \geq 0$, $Q(t) \geq 0$, $B(t) \geq 0$, and $p(t) \geq 0$, and one state, $G(t) \geq 0$. In the non-coordinated scenario, only three controls are used since $B(t) = 0$. From this point forward, the time argument is omitted. We solve the games by assuming that the players use a stationary feedback strategy, which is standard in differential games over the infinite time horizon (Dockner et al. 2000). Although the complex interfaces between marketing and operations that emerge in dynamic games suggest using an open-loop solution, we developed feedback strategies because they provide a time-consistent equilibrium. Moreover, the information obtained in the feedback strategy is much more appropriate from a managerial perspective in supply chain management studies due to the value of information that such strategies supply (He et al. 2007).

7.3 Equilibrium Results

To highlight the benefits that supply chain coordination provides, we have solved for two scenarios. In both cases players are aware of how total chain revenues will be split.

First, in the non-coordinated scenario, each player maximizes his payoff without taking into consideration the other player's strategies. The game is played *à la Stackelberg*, thus strategies are set up sequentially and non-collaboratively while $B(t) = 0$. Second, in a coordinated scenario, the retailer pays a fraction of the manufacturer's quality improvements expenses. Her motivations arise from the positive effect that quality improvements exerts on the state. Its contribution depends on the entire history of goodwill. Increasing production costs may discourage the manufacturer's investments. In a coordinated scenario, $B(t) \in (0, 1]$.

7.3.1 Non-Coordinated Scenario

We consider the scenario in which the retailer decides her pricing and advertising strategies without providing any support to the manufacturer. The manufacturer determines the investments in quality improvements. The fact that the retailer does not provide a support derives from marketing and operational motivations. On the one hand, quality improvements implies managing a trade-off involving higher state/higher production costs; thus, when its contribution to goodwill is low and the marginal production cost is high, providing a support to invest more in quality improvements becomes less important. On the other hand, the retailer has a marginal interest in the implementation of a support program if the contribution of advertising to goodwill is larger than the contribution of quality improvements to the stock. Under those conditions, providing a support program could damage the retailer's profits. The equilibrium strategies in the non-coordinated scenario have been identified through the superscript *NC* and displayed in the following proposition:

Proposition 7.1. *The equilibrium price, advertising, and quality improvements strategies in a non-coordinated scenario are given by:*

$$p^{NC} = \frac{[\theta(1-\phi) + m_3 bc\beta]\sqrt{G^{NC}}}{2\beta(1-\phi)} \tag{7.7}$$

$$Q^{NC} = \frac{\{(1-\phi)(2bm_1 - c\theta) + bc^2\beta m_3\}\sqrt{G^{NC}}}{2(1-\phi)} \tag{7.8}$$

$$A^{NC} = am_3 \tag{7.9}$$

while the payoff functions turn out to be:

$$V_M^{NC} = m_1 G^{NC} + m_2 \tag{7.10}$$

$$V_R^{NC} = m_3 G^{NC} + m_4 \tag{7.11}$$

where m_1, m_2, m_3, and m_4 are the constant parameters to be identified.

Proof. See the Appendix. □

Our results present novelties in the relationships between strategies and state. While a state-independent advertising strategy is well established in the marketing literature, state-dependent pricing and quality improvements strategies supply noteworthy managerial insights. The higher a company's capability to accumulate stock of goodwill, the higher the price that it is able to charge. This statement enhances a challenging link among marketing instruments: Companies can increase the retail price according to the level of goodwill without damaging profits. Nevertheless, price increases in a concave way, resulting $\frac{\partial p}{\partial G} = \frac{[\theta(1-\phi)+m_3 bc\beta]}{4\beta(1-\phi)\sqrt{G}} > 0$ and $\frac{\partial^2 p}{\partial G^2} = -\frac{[\theta(1-\phi)+m_3 bc\beta]}{8\beta(1-\phi)\sqrt{G^3}} < 0$.

A similar statement is also valid for a quality improvements strategy. This research presents the first attempt to model the quality improvements strategy as a state-dependent strategy. Previous research (e.g., De Giovanni 2011b; Nair and Narasimhan 2006) has developed quality improvements strategies that are state-independent, although they exert a positive influence on the state (e.g., goodwill dynamics). In the reality of business, manufacturers adjust their investments in quality according to the value of their brand. As Rao et al. (1999) observe, firms should carefully set their quality strategies because a brand name is already a signal of quality and, consequently, low quality has a strong negative impact on brand value. Our results show that a firm should establish quality improvements efforts according to the accumulated goodwill knowing that quality contributes to the accumulation of goodwill based on its entire history. Higher goodwill induces a manufacturer to spend more in quality improvements, as it results that $\frac{\partial Q}{\partial G} = \frac{\{(1-\phi)(2bm_1-c\theta)+bc^2\beta m_3\}}{4(1-\phi)\sqrt{G}} > 0$, although a quality improvements strategy depends on goodwill in a concave way (e.g., $\frac{\partial^2 Q}{\partial G^2} = -\frac{\{(1-\phi)(2bm_1-c\theta)+bc^2\beta m_3\}}{8(1-\phi)\sqrt{G^3}} < 0$).

Plugging (7.7) and (7.8) into (7.1), the goodwill at the steady state easily results in the following:

$$G_\infty^{NC} = \frac{a^2 m_3 2(1-\phi)}{2(1-\phi)\delta - (1-\phi)(2bm_1 - c\theta) - bc^2\beta m_3} \geq 0 \tag{7.12}$$

Because the numerator is nonnegative, for goodwill at the steady state to be positive it must result that $\frac{(2\delta+c\theta)(1-\phi)-bc^2\beta m_3}{2b(1-\phi)} > m_1$.

7.3.2 Coordinated Scenario

This section introduces a coordinated scenario in which the retailer pays a fraction of the manufacturer's quality improvements expenses. The support is intended to stimulate the manufacturer's quality improvements efforts. The game evolves according to the following sequence of events: The retailer announces a support strategy; then, the manufacturer establishes his quality improvements efforts; the retailer considers the manufacturer's strategy when deciding her pricing, advertising and support strategy. The retailer provides a support program because the quality improvements strategy exerts a positive influence on the stock of goodwill that depends on the stock itself. When $a = b$, which implies equal effectiveness of both advertising and quality improvements on goodwill, an initial stock $G(0) > 1$ will supply an incentive to the retailer to spend more to support an operational strategy (quality improvements) than to invest in marketing strategies (advertising). This result depends on the fact that quality improvements contribute to goodwill according to its square root: If the radicand is lower than 1, the square root function substantially penalizes the contribution of quality improvements. The issue of migrating from the use of an operational tool to a marketing tool and vice versa has been addressed by De Giovanni (2011b), who showed that a manufacturer is more willing to support an advertising campaign than to invest in quality improvements, depending on media effectiveness. In our game, the retailer's support is an attractive incentive for the manufacturer to invest more. Nevertheless, the increasing production cost can represent a serious barrier to invest more in quality improvements.

Proposition 7.2. *The equilibrium price, advertising, quality improvements, and support strategies in a coordinated scenario are expressed by:*

$$p = \frac{bc\beta l_3 + \theta(1-\phi)\left(1+\psi^2\right) - \{c\beta\left[b\left(l_1+l_3\right)-\theta c\right] + 2\theta(1-\phi)\}\,\psi}{2\beta(1-\phi)\left(1+\psi^2\right) + \beta\left(\beta c^2 - 4(1-\phi)\right)\psi}\sqrt{G^C} \tag{7.13}$$

$$Q^C = \frac{bl_1 - c(\theta-\beta\varphi)}{1-\psi}\sqrt{G^C} \tag{7.14}$$

$$A^C = al_3 \tag{7.15}$$

$$B = \frac{1}{3d_{13}}\left[\frac{\xi}{\sqrt[3]{2}} - \frac{\sqrt[3]{2}\left(3d_{13}d_{15}-d_{14}^2\right)}{\xi} - d_{14}\right] \tag{7.16}$$

while the value functions assume to form:

$$V_M^C = l_1G^C + l_2 \tag{7.17}$$

$$V_R^C = l_3G^C + l_4 \tag{7.18}$$

where l_1, l_2, l_3, *and* l_4 *are the parameters to be identified, while* $dj, j = 1 \ldots 14$ *are constants.*

Proof. See the Appendix. □

As for the non-coordinated scenario, there is a compensating effect between pricing and quality improvements; again, both turn out to be the only state dependent strategies. Nevertheless, the strategies are much more difficult to derive as support program development heavily influences strategies and payoffs. On the one hand, coordination requires higher efforts from a retailer: Increasing stocks of goodwill implies a higher retailer's value function and an increased incentive to provide more support while the manufacturer invests more in quality improvements efforts. On the other hand, coordination pushes the manufacturer to do more, while the positive impact of changes in goodwill on the manufacturer's value function discourages the implementation of a support program. Indeed, the manufacturer can push up its quality improvements efforts too much; thus, a supporting program may provide only marginal economic benefits for a retailer.

By inserting (7.14) and (7.15) into (7.1), the stock of goodwill at the steady state will assume the following form:

$$G_{\infty}^{C} = \frac{a^2 l_3 (1 - \psi)}{\delta (1 - \psi) - b [b l_1 - c(\theta - \beta\varphi)]} \tag{7.19}$$

For this stock to be positive, condition $\frac{\delta(1-\psi)+cb(\theta-\beta\varphi)}{b^2} > l_1$ must be satisfied.

7.4 Numerical Analysis

To address our research hypotheses and analyze strategies and outcomes of the games, we used a numerical analysis. The identified parameters for the scenarios are recursive although the value functions are linear. This is a peculiar result as in the domain of differential games when conjecturing linear value functions generally leads to identification of linear parameters (De Giovanni 2011a; Jørgensen et al. 2000, 2001, 2003). The different result reported in this paper is due to the interface between goodwill and a quality improvements strategy, which leads to nonlinear identified parameters. Finally, the network of relationships among parameters precludes any analytical solution.

Based on the results of Propositions 7.1 and 7.2, findings are derived for managers and practitioners by comparing profits and strategies in the two scenarios. We obtain a benchmark solution by setting the parameters according to values used in the marketing and operations literature (e.g., De Giovanni 2011a,b; De Giovanni and Zaccour 2013; El Ouardighi et al. 2008 as well as in Operations Research (e.g., Almeder et al. 2009); in particular:

Table 7.1 Sensitivity analysis

	p^{NC}	A^{NC}	Q^{NC}	G^{NC}	V_M^{NC}	V_R^{NC}	p^{NC}	A^{NC}	Q^{NC}	G^C	B	V_M^C	V_R^C
$\theta(1.1; 1.15; 1.2)$	+	+	+	+	+	+	+	+	+	+	+	+	+
$\beta(0.5; 0.55; 0.6)$	−	−	−	−	−	−	−	−	−	−	−	−	−
$a(1; 1.05; 1.1)$	+	+	+	+	+	+	+	+	+	+	+	+	+
$b(1; 1.05; 1.1)$	+	+	+	+	+	+	+	+	−	+	+	+	+
$c(0.2; 0.25; 0.3)$	−	−	−	−	−	−	−	−	−	−	−	−	−
$\phi(0.5; 0.55; 0.6)$	+	−	+	+	+	−	+	−	+	+	+	−	+
$\delta(0.2; 0.25; 0.3)$	−	−	−	−	−	−	−	−	−	−	−	−	−
$\rho(0.9; 0.95; 0.975)$	−	−	−	−	−	−	−	−	−	−	−	−	−

Demand parameters: $\beta = 0.5, \theta = 1, \phi = 0.5$
Operational parameters: $c = 0.2$
Goodwill parameters: $a = 1, b = 1, \delta = 0.2$
Dynamic parameters: $\rho = 0.9$

The above parameter values generate several solutions because the identified parameters m_1, m_3, l_1 and l_3 are not linear. Among the resulting solutions, only a limited number of roots led to a feasible solution, which means that $V_M^i \geq 0, V_R^i \geq 0, A^i \geq 0, Q^i \geq 0, p^i \geq 0, i = NC, C$, and $B \in (0, 1)$. Before comparing the strategies and payoffs, we run a sensitivity analysis to check that variations in the benchmark parameters do not violate these assumptions. The sensitivity analysis is reported in Table 7.1, starting from the benchmark values and evaluating changes in all parameters. A "+" ("–") in a cell should be interpreted as a positive (negative) change of a give element in the main row due to the change in a given parameter in the main colon. The sensitivity analysis shows that variations in any of the parameter model do not violate our assumptions; thus, we run a comparison of strategies and payoffs in the two scenarios. Starting from the benchmark solution, we found new results when comparing payoffs and strategies in coordinated and non-coordinated scenarios inside the space $\Omega\,(\phi, c, b)$, where $\phi \in (0.1, 0.6)$, $c \in (0.05, 0.3)$ and $b \in (0.3, 1)$. When running simulations, we keep the other parameters at the benchmark value as their variations supply little additional information.

Evaluation of the sharing parameter in that range provides noteworthy managerial insights, as a proper setting enhances coordination effectiveness. The sharing parameters is varied from low (e.g., $\phi = 0.1$) to high values (e.g., $\phi = 0.6$) to assess coordination effectiveness when most of the economic value created is retained in the upstream or downstream of the chain.

Considering the marginal production cost and the strategy effectiveness in the respective ranges, it is possible to identify four trade-off cases:

1. *Operational excellence*. This is the case in which quality effectiveness is really high (e.g., $b = 1$) while marginal production cost is low (e.g., $c = 0.05$). This combination pushes a decision maker through the implementation of a quality improvements strategy: It provides a considerable contribution to the state while it marginally affects production costs.

Fig. 7.1 Manufacturer's payoff comparison

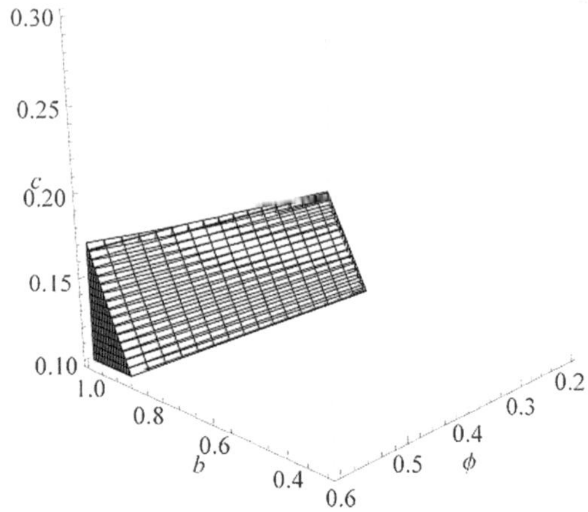

2. *Inefficient operations*. This case is characterized by high quality effectiveness (e.g., $b = 1$) and high marginal production cost (e.g., $c = 0.3$). The implementation of a quality improvements strategy is really appealing, but the high cost of quality imposes stringent barriers.
3. *Ineffective operations*. In this case, a firm faces low quality effectiveness (e.g., $b = 0.3$) along with low marginal production cost (e.g., $c = 0.05$). A quality improvements strategy is really attractive because it only slightly increases the cost of quality. Even so, the strategy has minor influence on the dynamics; thus, it supplies a low contribution to goodwill.
4. *Weak operations*. In this case, the marginal production cost is really high (e.g., $c = 0.3$) while quality improvements effectiveness is really low (e.g., $b = 0.3$). Both the implementation and the support of a quality improvements strategy are problematic when these conditions apply; thus, players depart from operational tools to exclusively espouse marketing strategies and pursue profit maximization.

Claim. Inside the region Ω, a manufacturer's preference for supply chain coordination depends on operational performance (see Fig. 7.1.)

A manufacturer's convenience in supply chain coordination slightly depends on the sharing parameter values. Higher sharing parameter values marginally increase the area inside which coordination will be preferred. Thus, the manufacturer should set his strategies mainly according to operational performance.

A necessary condition for a manufacturer to be economically better-off through coordination is that operational effectiveness is sufficiently high. In fact, when quality improvements exert only a marginal influence on the state, the operational strategy is not at all convenient, independently of both the sharing parameter values and the operational efficiency. Even so, high operational effectiveness is never a sufficient condition because operational efficiency plays an important role in the manufacturer's strategies. When the impact of quality on production cost is

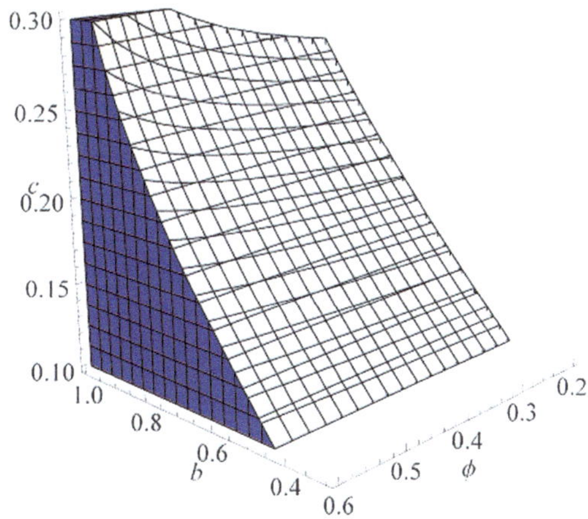

Fig. 7.2 Retailer's payoff comparison

marginal (high operational efficiency), the area in which the manufacturer is better-off through coordination is larger.

Finally, coordination directs the manufacturer through an improved payoff function only in the operational excellence case.

Claim. Inside Ω, a retailer always prefers supply chain coordination under an excellent operations case. She also prefers coordination under an inefficient operations case when the sharing parameter is sufficiently high (see Fig. 7.2).

The retailer, as chain leader, always announces coordination except in one case: when the cost of quality considerably increases according to quality improvements investments and, simultaneously, the contribution to the accumulation of goodwill is irrelevant. This is the scenario of low operational performance, which belongs to the *weak operations* case and discourages a retailer from announcing coordination. In all the other cases, the retailer can substantially increase her payoff, even when operations are inefficient or ineffective conditioned on a sufficiently high sharing parameter. This latter condition entails a manufacturer investing more in quality improvements, thus when this condition misses supply chain coordination is not attractive for a retailer because operations are inefficient. In fact, in the case of *inefficient operations*, the cost of quality is very high; thus, the retailer wants to support a quality improvements strategy because operational effectiveness is substantially relevant and the manufacturer will not be discouraged by higher marginal production cost when the share he receives is high enough. A sufficiently high sharing parameter eliminates the limitations created by operational inefficiency.

In contrast, in the case of *ineffective operations*, the cost of quality is very low; thus, the retailer will announce coordination because the marginal cost is of minor importance and the manufacturer finds it inexpensive to invest in this strategy. Nevertheless, the scarce contribution of quality improvements on goodwill discourages a retailer from coordinating the supply chain, independent of the

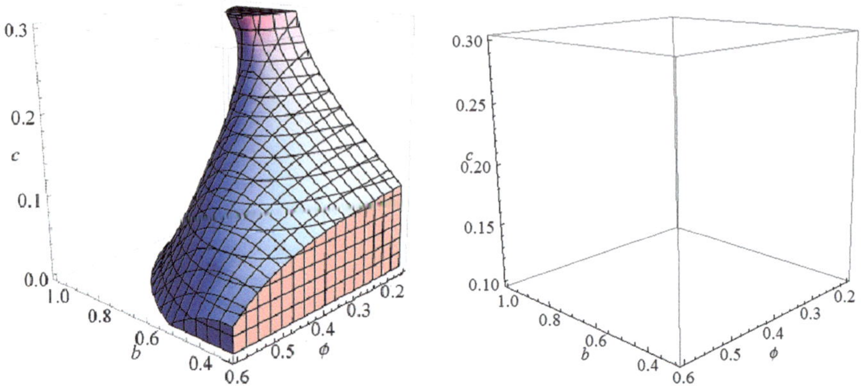

Fig. 7.3 Quality improvements (LHS) and advertising (RHS) strategies comparison

revenue sharing agreement. Thus, when quality is ineffective coordination is less important, even when the sharing parameter is sufficiently high.

Finally, the retailer will always announce a support program in an excellent operations case, and her decisions will no longer be influenced by the sharing parameter.

Claim. Inside Ω, the Pareto-improving region corresponds to the region inside which the manufacturer does prefer coordination (see Fig. 7.1).

Inside Ω, a retailer has a broader set of parameter values for being economically better-off through supply chain coordination. This mainly depends on the leader position that she covers in the game. According to Cachon's (2003) definition, coordination seems to be a difficult target. However, coordination will be an economically suitable option for both supply chain members conditioned on good operational performance and sufficiently high sharing parameter.

Claim. Inside Ω, the manufacturer always invests more in quality improvements only when the operational effectiveness is sufficiently low. In contrast, the retailer always advertises less under supply chain coordination (see Fig. 7.3).

Figure 7.3 interestingly shows that the retailer always advertises less under coordination. This result contrasts with several findings in marketing and operations (e.g., De Giovanni 2011b) that use a leader–follower structure. In such cases, the leader always does something more under coordination (e.g., higher advertising), independent of the amount of operational costs that she will pay. In our model, when coordination is an option, the leader migrates some economic resources from a marketing strategy (advertising) to an operational strategy (quality improvements) independent of the operational performance and revenue sharing agreements. This finding depends on the different ways through which advertising and quality improvements contribute to the accumulation of goodwill, as quality improvements' role depends on the stock of accumulated goodwill itself.

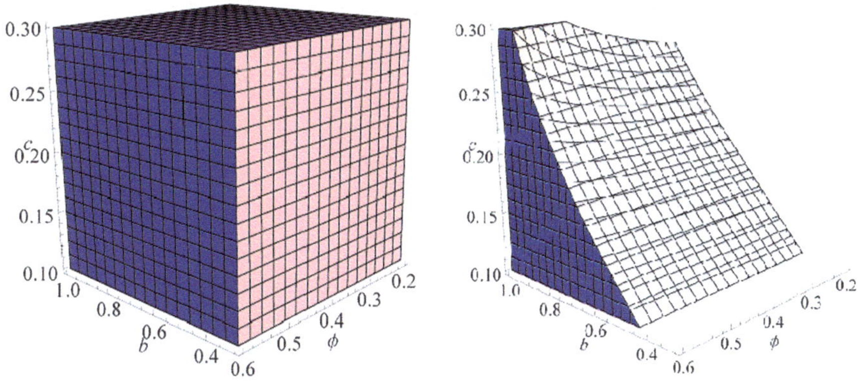

Fig. 7.4 Price strategy (LHS) and goodwill (RHS) comparison

In contrast, the manufacturer invests more in quality improvements efforts when the sharing parameter is sufficiently low. In fact, when the sharing parameter is low, the manufacturer sees coordination as an opportunity to increase his profits through higher investments in quality improvements because the sharing parameter cannot be modified over the course of the game. When the sharing parameter is sufficiently high the manufacturer does not invest more in quality improvements because it is in the interest of the retailer to boost the advertising to increase her profits. This result contrasts with El Ouardighi et al. (2008) and De Giovanni and Roselli (2012) who show that investments in quality increase the sharing parameter. These contrasting findings depend on the different game structure (Nash vs. Stackelberg) and the missing interface of quality in the dynamics of goodwill.

The manufacturer's strategy should also be interpreted with respect to the operational effectiveness. Indeed, increasing operational effectiveness leads to lower investments in quality improvements as the manufacturer can invest a lower amount of economic resources to obtain the same results. On the one hand, the manufacturer expects the retailer to invest more in advertising efforts; on the other hand, he does not want to increase quality improvements investments too much.

Claim. Inside Ω, the retailer always charges a higher price under coordination while the accumulated goodwill turns out to be lower under a weak operations case (see Fig. 7.4).

The findings displayed in Fig. 7.4 highlight an interesting interface between marketing and operations. The retailer always charges a higher price under coordination. This is because she always invests more in advertising and support of a quality improvements strategy may cover the inefficiency (lower sales) that a higher price may imply. This incremental increase in prices is not fully covered by a higher goodwill. Under a weak operations case, in fact, the accumulated goodwill under coordination no longer compensates for the decreasing demand due to the negative effect of higher prices. This compensation among marketing tools is not new in the literature of marketing and operations. Previous research shows that coordination allows a supply chain to reach higher levels of goodwill to overcome all inefficiency due to higher prices (e.g., De Giovanni 2011b; De Giovanni and Roselli 2012; El Ouardighi et al. 2008). In contrast, we demonstrate that supporting a quality improvements strategy that influences the stock of goodwill according to its entire history does not always lead to solving trade-offs among marketing strategies.

7.5 Conclusions

This paper introduced a supply chain dynamic game with operations and marketing interfaces. Our main research question related to the convenience of coordination in a supply chain, which is achieved through the implementation of a support program. Recent literature has demonstrated the higher effectiveness of such a program in coordinating a supply chain in relation to a traditional contract (De Giovanni and Roselli 2012), although it originally was a marketing strategy (e.g., Jørgensen et al. 2001). This way of coordinating the chain allows a retailer to use both marketing and operational devices to maximize her payoff function. The operational strategy quality improvements enhances an operational trade-off for the manufacturer: On the one hand, investing in quality boosts the goodwill dynamics, thus increasing the stock forever; on the other hand, a quality improvements strategy increases production costs, thus resulting in managerial concerns. Our main result relates to the leader's convenience in undertaking coordination. The operations management literature reports a clear result: A leader will always support an operational strategy (e.g., a quality improvements strategy) because the leader will always be economically better-off (e.g., De Giovanni 2011b). Our results contradict this statement when a quality improvements strategy contributes to the accumulation of goodwill based on its entire history. In particular, a leader should evaluate how profits are shared over the supply chain as well as the benefits and drawbacks that an operational strategy (e.g., a quality improvements strategy) implies.

One strength linked to quality improvements is its contribution to the state, which depends not only on the investment but also on the stock of goodwill itself. In this sense, investing in quality improvements could be more effective than investing in advertising, which also contributes to the state but disregards the history of goodwill. The players split the revenues based on a fixed sharing mechanism, whose application has been shown to be diffused in the marketing literature (Chintagunta and Jain 1992; Jørgensen and Zaccour 2003). As leader of the chain, the retailer announces her decision about coordination after evaluating supply chain performance (e.g., operational efficiency and effectiveness). Our findings, which are summarized in Table 7.2, suggest that the retailer is always in favor of supply chain coordination when a quality improvements strategy is efficient, while when operations is not efficient a sufficiently high sharing parameter is needed to involve the manufacturer to a greater extent. When this operational strategy does not provide

Table 7.2 Payoffs comparison

	Low sharing parameters	High sharing parameters
Operational excellence	$V_M^C > V_M^{NC}, V_R^C > V_R^{NC}$	$V_M^C > V_M^{NC}, V_R^C > V_R^{NC}$
Inefficient operations	$V_M^C < V_M^{NC}, V_R^C < V_R^{NC}$	$V_M^C < V_M^{NC}, V_R^C > V_R^{NC}$
Ineffective operations	$V_M^C < V_M^{NC}, V_R^C < V_R^{NC}$	$V_M^C < V_M^{NC}, V_R^C < V_R^{NC}$
Weak operations	$V_M^C < V_M^{NC}, V_R^C < V_R^{NC}$	$V_M^C < V_M^{NC}, V_R^C < V_R^{NC}$

sufficient contributions to goodwill and it damages the manufacturer's production cost too much, a retailer will not support a quality improvements strategy, with the result that coordination will not be an attractive option.

In the position of follower, the manufacturer hopes that operational performance is sufficiently high so that the leader will announce coordination and the trade-off of quality improvements will become a marginal concern.

Despite the novelty of the findings and the managerial implications, the conclusions present several limitations due to model assumptions. Windows for future research remain open for further development. While we have investigated coordination by means of a support program, other forms of coordination need to be investigated. For instance, supply chain coordination through a support program can be complemented either by the implementation of a formal contract (e.g., wholesale price contract, two-part tariff, revenue-sharing contract) or by the use of an incentive scheme based on strategies and/or state (e.g., joint maximization incentive (Jørgensen et al. 2006)). Moreover, future research could integrate other strategies beyond pricing, advertising, and quality improvements, such as product development, service, green investments, conformance quality, and durability. Consideration of competition or more players in the up- and downstream of the channel is another avenue to explore. Finally, it would be useful to test our results empirically with a case study and qualitative research.

Appendix

Proof of Proposition 7.1. Each player solves an optimal control problem to design its equilibrium strategies according to the Hamilton–Jacobi–Bellman (HJB) equations:

$$\rho V_M^{NC} = (\theta\sqrt{G^{NC}} - \beta p^{NC})(p^{NC}\phi - cQ^{NC}) - \frac{1}{2}Q^{NC^2} + V_M^{NC'}(aA^{NC} + bQ^{NC}\sqrt{G^{NC}} - \delta G^{NC}) \tag{7.20}$$

$$\rho V_R^{NC} = (\theta\sqrt{G^{NC}} - \beta p^{NC})p^{NC}(1-\phi) - \frac{1}{2}A^{NC^2} + V_R^{NC'}(aA^{NC} + bQ^{NC}\sqrt{G^{NC}} - \delta G^{NC}) \tag{7.21}$$

As the game is played *à la Stackelberg* where the retailer is the leader, we first solved for the manufacturer's strategy:

$$Q^{NC} = (bV_M^{NC'} - c\theta)\sqrt{G^{NC}} + c\beta p^{NC} \tag{7.22}$$

Then, substituting (7.22) in the retailer's HJB functional we obtain

$$\begin{aligned} \rho V_R^{NC} &= (\theta\sqrt{G^{NC}} - \beta p^{NC}) p^{NC}(1-\phi) - \frac{1}{2}A^{NC^2} + V_R^{NC'}(aA^{NC} \\ &\quad + b\left[(bV_M^{NC'} - c\theta)\sqrt{G^{NC}} + c\beta p^{NC}\right]\sqrt{G^{NC}} - \delta G^{NC}) \end{aligned} \tag{7.23}$$

Optimizing the leader's functional with respect to her controls, the strategies turn out to be:

$$p^{NC} = \frac{\left[\theta(1-\phi) + V_R^{NC'}bc\beta\right]\sqrt{G^{NC}}}{2\beta(1-\phi)} \tag{7.24}$$

$$A^{NC} = aV_R^{NC'} \tag{7.25}$$

The quality improvements strategy finally becomes:

$$Q^{NC} = \frac{\left\{(1-\phi)(2bV_M^{NC'} - c\theta) + bc^2\beta V_R^{NC'}\right\}\sqrt{G^{NC}}}{2(1-\phi)} \tag{7.26}$$

Substituting (7.24), (7.25), and (7.26) inside (7.20) and (7.23), after several manipulations we obtain:

$$\begin{aligned} 8\beta(1-\phi)^2\rho V_M^{NC} &= 2G^{NC}\left[\theta(1-\phi) - V_R^{NC'}bc\beta\right] \\ &\quad \left\{(1-\phi)\left[\theta\phi - c\beta(2bV_M^{NC'} - c\theta)\right] + V_R^{NC'}bc\beta\left(\phi - c^2\beta\right)\right\} \\ &\quad + \left[(1-\phi)(2bV_M^{NC'} + c\theta) - bc^2\beta V_R^{NC'}\right] \\ &\quad \beta\left\{(1-\phi)(2bV_M^{NC'} - c\theta) + bc^2\beta V_R^{NC'}\right\}G^{NC} \\ &\quad + 8\beta(1-\phi)^2 V_M^{NC'}(a^2V_R^{NC'} - \delta G^{NC}) \end{aligned} \tag{7.27}$$

$$\begin{aligned} 4\beta(1-\phi)\rho V_R^{NC} &= \left[\theta^2(1-\phi)^2 - \left(V_R^{NC'}bc\beta\right)^2\right]G^{NC} + 2\beta V_R^{NC'}(a^2(1-\phi)V_R^{NC'} \\ &\quad + b\left[(1-\phi)(2bV_M^{NC'} - c\theta) + V_R^{NC'}bc^2\beta\right]G^{NC} - 2(1-\phi)\delta G^{NC}) \end{aligned} \tag{7.28}$$

Because both (7.27) and (7.28) are linear in the state, it is possible to conjecture linear value functions, $V_M^{NC} = m_1 G + m_2$ and $V_R^{NC} = m_3 G + m_4$. Substituting our conjectures and their derivatives in (7.27) and (7.28) we get:

$$\begin{aligned}
8\beta(1-\phi)^2\rho\left(m_1 G^{NC} + m_2\right) &= 2G^{NC}\left[\theta(1-\phi) - m_3 bc\beta\right] \\
&\quad \left\{(1-\phi)\left[\theta\phi - c\beta(2bm_1 - c\theta)\right] + m_3 bc\beta\left(\phi - c^2\beta\right)\right\} \\
&\quad + \left[(1-\phi)(2bm_1 + c\theta) - bc^2\beta m_3\right]\beta\left\{(1-\phi)(2bm_1 - c\theta) + bc^2\beta m_3\right\}G^{NC} \\
&\quad + 8\beta(1-\phi)^2 m_1(a^2 m_3 - \delta G^{NC})
\end{aligned} \tag{7.29}$$

$$\begin{aligned}
4\beta(1-\phi)\rho\left(m_3 G + m_4\right) &= \left[\theta^2(1-\phi)^2 - (m_3 bc\beta)^2\right]G^{NC} + 2\beta m_3(a^2(1-\phi)m_3 \\
&\quad + b\left[(1-\phi)(2bm_1 - c\theta) + m_3 bc^2\beta\right]G^{NC} - 2(1-\phi)\delta G^{NC})
\end{aligned} \tag{7.30}$$

By identification, it is possible to derive the constant parameters m_1, m_2, m_3, and m_4 as follows,

$$\left\{\begin{array}{c}
2\left[\theta(1-\phi) - m_3 bc\beta\right] \\
\left\{(1-\phi)\left[\theta\phi - c\beta(2bm_1 - c\theta)\right] + m_3 bc\beta\left(\phi - c^2\beta\right)\right\} \\
+\left[(1-\phi)(2bm_1 + c\theta) - bc^2\beta m_3\right] \\
\beta\left\{(1-\phi)(2bm_1 - c\theta) + bc^2\beta m_3\right\} - 8\beta(1-\phi)^2(\rho+\delta)m_1
\end{array}\right\} = 0 \tag{7.31}$$

$$m_1 m_3 a^2 - \rho m_2 = 0 \tag{7.32}$$

$$m_3^2 b^2 c^2 \beta^2 + 2\beta(1-\phi)\left[b(2bm_1 - c\theta) - 2(\rho+\delta)\right]m_3 + \theta^2(1-\phi)^2 = 0 \tag{7.33}$$

$$2\rho m_4 - m_3^2 a^2 = 0 \tag{7.34}$$

□

Proof of Proposition 7.2. Each player solves an optimal control problem where equilibrium strategies are derived from the HJB equations:

$$\begin{aligned}
\rho V_M^C &= (\theta\sqrt{G^C} - \beta p^C)(p^C\phi - cQ^C) - \frac{(1-B)}{2}Q^{C^2} \\
&\quad + V_M^{C'}(aA^C + bQ^C\sqrt{G^C} - \delta G^C)
\end{aligned} \tag{7.35}$$

$$\begin{aligned}
\rho V_R^C &= (\theta\sqrt{G^C} - \beta p^C)p^C(1-\phi) - \frac{1}{2}A^{C^2} - \frac{1}{2}BQ^{C^2} \\
&\quad + V_R^{C'}(aA^C + bQ^C\sqrt{G^C} - \delta G^C)
\end{aligned} \tag{7.36}$$

As the game is played *à la Stackelberg* where the retailer is the leader, we first solved for the manufacturer's strategy:

$$Q^C = \frac{\left(bV_M^{C'} - c\theta\right)\sqrt{G^C} + \beta c p^C}{1 - B} \tag{7.37}$$

Therefore, we substituted (7.37) in the retailer's HJB that assumes the following form:

$$\begin{aligned} \rho V_R^C = {} & (\theta\sqrt{G^C} - \beta p^C)p^C(1-\phi) - \frac{1}{2}A^{C^2} \\ & - \frac{B}{2(1-B)^2}\Big[b^2 V_M^{C'2} G^C + c^2\theta^2 G^C + \beta^2 c^2 p^2 - 2bc\theta V_M^{C'} G^C \\ & + 2b\beta c V_M^{C'} p^C \sqrt{G^C} - 2c^2\theta\beta c p^C \sqrt{G^C}\Big] \\ & + V_R^{C'}\left(aA^C + b\frac{\left(bV_M^{C'} - c\theta\right)\sqrt{G^C} + \beta c p^C}{1-B}\sqrt{G^C} - \delta G^C\right) \end{aligned} \tag{7.38}$$

The retailer's strategies turn out to be:

$$p = \frac{\begin{array}{l} bc\beta V_R^{C'} + \theta(1-\phi)\left(1 + B*^2\right) \\ \quad - \left\{c\beta\left[b\left(V_M^{C'} + V_R^{C'}\right) - \theta c\right] + 2\theta(1-\phi)\right\} B* \end{array}}{2\beta(1-\phi)(1 + B*^2) + \beta\left(\beta c^2 - 4(1-\phi)\right) B*}\sqrt{G^C} \tag{7.39}$$

$$B = B^*\left(V_M^{C'}, V_M^{C'}, \Sigma\right) : B* \in (0,1) \tag{7.40}$$

$$A^C = aV_R^{C'} \tag{7.41}$$

where $\Sigma \in [\theta, \beta, \phi, a, b, c, \delta, \rho]$ and B^* is obtained by solving a polynomial equation of third degree $d_{14}B + d_{13}B^2 + d_{10}B^3 - d_7 = 0$, where the constant parameters $d_j, j = 0 \ldots 14$ are given by:

$$d_0 = \left[\left(2V_R^{C'} - V_M^{C'}\right)b + c\theta\right]$$

$$d_1 = \left[\left(2V_R^{C'} + V_M^{C'}\right)b - c\theta\right]$$

$$d_2 = \theta(1-\phi)c\beta$$

$$d_3 = \left[bV_M^{C'} - \theta c\right]c^2\beta^2$$

$$d_4 = V_R^{C'} c^2\beta^2 b$$

$$
\begin{aligned}
d_5 &= 2\beta(1-\phi)\\
d_6 &= \beta^2 c^2\\
d_7 &= d_1 d_5 - d_2 - d_4\\
d_8 &= d_0 d_5 + d_2 + d_4\\
d_9 &= (d_1 d_5 - d_2)\\
d_{10} &= (d_0 d_5 + d_2)\\
d_{11} &= (d_1 (d_6 - 2d_5) + 2d_2 + d_4 + d_3)\\
d_{12} &= (d_0 (d_6 - 2d_5) - 2d_2 - d_4 - d_3)\\
d_{13} &= d_{12} - d_9\\
d_{14} &= d_8 - d_{11}
\end{aligned}
$$

Among the three solutions obtained, only one solution satisfy our assumptions, that is $B \in (0, 1)$, and it is given by:

$$
B = \frac{1}{3d_{10}}\left[\frac{\xi}{\sqrt[3]{2}} - \frac{\sqrt[3]{2}\left(3d_{10}d_{14} - d_{13}^2\right)}{\xi} - d_{\_13}\right]
$$

where

$$
\xi = \sqrt[3]{27d_{10}^2 d_7 + 9d_{10}d_{13}d_{14} - 2d_{13}^3 + \sqrt{\left[\begin{array}{l} 4\left(3d_{10}d_{14} - d_{13}^2\right)^3 \\ +\left(27d_{10}^7 d_7 + 9d_{10}d_{13}d_{14} - 2d_{13}^3\right)^2 \end{array}\right]}}
$$

To save notations, we write the pricing and support strategies such as:

$$
p = \varphi\sqrt{G^C} \tag{7.42}
$$

$$
B = \psi \tag{7.43}
$$

where

$$
\varphi = \frac{bc\beta V_R^{C'} + \theta(1-\phi)\left(1+\psi^2\right) - \left\{c\beta\left[b\left(V_M^{C'} + V_R^{C'}\right) - \theta c\right] + 2\theta(1-\phi)\right\}\psi}{2\beta(1-\phi)\left(1+\psi^2\right) + \beta\left(\beta c^2 - 4(1-\phi)\right)\psi} \tag{7.44}
$$

and

$$
\psi = \frac{1}{3d_{10}}\left[\frac{\xi}{\sqrt[3]{2}} - \frac{\sqrt[3]{2}\left(3d_{10}d_{14} - d_{13}^2\right)}{\xi} - d_{13}\right] \tag{7.45}
$$

Plugging (7.42) and (7.43) into (7.37), the quality improvements strategy turns out to be

$$Q^C = \frac{bV_M^{C'} - c(\theta - \beta\varphi)}{1-\psi}\sqrt{G^C} \tag{7.46}$$

Substituting (7.42), (7.43), and (7.46) inside (7.35) and (7.38) we obtain:

$$\begin{aligned} 2(1-\psi)\rho V_M^C &= 2(\theta-\beta\varphi)\left\{\varphi\phi(1-\psi) - c\left[bV_M^{C'} - c(\theta-\beta\varphi)\right]\right\}G^C \\ &\quad + \left[bV_M^{C'} - c(\theta-\beta\varphi)\right]^2 G^C + 2V_M^{C'}(1-\psi)\left[a^2V_R^{C'} - \delta G^C\right] \end{aligned} \tag{7.47}$$

$$\begin{aligned} 2(1-\psi)^2\rho V_R^C &= 2(1-\psi)^2(\theta-\beta\varphi)\varphi(1-\phi)G^C \\ &\quad + \left\{2bV_R^{C'} - \psi\left[\left(2V_R^{C'} + V_M^{C'}\right)b - c(\theta-\beta\varphi)\right]\right\}\left[bV_M^{C'} - c(\theta-\beta\varphi)\right]G^C \\ &\quad + (1-\psi)^2 V_R^{C'}(a^2V_R^{C'} - 2\delta G^C) \end{aligned} \tag{7.48}$$

Because also in the coordination scenario both HJBs are linear in the state, we conjecture linear value functions, $V_M^C = l_1G^C + l_2$ and $V_R^C = l_3G^C + l_4$. Substituting our conjectures and their derivatives in (7.47) and (7.48), we obtain:

$$\begin{aligned} 2(1-\psi)\rho\left(l_1G^C + l_2\right) &= 2(\theta-\beta\varphi)\left[\varphi\phi(1-\psi) - bl_1 + c^2(\theta-\beta\varphi)\right]G^C \\ &\quad + \left[bV_M^{C'} + c(\theta-\beta\varphi)\right]\left[bV_M^{C'} - c(\theta-\beta\varphi)\right]G^C \\ &\quad + 2l_1(1-\psi)\left\{a^2l_3 - \delta G^C\right\} \end{aligned} \tag{7.49}$$

$$\begin{aligned} 2(1-\psi)^2\rho\left(l_3G^C + l_4\right) &= 2(1-\psi)^2(\theta-\beta\varphi)\varphi(1-\phi)G^C \\ &\quad + \{2bl_3 - \psi\left[(2l_3 + l_1)b - c(\theta-\beta\varphi)\right]\}\left[bl_1 - c(\theta-\beta\varphi)\right]G^C \\ &\quad + (1-\psi)^2 l_3(a^2l_3 - 2\delta G^C) \end{aligned} \tag{7.50}$$

By identification it is possible to derive the constant parameters to be identified:

$$\left\{\begin{array}{c} 2(\theta-\beta\varphi)\left[\varphi\phi(1-\psi) - bl_1 + c^2(\theta-\beta\varphi)\right] \\ + \left[bV_M^{C'} + c(\theta-\beta\varphi)\right]\left[bV_M^{C'} - c(\theta-\beta\varphi)\right] - (1-\psi)(\rho+\delta)l_1 \end{array}\right\} = 0 \tag{7.51}$$

$$\rho l_2 - a^2l_1l_3 = 0 \tag{7.52}$$

$$\left\{\begin{array}{c} 2(1-\psi)^2(\theta-\beta\varphi)\varphi(1-\phi) - 2(1-\psi)^2(\rho+\delta)l_3 \\ + \{2bl_3 - \psi\left[(2l_3+l_1)b - c(\theta-\beta\varphi)\right]\}\left[bl_1 - c(\theta-\beta\varphi)\right] \end{array}\right\} = 0 \tag{7.53}$$

$$2\rho l_4 - a^2l_3^2 = 0 \tag{7.54}$$

□

References

Almeder C, Preusser M, Hartl R (2009) Simulation and optimization of supply chains: alternative or complementary approaches? OR Spectr 31(1):95–119

Cachon G (2003) Supply chain coordination with contracts. In: Graves S, de Kok T (eds) Handbooks in operations research and management science: Supply chain management. North-Holland, Amsterdam

Chand S, Moskowitz H, Novak A, Rekhi I, Sorger G (1996) Capacity allocation for dynamic process improvement with quality and demand considerations. Oper Res 44(6):964–976

Chintagunta P, Jain D (1992) A dynamic model of channel member strategies for marketing expenditures. Mark Sci 11:168–188

De Giovanni P (2011a) Environmental collaboration in a closed loop supply chain with a reverse revenue sharing contract. Ann Oper Res Online First Articles:23 pages. doi:10.1007/s10479-011-0912-5

De Giovanni P (2011b) Quality improvements vs. advertising support: Which strategy works better for a manufacturer? Eur J Oper Res 208(2):119–130

De Giovanni P, Roselli M (2012) Overcoming the drawbacks of a revenue sharing contract through a support program. Ann Oper Res 196:201–222

De Giovanni P, Zaccour G (2013) Cost-revenue sharing contract in closed-loop supply chains. In: Cardaliaguet P, Cressman R (eds) Advances in dynamic games. Annals of the international society of dynamic games, vol 12. Birkhäuser, Boston, pp 395–421

Dockner E, Jørgensen S, Long N, Sorger G (2000) Differential games in economics and management science. Cambridge University Press, Cambridge

Doganoglu T, Klapper D (2006) Goodwill and dynamic advertising strategies. Quant Market Econ 4:5–29

El Ouardighi F, Kim B (2010) Supply quality management with wholesale price and revenue-sharing contracts under horizontal competition. Eur J Oper Res 206:329–340

El Ouardighi F, Jørgensen S, Pasin F (2008) A dynamic game of operations and marketing management in a supply chain. Int J Game Theory Rev 34:59–77

Erickson G (2011) A differential game model of the marketing-operations interface. Eur J Oper Res 211(2):394–402

Fine C (1986) Quality improvements and learning in productive systems. Manag Sci 10:1301–1315

Fine C (1988) A quality control model with learning effects. Oper Res 36:437–444

He X, Prasad A, Sethi S, Gutierrez G (2007) A survey of Stackelberg differential game models in supply and marketing channels. J Syst Sci Syst Eng 16:385–413

He X, Prasad A, Sethi S (2009) Coordinated advertising and pricing in a dynamic stochastic supply chain: Feedback Stackelberg strategies. Prod Oper Manag 18:78–94

Jørgensen S (1986) Optimal production, purchasing and pricing: A differential game approach. Eur J Oper Res 24:64–76

Jørgensen S, Zaccour G (2003) A differential game of retailer promotions. Automatica 39: 1145–1155

Jørgensen S, Sigué S, Zaccour G (2000) Dynamic coordinated advertising in a channel. J Retailing 76:71–92

Jørgensen S, Sigué S, Zaccour G (2001) Stackelberg leadership in a marketing channel. Int Game Theory Rev 3:13–26

Jørgensen S, Taboubi S, Zaccour G (2003) Retail promotions with negative brand image effects: Is cooperation possible? Eur J Oper Res 150:395–405

Jørgensen S, Taboubi S, Zaccour G (2006) Incentives for retailer promotion in a marketing channel. In: Annals of the International Society of Dynamic Games, vol 8. Birkhäuser, Boston, pp 365–378

Karray S, Zaccour G (2006) Could co-op advertising be a manufacturer's counterstrategy to store brands? J Bus Res 59:1008–1015

Kim B (1998) Optimal development of production technology when autonomous and induced learning are present. Int J Prod Econ 55:39–52

Kim B (2003) Dynamic outsourcing to contract manufacturers with different capabilities of reducing the supply cost. Int J Prod Econ 86:63–80

von Lanzenauer C, Pilz-Glombik K (2002) Coordinating supply chain decisions: An optimization model. OR Spectr 24(1):59–78

Martín-Herrán G, Taboubi S, Zaccour G (2012) Dual role of price and myopia in a marketing channel. Eur J Oper Res 219:284–295

Mentzer J, de Witt W, Keebler J, Min S, Nix N, Smith C, Zacham Z (2001) Defining supply chain management. J Bus Logist 22(2):1–25

Nair A, Narasimhan N (2006) Dynamics of competing with quality- and advertising-based goodwill. Eur J Oper Res 175:462–474

Rao A, Qul L, Ruekert R (1999) Signalling unobservable product quality through a brand ally. J Mark Res 36:258–268

Roselli M, De Giovanni P (2012) Recent developments in the differential equations of design and conformance quality. Int J Decis Sci Risk Manag 4(1–2):163–174

Tapiero C (1987) Production learning and quality control. IIE Trans 19(4):362–370

Vörös J (2006) The dynamics of price, quality and productivity improvement decisions. Eur J Oper Res 170:809–823

Chapter 8
A Large Population Parental Care Game with Asynchronous Moves

David M. Ramsey

Abstract This article considers two game-theoretic models of parental care which take into account the feedback between patterns of care and the operational sex ratio. Attention is paid to fish species which care for their young by mouthbrooding, in particular to St. Peter's Fish. It is assumed here that individuals can be in one of the two states: searching for a mate or breeding (including caring for their offspring). However, the sets of states can be adapted to the physiology of a particular species. The length of time an individual remains in the breeding state depends on the level of care he/she gives. According to one model, parents make their decision regarding the amount of care they give simultaneously. Under the second model, one individual in a pair (for convenience, the female) makes her decision before the male makes his decision. When in the searching state, individuals find partners at a rate dependent on the proportion of members of the opposite sex searching. These rates are defined to satisfy the Fisher condition that the total number of offspring of males equals the total number of offspring of females. The operational sex ratio is not defined exogenously, but can be derived from the adult sex ratio and the pattern of parental care. The results obtained go some way to explain the variety of parental care behaviour observed in fish, in particular the high frequency of male care, although further work is required to explain the exact patterns observed.

Keywords Evolutionarily stable strategy • Neutrally stable strategy • Strongly stable strategy • Polymorphic equilibrium • Large population game • Parental care • Asynchronous moves

---

D.M. Ramsey (✉)
Department of Computer Science and Management, Wrocław University of Technology, Wybrzeże Wyspiańskiego 27, Wrocław, Poland
e-mail: david.ramsey@pwr.wroc.pl

V. Křivan and G. Zaccour (eds.), *Advances in Dynamic Games*, Annals of the International Society of Dynamic Games 13, DOI 10.1007/978-3-319-02690-9_8,

8.1 Introduction

Research on the evolution of patterns of parental care has indicated the complex nature of this process. Trivers (1972) gave an in-depth exposition of the then current state of research into patterns of parental care. By definition females invest more in gametes. He argues that females have more to lose than males if the offspring die and should thus invest more in caring. Males can potentially reproduce at a much higher rate than females. However, due to the physiological constraints of breeding, the ratio of the number of males searching for a mate to the number of such females (the operational sex ratio, OSR) may be much greater than one, i.e. males face strong competition from other males when looking for a mate. Hence, it is argued that males should attempt to maximise the number of females they breed with by being attractive to females and/or outcompeting other males, rather than investing in parental care.

Emlen and Oring (1977) make an excellent review on the evolution of mating systems. They define the OSR and its relation with sexual selection. The feedback between patterns of parental care and the mating system is noted (e.g., mutual mate choice is normally associated with biparental care). In addition, they state that parental care in the form of egg incubation among birds will affect the OSR (the more males care, the less male-biased the OSR is). Kokko and Jennions (2008) argue that if males desert, then it is difficult for them to find a partner (since the OSR is male biased). Hence, if the level of male desertion increases, paternal care may become a more successful strategy, i.e. parental care is subject to frequency-dependent selection.

Dawkins and Carlisle (1976) state that Trivers' argument is a type of "Concorde fallacy", i.e. if one has invested heavily in a project, then one should continue even if losses are expected. Maynard Smith (1977) defends Trivers' approach by stating that in calculating the expected number of future offspring one needs to take into account the investment that has to be made. In that paper Maynard Smith describes three models of parental care. The first two are matrix games in which deserting males find another partner with probability p. However, he recognized that this probability depends on the behaviour of the population as a whole. The third model is more realistic, since it takes into account the Fisher condition that the total number of offspring of males equals the total number of offspring of females (see Houston and McNamara 2002, 2005; Kokko and Jennions 2003). This third model is a so-called time in/time out model in which individuals spend some time breeding and caring for their young and some time searching for a partner (between breeding attempts). The cycle time of an individual is the mean time between his/her breeding attempts. It is assumed that individuals maximise the rate of producing offspring that survive to adulthood. Grafen and Sibly (1978) develop this approach. However, these models assume that there is a pure equilibrium and so do not investigate the possibility of stable polymorphisms or mixed strategies. Yamamura and Tsuji (1993) adapt the model of Maynard Smith (1977). They assume that members of the less common sex in the mating pool immediately find a mate. It is assumed that parents can only make one of the two decisions: care or desert. This model was adapted to the life cycle of the St. Peter's Fish, *Sarotherodon galilaeus*, by Balshine-Earn and Earn (1997). This model was extended by Ramsey (2010) to take into account the fact that searchers spend some time in the mating pool. The rate at which mates are found depends on an exogenously defined interaction rate, λ_1. This parameter can be thought of as a measure of the density and mobility of the population. As Kokko and Rankin (2006) argue, density effects may be very important in the evolution of behaviour. This allows us to model the feedback between patterns of parental care and the OSR. By defining λ_1 to be arbitrarily large, we essentially obtain the model of Yamamura and Tsuji (1993). This paper further extends the model by assuming that one of the members of a mating pair makes its decision about whether to care or desert before the other. Since fish reproduce via external fertilisation, it is assumed that the female is the first to make such a choice.

The model presented can be adapted to the particular nature of physiological processes involved in reproduction (e.g. by assuming that females can be receptive or non-receptive to model œstrus cycles in mammals). However, it should be noted that these processes are assumed to be given (i.e. the model cannot explain why these processes evolved in the first place).

As in Kokko and Jennions (2008), members of the less common sex in the mating pool find mates at a faster rate than members of the other sex in such a way that ensures each female mating corresponds to a male mating. Hence, the ASR (the ratio of the number of fertile males to the number of fertile females) is fixed, but the OSR results from the ASR and the observed pattern of parental care, rather than being given as an exogenous parameter. In this way, the OSR and the pattern of parental-care co-evolve as argued in Jennions and Kokko (2010).

Clutton-Brock and Parker (1992) consider a similar time-in/time-out model to derive the OSR given the ASR as well as, patterns of parental investment (including both gamete production and parental care). They include a parameter describing the level of interaction between the sexes in a population, which in turn defines the mean time individuals spend looking for a mate. However, they assume that the amount of parental care given is fixed, since their goal is to derive the OSR and thus predict which sex will compete most strongly for mates (assumed to be the most common sex in the mating pool). As such, this model does not give us any insight into why a particular pattern of parental care evolves. As well as giving an excellent review of the research on patterns of mate choice and parental care, Kokko and Jennions (2008) extend this model by allowing the level of parental care to evolve. Offspring survival is increasing in the level of care from a parent, given the level of care from the other parent. At the time of fertilisation, parents simultaneously choose the amount of time for which they care from a continuous range. The minimum time females can choose is assumed to be larger than the minimum time a male can choose, since males can replenish their gametes more quickly than females. Sexual selection is incorporated into the model by assuming that only a fraction of the members of a particular sex mate. Due to the complexity of the model, they assume that the level of sexual selection is fixed, although they admit that in reality it evolves along with the parental care strategies.

All these models assume that breeding is non-seasonal and the population size is large. In such a case, at equilibrium the OSR will be constant over time. In the case of seasonal breeders, the strategies used by individuals will change over the breeding season and this is associated with temporal fluctuations in the OSR (see Webb et al. 1999 and McNamara et al. 2005).

The models described above implicitly assume that the decisions are made simultaneously. Asynchronous moves have been considered in 2-player games (see Maynard Smith 1982). This paper extends this approach to large population parental games. The results indicate that, along with physiological constraints, the order in which players move has an important influence on patterns of parental care in agreement with the results from two-player games, i.e. the sex that has the first opportunity to desert will desert when there is single-parent care.

The evolution of mating systems depends on many interacting factors. Verbal explanations of such evolution cannot realistically take these interactions into account. The development of mathematical models that take such interactions into account will prove useful in explaining behaviour and predicting the reaction of mating systems to changes in the environment.

Section 8.2 presents the basic model. Section 8.3 describes the derivation of pure stable profiles in which all the members of a particular sex always use the same strategy. For example, in the game with simultaneous moves, four pure stable profiles are possible: no parental care, maternal care, paternal care and biparental care. Section 8.4 considers stable polymorphisms, where there is variation between the behaviour of individuals of the same sex. Some analytic results are given and a procedure for estimating such stable profiles, based on replicator dynamics, is described. Section 8.5 gives a brief conclusion and directions for future research.

8.2 The Model

This model is adapted from the one presented by Ramsey (2010). Consider a large population with no variation in the quality of mates and individuals only decide whether to care for their young or desert. The model can be adapted so individuals choose the level of care they give from a continuous interval. However, discrete choices seem reasonable, e.g. in the case of the St. Peter's Fish, where care consists of mouthbrooding the young until they hatch.

There is no breeding season. Individuals may be in one of the two states: searching or breeding. For simplicity, it is assumed that individuals in the breeding state do not attempt to (or cannot) breed with other partners. The ASR is denoted by r. Denote the proportions of males in the two male states, searching and breeding, as p_1 and $p_2 = 1 - p_1$, respectively. The proportions of females in these states are denoted as q_1 and $q_2 = 1 - q_1$, respectively.

Males in the searching state find a mate at a rate proportional to the number of searching females, namely at rate $\lambda_1 q_1$. Hence, in a small interval of time of length δ units, a proportion $\lambda_1 q_1 \delta$ of searching males will find a mate. Similarly, searching

Table 8.1 Glossary of the notation used (ratios and relative nos.)

r	adult sex ratio	k_m	no. of offspring with male care
k_f	no. of offspring with female care	k_b	no. of offspring with biparental care

Table 8.2 Glossary of the notation used (proportions)

p_1	prop. of all males searching	$p_{1,C}$	of caring males searching
$p_{1,D}$	of deserting males searching	q_1	of all females searching
$q_{1,C}$	of caring females searching	$q_{1,D}$	of deserting females searching
s_m	of males who care	s_f	of females who care

Table 8.3 Glossary of the notation used (rates)

λ_m^D	deserting males return to searching	λ_m^C	caring males return to searching
λ_f^D	deserting females return to searching	λ_f^C	caring females return to searching
λ_1	interaction rate		

females find a mate at a rate proportional to the number of searching males, i.e. at rate $\lambda_1 p_1 r$. Note that these assumptions satisfy the condition that a male entering the breeding state corresponds to a female going into the breeding state, i.e. the Fisher condition is satisfied. Also, it is assumed that the population is freely mixing, e.g. when a male mates with a female then the strategy used by the female is chosen at random from the distribution of strategies used by the females in the mating pool.

The rate at which individuals return to the mating pool depends on their sex and level of care given. If they do not care for their young, males return to the mating pool at rate λ_m^D, i.e. on average the mating process and time to replenish sperm supplies together occupy on average $\frac{1}{\lambda_m^D}$ units of time. Similarly, if females do not care for their young, they return to the mating pool at rate λ_f^D. It is assumed that λ_m^D is larger than λ_f^D, i.e. male deserters return to searching for a new mate faster than female deserters. Since fish breed via external fertilization, it seems reasonable that the time deserters require to return to the mating pool is proportional to the amount of energy invested in gamete production. According to Hayward and Gillooly (2011), females tend to invest between twice and four times as much as males, depending on the species.

When they care for offspring, males and females return to the mating pool at rates λ_m^C and λ_f^C, respectively. In the case of St. Peter's Fish, we may assume that $\lambda_m^C = \lambda_f^C$. The transitions between states are illustrated in Fig. 8.1. The notation used is summarized in Tables 8.1–8.3.

The number of young surviving to maturity per brood is measured in relation to the number surviving when no parental care is given. Suppose the relative number of young surviving to maturity when (a) just the female cares, (b) just the male cares and (c) both parents care are k_f, k_m and k_b, respectively. It is assumed that $1 < k_f < k_b$ and $1 < k_m < k_b$, i.e. the greater the number of caring

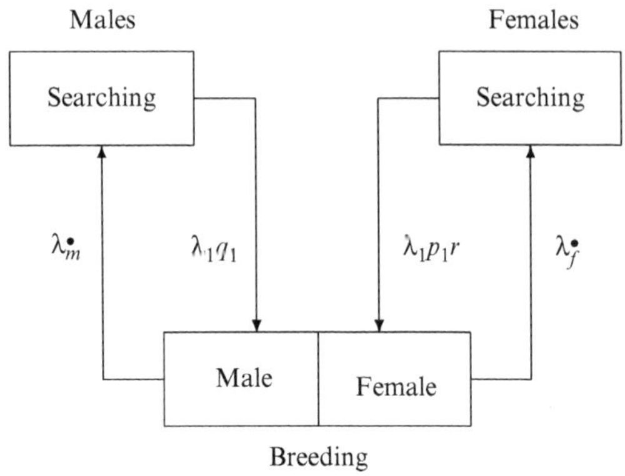

Fig. 8.1 Transition rates between states

• represents C or D according to whether a given sex cares for their offspring or not

parents, the greater the number of surviving offspring per brood. See Gubernick and Teferi (2000) and Wright (2006) for examples of parental care increasing offspring survival. In the case of St. Peter's Fish, since parental care consists solely of mouthbrooding and there is no significant size dimorphism, it may be assumed that $k_m = k_f$ and $k_b = 2k_f$ (see also Balshine-Earn 1997). It is assumed that the goal of each individual is to maximise the rate of producing offspring that survive until maturity. For simplicity this is referred to as the reproduction rate. This implicitly assumes that the mortality rate is independent of the strategy used.

For other game-theoretic models of large population games with state transitions, see Broom and Ruxton (1998) and Eriksson et al. (2004).

8.3 Derivation of Pure Stable Strategies

8.3.1 When Moves Are Made Simultaneously

When moves are made simultaneously, players only have two possible pure strategies: C—care and D—defect. In order to investigate the pure ESSes of such a system, we must first derive the "steady-state" proportions of individuals in each state given the strategy profile used. A strategy profile is defined by a description of both the strategies used by the males and the strategies used by the females. In this section, it is assumed that all individuals of a particular sex use the same strategy. Note that the term "steady-state" will only be used to describe the values q_1 and p_1 tend to, given that the strategy profile used does not change over time. One important aspect to note is the feedback between selection and these steady-state proportions. Selection changes the proportions of males and females who care for their offspring, which in turn changes these steady-state proportions.

At the steady-state proportions, the number of individuals moving from state A to state B per unit time must equal the number of individuals moving from state B to state A. Considering the transition of females from searching to breeding, the relative number of females finding partners per unit time is the proportion of females searching times the rate at which a female finds mates, i.e. $\lambda_1 p_1 q_1 r$. This rate is called the female population rate of transition from searching to breeding. Similarly, the population rate of females returning to the mating pool is $\lambda_f^{\bullet}(1-q_1)$, where $\bullet \in \{C, D\}$ denotes the action taken by females. Hence, at the steady-state proportions

$$\lambda_1 p_1 q_1 r = \lambda_f^{\bullet}(1 - q_1). \tag{8.1}$$

Equating the male population rate of transition from searching to breeding to the male population rate of transition from breeding to searching, it follows that

$$\lambda_1 p_1 q_1 = \lambda_m^{\bullet}(1 - p_1). \tag{8.2}$$

Equations (8.1) and (8.2) together lead to $aq_1^2 + bq_1 + c = 0$, where

$$\begin{aligned} a &= \lambda_f^{\bullet}\lambda_1 \\ b &= \lambda_m^{\bullet}\lambda_f^{\bullet} + \lambda_1 r \lambda_m^{\bullet} - \lambda_f^{\bullet}\lambda_1 \\ c &= -\lambda_m^{\bullet}\lambda_f^{\bullet}. \end{aligned}$$

The unique solution to this equation between 0 and 1 is

$$q_1 = \frac{-b + \sqrt{b^2 - 4ac}}{2a}. \tag{8.3}$$

From (8.2), it follows that

$$p_1 = \frac{\lambda_m^{\bullet}}{\lambda_m^{\bullet} + \lambda_1 q_1}. \tag{8.4}$$

Since these steady-state proportions depend on the strategy profiles adopted, the strategy profile will be denoted using superscripts indicating firstly the strategy used by males and secondly the strategy used by females. For example, p_1^{CD} denotes the equilibrium proportion of males searching when males care for their offspring, but females do not.

First, consider the conditions for no parental care to be an ESS. To find the corresponding steady-state frequencies, set $\lambda_f^{\bullet} = \lambda_f^D$ and $\lambda_m^{\bullet} = \lambda_m^D$.

There are two ways to calculate the reproduction rate of males. Firstly, it is the number of offspring surviving from a breeding attempt divided by the mean cycle time, which is the mean time required to move from the searching state to the breeding state and back again. Assume that neither parent cares for the offspring. Denote this mean cycle time by T^{DD}. It follows that

$$T^{DD} = \frac{1}{\lambda_1 q_1^{DD}} + \frac{1}{\lambda_m^D} = \frac{\lambda_m^D + \lambda_1 q_1^{DD}}{\lambda_m^D \lambda_1 q_1^{DD}}.$$

Since the relative number of offspring surviving per breeding attempt is 1, the mean reproduction rate of males is given by R^{DD}, where

$$R^{DD} = \frac{1}{T^{DD}} = \frac{\lambda_m^D \lambda_1 q_1^{DD}}{\lambda_m^D + \lambda_1 q_1^{DD}}. \tag{8.5}$$

The second way to calculate the reproduction rate of males is by noting that it must be the population rate of males entering the breeding state multiplied by the relative number of surviving offspring per breeding attempt. Hence, the mean reproduction rate of males is given by

$$R^{DD} = \lambda_1 q_1^{DD} p_1^{DD}. \tag{8.6}$$

From the Fisher condition, the average reproduction rate of females must be r times the average reproduction rate of males.

For no parental care to be an ESS, R^{DD} must be greater than the reproduction rate of a male mutant who cares for his offspring. Since this is a large population game, such a mutant does not affect the steady-state frequencies or the population's reproduction rate. The reproduction rates of mutants are calculated by considering their mean cycle time. The mean cycle time of a male mutant who cares for offspring when the rest of the population desert, denoted T_m^{DD}, is

$$T_m^{DD} = \frac{1}{\lambda_1 q_1^{DD}} + \frac{1}{\lambda_m^C} = \frac{\lambda_m^C + \lambda_1 q_1^{DD}}{\lambda_m^C \lambda_1 q_1^{DD}}.$$

Since the relative number of surviving offspring of such a male per brood is k_m, it follows that a male mutant cannot invade if

$$\lambda_1 p_1^{DD} q_1^{DD} > \frac{k_m \lambda_m^C \lambda_1 q_1^{DD}}{\lambda_m^C + \lambda_1 q_1^{DD}}. \tag{8.7}$$

Arguing similarly, a female mutant who cares cannot invade if

$$\lambda_1 p_1^{DD} q_1^{DD} > \frac{k_f \lambda_f^C \lambda_1 p_1^{DD}}{\lambda_f^C + \lambda_1 r p_1^{DD}}. \tag{8.8}$$

It should be noted that the OSR at such an equilibrium, denoted S^{DD}, is given by

$$S^{DD} = \frac{r p_1^{DD}}{q_1^{DD}}.$$

The OSR at other equilibria can be calculated in an analogous way.

The derivation of the stability conditions for the remaining three possible pure equilibria are analogous. Therefore, the equilibrium conditions are presented in Appendix 1. Suppose that $\lambda_1 = 20$, $\lambda_m^C = \lambda_f^C = 0.05$, $\lambda_m^D = 5$, $\lambda_f^D = 2$, $r = 1$. Assuming that $k_m = k_f$, the conditions for the existence of a pure ESS profile are given below:

1. No parental care is an ESS when $k_f < 34.8215$.
2. Just female parental care is an ESS when $k_f > 36.4221$ and $\frac{k_b}{k_f} < 1.9875$.
3. Just male parental care is an ESS when $k_f > 79.7970$ and $\frac{k_b}{k_f} < 1.9725$.
4. Biparental care is an ESS when $\frac{k_b}{k_f} > 17.1591$.

Now consider the problem in which $\lambda_1 = 0.5$, but the values of the remaining parameters are unchanged. These parameters represent a case in which there is less opportunity to find future partners. The conditions for the existence of a pure ESS profile are given below:

1. No parental care is an ESS when $k_f < 8.3241$.
2. Just female parental care is an ESS when $k_f > 8.7432$ and $\frac{k_b}{k_f} < 1.8993$.
3. Just male parental care is an ESS when $k_f > 9.8141$ and $\frac{k_b}{k_f} < 1.8845$.
4. Biparental care is an ESS when $\frac{k_b}{k_f} > 3.6042$.

From Ramsey (2010), when k_f is relatively large and $k_b/k_f = 2$, there is no pure equilibrium for a wide range of parameter sets. In the case of St. Peter's Fish, such values of k_f and k_b seem reasonable. Firstly, if neither parent cares, then the expected number of offspring surviving will be very low. Secondly, due to the lack of size dimorphism, about twice as many offspring survive when both parents mouthbrood their offspring, compared to uniparental care. It is thus unsurprising that a wide range of patterns of parental care has been observed in populations of St. Peter's Fish (see Fishelson and Hilzerman 2002).

8.3.2 *When Females Move First*

In the case of fish, females first lay their eggs and then males deposit sperm. Thus the female has the first opportunity to defect. Hence, we consider a version of the game where the female decides which action to take and then the male chooses his action. As before, females have two possible pure strategies C and D. Males may condition their action on the action taken by the female. The four possible pure strategies of the male are denoted $(C, C), (C, D), (D, C), (D, D)$ where the first component is the action taken by a male when the female cares and the second is the action taken by a male when the female deserts.

We now look for a pure equilibrium in this game. Given the strategy profile played by the population as a whole, the actions of each individual are well defined. For example, suppose the strategy profile used is $[(D, C), D]$, i.e. females play D

and males respond to C by playing D and respond to D by playing C. In this case, females defect and males care. Hence, we can derive the steady-state distributions as in the game with simultaneous moves.

The optimal actions of a male/female pair in a mating subgame (played when two individuals mate) against this background can be derived by recursion as for a standard game in extensive form. Also, it should be noted that since the payoffs of individuals involved in such a subgame depend on the actions taken in the population as a whole, then the equilibrium of the population game can be mixed or polymorphic (see Ramsey 2010).

We first consider the stability of $[(D, C), D]$. Suppose all males play (D, C) and all females play D. When all males care and females defect, the subgame perfect equilibrium of the induced game faced by a male/female pair is $[(D, C), D]$ when the following three conditions are satisfied:

1. When the female cares, the male prefers deserting to caring.
2. When the female deserts, the male prefers caring to deserting.
3. The female must prefer the action pair CD (which results when she defects) to the action pair DC (which results when she cares).

However, when all the population follow the appropriate strategy from the strategy pair $[(D, C), D]$, there is no selection pressure on the response of males to females caring for their young. Hence, it is expected that when all females defect, then the proportions of males using (D, C) and (C, C) will be subject to drift. It is possible that the proportion of males using (C, C) will rise to a level where it would start paying a female to change her strategy to C. Thus when the three conditions above are satisfied, then we say that $[(D, C), D]$ is neutrally stable. If in addition the following condition is satisfied, then $[(D, C), D]$ is said to be strongly stable.

4. When all males care and females defect, the female must prefer the action pair CD to CC [which results when she plays C and the male plays (C, C)].

Note that when this condition is satisfied, it does not pay a female to switch to C whatever the proportion of males playing (C, C) is.

It will be assumed henceforth that $k_m = k_f$. Suppose the population follows a strategy profile which results in male only care, but the male meets a female mutant who cares. The expected cycle length for a male deserting in this case is given by

$$\frac{1}{\lambda_m^D} + \frac{1}{\lambda_1 q_1^{CD}} = \frac{\lambda_1 q_1^{CD} + \lambda_m^D}{\lambda_m^D \lambda_1 q_1^{CD}}.$$

Since the relative number of offspring surviving in this cycle is k_f, the reproductive rate of the male deserter is

$$\frac{k_f \lambda_m^D \lambda_1 q_1^{CD}}{\lambda_1 q_1^{CD} + \lambda_m^D}.$$

Arguing similarly, the reproductive rate of a male carer in this case is

$$\frac{k_b \lambda_m^C \lambda_1 q_1^{CD}}{\lambda_1 q_1^{CD} + \lambda_m^C}.$$

It follows that Condition 1 is satisfied if and only if

$$\frac{k_b}{k_f} < \frac{\lambda_m^D(\lambda_m^C + \lambda_1 q_1^{CD})}{\lambda_m^C(\lambda_m^D + \lambda_1 q_1^{CD})}. \tag{8.9}$$

Similarly, considering the reproduction rates when males care and females defect, Condition 2 leads to

$$k_f > \frac{\lambda_m^D(\lambda_m^C + \lambda_1 q_1^{CD})}{\lambda_m^C(\lambda_m^D + \lambda_1 q_1^{CD})}. \tag{8.10}$$

It can be shown that the third condition is always satisfied when $k_m = k_f$. Hence, $[(D, C), D]$ is neutrally stable when (8.9) and (8.10) are satisfied. Considering the female's reproduction rate when both parents care, $[(D, C), D]$ is strongly stable when, in addition,

$$\frac{k_b}{k_f} < \frac{\lambda_f^D(\lambda_1 p_1^{CD} r + \lambda_f^C)}{\lambda_f^C(\lambda_1 p_1^{CD} r + \lambda_f^D)}.$$

Using a similar argument, it can be shown that when females can desert first, just female care, i.e. $[(D, D), C]$ or $[(D, C), C]$, cannot be neutrally stable. In addition, neither can $[(C, D), D]$ nor $[(C, C), D]$ be neutrally stable profiles. The conditions for the other strategy profiles to be neutrally or strongly stable are given in Appendix 2.

As before, consider the problem with $\lambda_1 = 20$, $\lambda_m^C = \lambda_f^C = 0.05$, $\lambda_m^D = 5$, $\lambda_f^D = 2$, $r = 1$. The stability conditions are given below:

1. $[(D, D), D]$ is neutrally stable when $k_f < 34.8215$, $k_b/k_f < 35.3418$. It is strongly stable if, in addition, $k_b < 7.0830$. Under such a profile, there is no parental care.
2. $[(D, C), D]$ is neutrally stable when $k_f > 79.7970$ and $k_b/k_f < 79.7970$. It is strongly stable if, in addition, $k_b/k_f < 1.9725$. Under such a profile, just males care.
3. $[(C, C), C]$ is strongly (and thus neutrally) stable when $k_f > 17.1591$ and $k_b/k_f > 17.1591$. Under such a profile, there is biparental care.
4. $[(C, D), C]$ is strongly (and thus neutrally) stable when $k_f < 17.1591$ and $k_b/k_f > 17.1591$. At such an equilibrium there is biparental care.

When the interaction rate is reduced to $\lambda_1 = 0.5$, the stability conditions are given below:

1. $[(D, D), D]$ is neutrally stable when $k_f < 8.3241$, $k_b/k_f < 8.4368$. It is strongly stable when $k_b < 2.0403$. Under such a profile, there is no parental care.
2. $[(D, C), D]$ is neutrally stable when $k_f > 9.8141$ and $k_b/k_f < 9.8141$. It is strongly stable when $k_b/k_f < 1.8845$. Under such a profile, just males care.
3. $[(C, C), C]$ is strongly (and thus neutrally) stable when $k_f > 3.6042$ and $k_b/k_f > 3.6042$. Under such a profile, there is biparental care.
4. $[(C, D), C]$ is strongly (and thus neutrally) stable when $k_f < 3.6042$ and $k_b/k_f > 3.6042$. Under such a profile, there is biparental care.

From the stability conditions, polymorphic equilibria are more likely when the interaction rate is high. In such a case, selection will be highly frequency dependent. When all individuals care, although there are not many prospective partners in the mating pool, a mutant deserter will find a partner reasonably quickly and will thus be selected for. On the other hand, when no individuals care, males return to the mating pool more quickly than females do. Hence, it may be relatively hard for males to find a new partner and thus caring males will be selected for.

8.3.3 Effect of the Asynchronocity of Moves on Behaviour

In the examples presented, there is at most one strongly stable pure profile. It is assumed that if a strongly stable profile exists, then the population will evolve to such a profile. If there is also a neutrally stable profile, then initially the population may evolve towards such a profile. However, it is expected that genetic drift will cause the population to eventually evolve to the strongly stable profile.

The conditions for biparental care to be stable (evolutionarily or strongly, as appropriate) are identical in both games, while the region in which no parental care is stable is smaller in the game with asynchronous moves (it is never stable when the gains from biparental care are large). Hence, when the gains from uniparental care are small and the gains from biparental care are large, it seems more likely that biparental care will evolve when the decisions of the parents are made in sequence.

Just female care is never stable in the game with asynchronous moves. The region in which $[(D, C), D]$ (i.e. just male care) is strongly stable in the game with synchronous moves corresponds to the region in which just male care is an ESS in the game with synchronous moves (i.e. k_f is large and k_b/k_f is small). However, in the game with synchronous moves, just female care is also an ESS and is expected to evolve, since it has a larger basin of attraction.

For large values of k_f and intermediate values of k_b/k_f, there is no ESS in the game with synchronous moves. However, $[(D, C), D]$ is neutrally stable in the game with asynchronous moves. Since there is no other stable strategy profile, it is expected that predominantly male care will evolve. When the gains from biparental care relative to uniparental care increase, i.e. k_b/k_f increases, biparental care is the only strongly stable profile and thus is predicted to evolve.

In general, biparental care or, in particular, just male care are more likely to evolve when moves are made in sequence rather than simultaneously.

In the game played by mouthbrooding fish, k_f will be large and $k_b/k_f \approx 2$. In the example with $\lambda_1 = 20$, when k_f is moderately large ($34.8215 < k_f < 79.7970$), then there is neither a pure ESS in the game with simultaneous moves nor a stable strategy profile in the game with asynchronous moves. In the case where $k_f > 79.7970$, then $[(D, C), D]$ is neutrally stable in the game with asynchronous moves, but there is no pure ESS in the game with simultaneous moves. Hence, the new model seems to predict predominantly male care in St. Peter's Fish.

It should be noted that the existence of a pure ESS does not necessarily mean that there is no mixed or polymorphic equilibrium, since selection is frequency dependent.

8.4 Polymorphic Equilibria

In this section we concentrate our attention on the type of games played by mouthbrooding fish, i.e. k_f is large and $k_b/k_f = 2$. Suppose that $\lambda_m^C = \lambda_f^C = 0.05$, $\lambda_m^D = 5$, $\lambda_f^D = 2$, $r = 1$. These parameters are chosen to reflect a fish species in which the adult sex ratio is one and females invest 2.5 times as much energy in producing gametes as males.

8.4.1 The Game with Simultaneous Moves

Ramsey (2010) shows that when k_f is large and k_b/k_f is around 2, then we expect an equilibrium where all females care and males show varied behaviour. That article differentiates between polymorphisms (where each male always uses the same action, but some males act differently from others) and mixed equilibria (where each male chooses each action with an appropriate probability after mating). This article only considers polymorphisms. Suppose that all females and a proportion s^{PC} of males care at such a polymorphic equilibrium. The superscript PC shows that males are polymorphic while females care. The remaining notation is adapted accordingly. Let $p_{1,C}$ and $p_{1,D}$ be the proportion of male carers and male deserters, respectively, who are searching. The steady-state equations are

$$\lambda_1 q_1^{PC} p_{1,C}^{PC} = \lambda_m^C (1 - p_{1,C}^{PC}) \tag{8.11}$$

$$\lambda_1 q_1^{PC} p_{1,D}^{PC} = \lambda_m^D (1 - p_{1,D}^{PC}) \tag{8.12}$$

$$\lambda_1 [s^{PC} p_{1,C}^{PC} + (1 - s^{PC}) p_{1,D}^{PC}] r q_1^{PC} = \lambda_f^C (1 - q_1^{PC}). \tag{8.13}$$

These expressions equate the population transition rates between the searching and breeding states for (a) male carers, (b) male deserters, and (c) females, respectively. In addition, the equilibrium conditions state that (a) the reproductive rates of male carers must equal the reproductive rate of male deserters and (b) the reproductive rate of female carers must be greater than the reproductive rate of female deserters. These conditions lead to (see Ramsey 2010)

$$k_b p_{1,C}^{PC} = k_f p_{1,D}^{PC} \tag{8.14}$$

$$\frac{\lambda_f^D[s^{PC} p_{1,C}^{PC} k_f + (1-s^{PC}) p_{1,D}^{PC}]}{\lambda_f^D + \lambda_1[s^{PC} p_{1,C}^{PC} + (1-s^{PC}) p_{1,D}^{PC}]} < \frac{\lambda_f^C[s^{PC} p_{1,C}^{PC} k_b + (1-s^{PC}) p_{1,D}^{PC} k_f]}{\lambda_f^C + \lambda_1[s^{PC} p_{1,C}^{PC} + (1-s^{PC}) p_{1,D}^{PC}]}. \tag{8.15}$$

Setting $\gamma = k_b/k_f$ and solving (8.11)–(8.14), we obtain

$$p_{1,C}^{PC} = \frac{\lambda_m^D - \gamma\lambda_m^C}{\gamma(\lambda_m^D - \lambda_m^C)} \tag{8.16}$$

$$p_{1,D}^{PC} = \frac{\lambda_m^D - \gamma\lambda_m^C}{\lambda_m^D - \lambda_m^C} \tag{8.17}$$

$$q_1^{PC} = \frac{(\gamma-1)\lambda_m^C\lambda_m^D}{\lambda_1(\lambda_m^D - \gamma\lambda_m^C)} \tag{8.18}$$

$$s^{PC} = \frac{\gamma}{\gamma-1} - \frac{\gamma\lambda_f^C(\lambda_m^D - \lambda_m^C)[\lambda_1(\lambda_m^D - \gamma\lambda_m^C) - (\gamma-1)\lambda_m^C\lambda_m^D]}{\lambda_1 r(\gamma-1)^2\lambda_m^C\lambda_m^D(\lambda_m^D - \gamma\lambda_m^C)}. \tag{8.19}$$

When $\lambda_1 = 20$, from (8.19) approximately 2.5% of males care at equilibrium. From Condition (8.15), such an equilibrium exists if $k_f > 65.1628$. When $\lambda_1 = 0.5$, approximately 22.2% of males care at equilibrium and such an equilibrium exists if $k_f > 57.7576$. Unsurprisingly, as it becomes harder to find a future partner, males are more likely to care for their offspring.

Note also that given the physiological parameters, this equilibrium depends on the ratio between k_b and k_f rather than their individual values.

8.4.1.1 Estimating Fully Polymorphic Equilibria Using Replicator Dynamics

When $\lambda_1 = 20$, $34.8215 < k_f < 65.1628$ and $\gamma = 2$, there is no pure equilibrium and the polymorphism described above is unstable. It is expected that some individuals of either sex care and some desert. We denote the proportion of females caring and the proportion of males caring by s_f and s_m, respectively. Let $p_{1,C}$, $p_{1,D}$, $q_{1,C}$ and $q_{1,D}$ denote the proportion of caring males who are searching, the proportion of deserting males who are searching, the proportion of caring females who are searching and the proportion of deserting females who are searching, respectively. Consider the rate at which: (a) caring males, (b) deserting males, (c) caring females and (d) deserting females, move between the mating pool and breeding state. The steady-state equations are given by

$$\lambda_1 p_{1,C}[s_f q_{1,C} + (1 - s_f) q_{1,D}] = \lambda_m^C (1 - p_{1,C}) \tag{8.20}$$

$$\lambda_1 p_{1,D}[s_f q_{1,C} + (1 - s_f) q_{1,D}] = \lambda_m^D (1 - p_{1,D}) \tag{8.21}$$

$$\lambda_1 q_{1,C}[s_m p_{1,C} + (1 - s_m) p_{1,D}] = \lambda_f^C (1 - q_{1,C}) \tag{8.22}$$

$$\lambda_1 q_{1,D}[s_m p_{1,C} + (1 - s_m) p_{1,D}] = \lambda_f^D (1 - q_{1,D}). \tag{8.23}$$

Assume that s_m and s_f are fixed, i.e. we treat (8.20)–(8.23) as a system of equations for $p_{1,C}, p_{1,D}, q_{1,C}$ and $q_{1,D}$. This leads to

$$q_{1,D} = \frac{\lambda_f^D q_{1,C}}{\lambda_f^C + (\lambda_f^D - \lambda_f^C) q_{1,C}}$$

$$p_{1,C} = \frac{\lambda_m^C [\lambda_f^C + (\lambda_f^D - \lambda_f^C) q_{1,C}]}{A}$$

$$p_{1,D} = \frac{\lambda_m^D \lambda_m^C [\lambda_f^C + (\lambda_f^D - \lambda_f^C) q_{1,C}]}{(\lambda_m^D - \lambda_m^C) \lambda_m^C [\lambda_f^C + (\lambda_f^D - \lambda_f^C) q_{1,C}] + \lambda_m^C A},$$

where

$$A = \lambda_m^C \lambda_f^C + q_{1,C}[\lambda_m^C(\lambda_f^D - \lambda_f^C) + \lambda_1 \lambda_f^C s_f + \lambda_1 (1 - s_f) \lambda_f^D] + \lambda_1 s_f (\lambda_f^D - \lambda_f^C) q_{1,C}^2.$$

Substituting the expressions for $p_{1,C}$ and $p_{1,D}$ into (8.22), we obtain an equation for $q_{1,C}$ of the form $f(q_{1,C}) = g(q_{1,C})$, where $g(q_{1,C}) = \lambda_f^C (1 - q_{1,C})$. We have $f(0) = 0 < g(0)$ and $g(1) = 0 < f(1)$. The right-hand side of this equation is clearly decreasing in $q_{1,C}$, whilst it can be shown by differentiation that the left-hand side is increasing in $q_{1,C}$ for $0 < q_{1,C} < 1$. Hence, there is a unique solution of this equation in the interval $(0, 1)$. Also, given a value of $q_{1,C}$ in $(0, 1)$, it can be shown that the remaining values of the unknowns in this system of equations take unique values in the interval $(0, 1)$. Hence, there is a unique solution to the system of equations given by (8.20)–(8.23). This system of equations was solved numerically.

We can now find the reproduction rates of individuals according to sex and the strategy they follow. For example, consider a male deserter. The proportion of females who care and are searching is $q_{1,C} s_f$ and the proportion of females who desert and are searching is $q_{1,D}(1 - s_f)$. Hence, the expected number of offspring that a male deserter produces per cycle is given by

$$\frac{q_{1,C} s_f k_f + (1 - s_f) q_{1,D}}{q_{1,C} s_f + (1 - s_f) q_{1,D}}.$$

Also, the expected length of a cycle is given by

$$\frac{1}{\lambda_m^D} + \frac{1}{\lambda_1[q_{1,C}s_f + q_{1,D}(1-s_f)]} = \frac{\lambda_m^D + \lambda_1[q_{1,C}s_f + q_{1,D}(1-s_f)]}{\lambda_m^D\lambda_1[q_{1,C}s_f + q_{1,D}(1-s_f)]}.$$

Dividing the expected number of offspring by the expected length of a cycle, we obtain the reproduction rate of male deserters, denoted R_m^D.

$$R_m^D = \frac{\lambda_m^D\lambda_1[q_{1,C}s_fk_f + q_{1,D}(1-s_f)]}{\lambda_m^D + \lambda_1[q_{1,C}s_f + q_{1,D}(1-s_f)]}. \tag{8.24}$$

Arguing similarly, the reproduction rates of male carers, female deserters and female carers (R_m^C, R_f^D and R_f^C, respectively) are given by

$$R_m^C = \frac{\lambda_m^C\lambda_1[q_{1,C}s_fk_b + q_{1,D}(1-s_f)k_f]}{\lambda_m^C + \lambda_1[q_{1,C}s_f + q_{1,D}(1-s_f)]} \tag{8.25}$$

$$R_f^D = \frac{\lambda_f^D\lambda_1[p_{1,C}s_mk_f + p_{1,D}(1-s_m)]}{\lambda_f^D + \lambda_1[p_{1,C}s_m + p_{1,D}(1-s_m)]} \tag{8.26}$$

$$R_f^C = \frac{\lambda_f^C\lambda_1[p_{1,C}s_mk_b + p_{1,D}(1-s_m)k_f]}{\lambda_f^C + \lambda_1[p_{1,C}s_m + p_{1,D}(1-s_m)]}. \tag{8.27}$$

By assuming that evolution acts slowly relative to the speed with which the steady state is attained under a given strategy profile, we may use (8.24)–(8.27) to define the replicator dynamics of such a population. Let $\tilde{s}_m$ and $\tilde{s}_f$ be the updated proportions of male and females, respectively, caring in the next generation. It follows that

$$\tilde{s}_m = \frac{s_m R_m^C}{s_m R_m^C + (1-s_m)R_m^D} \tag{8.28}$$

$$\tilde{s}_f = \frac{s_f R_f^C}{s_f R_f^C + (1-s_f)R_f^D}. \tag{8.29}$$

Suppose $\lambda_1 = 20$, $\lambda_m^C = \lambda_f^C = 0.05$, $\lambda_m^D = 5$, $\lambda_f^D = 2$, $r = 1$, $k_f = 50$ and $k_b = 100$. Using these replicator dynamics, at the equilibrium about 1.79% of males and 99.15% of females care. By varying the initial frequencies with which males and females care, it seems that this is the only attractor for this problem.

8.4.2 The Game with Asynchronous Moves

In the example with the high interaction rate, when $k_f > 79.7970$, there is a unique weakly stable strategy profile where only males care. Consider the realisation of the problem in which $k_f = 50$, $k_b = 100$. We expect that at least one of the sexes will be polymorphic at equilibrium. In order to investigate this, we define the replicator equations for this game. As before, s_f, $q_{1,C}$ and $q_{1,D}$ denote the proportion of females who care, the proportion of caring females in the mating pool and the proportion of deserting females in the mating pool, respectively. Let $r_{D,D}, r_{D,C}, r_{C,D}$ and $r_{C,C}$ denote the proportion of males who play the strategies (D,D), (D,C), (C,D) and (C,C), respectively. Let $p_{1,\bullet}$ be the proportion of $\bullet$ playing males in the mating pool, where $\bullet \in \{(D,D),(D,C),(C,D),(C,C)\}$.

It should be noted that males using the strategy (D,D) always return to the mating pool at rate λ_m^D. Similarly, males using the strategy (C,C) always return to the mating pool at rate λ_m^C. Males using the strategy (D,C) return to the mating pool at rate λ_m^D when the female cares and at rate λ_m^C when the female deserts. Considering the probability that a female in the mating pool is a carer, the mean time spent breeding by such a male is given by

$$\frac{s_f q_{1,C}}{\lambda_m^D[s_f q_{1,C} + (1-s_f)q_{1,D}]} + \frac{(1-s_f)q_{1,D}}{\lambda_m^C[s_f q_{1,C} + (1-s_f)q_{1,D}]} = \frac{\lambda_m^C s_f q_{1,C} + \lambda_m^D(1-s_f)q_{1,D}}{\lambda_m^D\lambda_m^C[s_f q_{1,C} + (1-s_f)q_{1,D}]}.$$

The rate at which such males return to the mating pool is the reciprocal of this expression. The rate at which males using the strategy (C,D) return to the mating pool can be derived in a similar way. Hence, the steady-state equations for this game are given by

$$\lambda_1 p_{1,(D,D)}[s_f q_{1,C} + (1-s_f)q_{1,D}] = \lambda_m^D(1-p_{1,(D,D)}) \tag{8.30}$$

$$\lambda_1 p_{1,(D,C)}[s_f q_{1,C}+(1-s_f)q_{1,D}] = \frac{\lambda_m^D\lambda_m^C[s_f q_{1,C}+(1-s_f)q_{1,D}](1-p_{1,(D,C)})}{s_f q_{1,C}\lambda_m^C+(1-s_f)q_{1,D}\lambda_m^D} \tag{8.31}$$

$$\lambda_1 p_{1,(C,D)}[s_f q_{1,C}+(1-s_f)q_{1,D}] = \frac{\lambda_m^D\lambda_m^C[s_f q_{1,C}+(1-s_f)q_{1,D}](1-p_{1,(C,D)})}{s_f q_{1,C}\lambda_m^D+(1-s_f)q_{1,D}\lambda_m^C} \tag{8.32}$$

$$\lambda_1 p_{1,(C,C)}[s_f q_{1,C} + (1-s_f)q_{1,D}] = \lambda_m^C(1-p_{1,(C,C)}) \tag{8.33}$$

$$\lambda_1 q_{1,C}[r_{D,D}p_{1,(D,D)} + r_{D,C}p_{1,(D,C)} + r_{C,D}p_{1,(C,D)} + r_{C,C}p_{1,(C,C)}] = \lambda_f^C(1-q_{1,C}) \tag{8.34}$$

$$\lambda_1 q_{1,D}[r_{D,D}p_{1,(D,D)} + r_{D,C}p_{1,(D,C)} + r_{C,D}p_{1,(C,D)} + r_{C,C}p_{1,(C,C)}] = \lambda_f^D(1-q_{1,D}). \tag{8.35}$$

Assuming that the strategy profile [given by s_f, $r_{D,D}$, $r_{D,C}$, $r_{C,D}$ and $r_{C,C}$] is fixed, we can determine the steady state [given by $q_{1,C}$, $q_{1,D}$, $p_{1,(D,D)}$, $p_{1,(D,C)}$, $p_{1,(C,D)}$ and $p_{1,(C,C)}$] from (8.30)–(8.35). Using a similar argument to the one used for the game with simultaneous moves, it can be shown that there is a unique solution to this system of equations.

We now derive the reproduction rates of individuals given the strategy profile used. First consider female carers. The cycle time of such a female is given by

$$T_f^C = \frac{1}{\lambda_f^C} + \frac{1}{\lambda_1[r_{D,D}p_{1,(D,D)} + r_{D,C}p_{1,(D,C)} + r_{C,D}p_{1,(C,D)} + r_{C,C}p_{1,(C,C)}]}.$$

When the female cares, there is biparental care when the male plays (C, D) or (C, C), otherwise only the female cares. Considering the probability that a male in the mating pool plays (C, D) or (C, C), the expected number of offspring of a female carer per cycle is given by

$$N_f^C = \frac{k_b(r_{C,D}p_{1,(C,D)} + r_{C,C}p_{1,(C,C)}) + k_f(r_{D,C}p_{1,(D,C)} + r_{D,D}p_{1,(D,D)})}{r_{C,D}p_{1,(C,D)} + r_{C,C}p_{1,(C,C)} + r_{D,C}p_{1,(D,C)} + r_{D,D}p_{1,(D,D)}}.$$

It follows that the reproduction rate of female carers is given by

$$R_f^C = \frac{\lambda_f^C\lambda_1[k_b(r_{C,D}p_{1,(C,D)} + r_{C,C}p_{1,(C,C)}) + k_f(r_{D,C}p_{1,(D,C)} + r_{D,D}p_{1,(D,D)})]}{\lambda_f^C + \lambda_1[r_{D,D}p_{1,(D,D)} + r_{D,C}p_{1,(D,C)} + r_{C,D}p_{1,(C,D)} + r_{C,C}p_{1,(C,C)}]}. \quad (8.36)$$

Arguing similarly, the reproduction rate of female deserters is given by

$$R_f^D = \frac{\lambda_f^D\lambda_1[k_f(r_{D,C}p_{1,(D,C)} + r_{C,C}p_{1,(C,C)}) + r_{C,D}p_{1,(C,D)} + r_{D,D}p_{1,(D,D)}]}{\lambda_f^D + \lambda_1[r_{D,D}p_{1,(D,D)} + r_{D,C}p_{1,(D,C)} + r_{C,D}p_{1,(C,D)} + r_{C,C}p_{1,(C,C)}]}. \quad (8.37)$$

Also, the reproduction rates of males according to their strategy are given by

$$R_m^{D,D} = \frac{\lambda_1\lambda_m^D[k_f s_f q_{1,C} + (1-s_f)q_{1,D}]}{\lambda_m^D + \lambda_1[s_f q_{1,C} + (1-s_f)q_{1,D}]} \quad (8.38)$$

$$R_m^{D,C} = \frac{k_f\lambda_m^D\lambda_m^C\lambda_1[s_f q_{1,C} + (1-s_f)q_{1,D}]}{\lambda_m^C\lambda_m^D + \lambda_1[s_f\lambda_m^C + (1-s_f)\lambda_m^D][s_f q_{1,C} + (1-s_f)q_{1,D}]} \quad (8.39)$$

$$R_m^{C,D} = \frac{\lambda_m^D\lambda_m^C\lambda_1[k_b s_f q_{1,C} + (1-s_f)q_{1,D}]}{\lambda_m^C\lambda_m^D + \lambda_1[s_f\lambda_m^D + (1-s_f)\lambda_m^C][s_f q_{1,C} + (1-s_f)q_{1,D}]} \quad (8.40)$$

$$R_m^{C,C} = \frac{\lambda_1\lambda_m^C[k_b s_f q_{1,C} + k_f(1-s_f)q_{1,D}]}{\lambda_m^C + \lambda_1[s_f q_{1,C} + (1-s_f)q_{1,D}]}. \quad (8.41)$$

As before, we may use (8.36)–(8.41) to define the replicator dynamics of such a population. Let $\tilde{s}_f$ be the updated proportion of females caring and $\tilde{r}_{D,D}$, $\tilde{r}_{D,C}$, $\tilde{r}_{C,D}$ and $\tilde{r}_{C,C}$ be the updated proportions of males playing (D, D), (D, C), (C, D) and (C, C), respectively. It follows that

$$\tilde{s}_f = \frac{s_f R_f^C}{s_f R_f^C + (1 - s_f) R_f^D} \tag{8.42}$$

$$\tilde{r}_\bullet = \frac{r_\bullet R_m^\bullet}{r_{D,D} R_m^{D,D} + r_{D,C} R_m^{D,C} + r_{C,D} R_m^{C,D} + r_{C,C} R_m^{C,C}}, \tag{8.43}$$

where $\bullet \in \{(D, D), (D, C), (C, D), (C, C)\}$.

Suppose $\lambda_1 = 20$, $\lambda_m^C = \lambda_f^C = 0.05$, $\lambda_m^D = 5$, $\lambda_f^D = 2$, $r = 1$, $k_f = 100$ and $k_b = 200$. The unique neutrally stable strategy profile is $[(D, C), D]$. The replicator dynamics were used with initially 50% of females caring and each of the male strategies used by 25% of the males. The population evolved so that no females cared, 99.89% of males used (D, C) and 0.11% of males used (C, C). Note that when all females desert, in the absence of mutation, the only selection pressure on males is to make the optimal response when a female deserts, i.e. to care. Hence, when caring females and males responding to desertion with desertion have died out, the fraction of males using (D, C) remains fixed. This fraction depends on the starting point used.

Now consider the game where $k_f = 50$ and $k_b = 100$. At the equilibrium about 60.17% of males use (D, D) and 39.83% use (D, C). Approximately 0.36% of females care. At such an equilibrium, the probability of no parental care is $0.9964 \times 0.6017 \approx 0.5995$. The probability of just male care is $0.9964 \times 0.3983 \approx 0.3969$. The probability of just female care is 0.0036. Biparental care never exists at such an equilibrium. The replicator dynamics converge to the equilibrium profile, since both types of female behaviour are present.

8.5 Conclusion

As far as the author knows, this is the first model of a parental care game, which combines the time-in/time-out approach with asynchronous moves. The model was used to predict behaviour in a species of mouth brooding fish.

As predicted by the model, there is a variety of parental care strategies observed in St. Peter's Fish. However, observations suggest that in reality the level of male care lies in between the level predicted by the game with simultaneous moves (almost all males desert) and the model where females decide first (many males care). Similarly, the level of female care observed seems to lie in between the levels predicted by these two models (see Fishelson and Hilzerman 2002). The following factors may explain the differences to some degree.

1. A pair seem to play a war of attrition game before deciding whether they will brood. This occurs due to variation in the size of partners and, particularly, in the number of eggs laid and their state (see Jennions and Polakow 2001). Players should gain information on the state of the eggs, but balance this against losses

due to mortality when offspring are not brooded immediately. Yaniv and Motro (2004) consider the game played by a mating pair as a war of attrition. Unlike the large population game considered here, they consider a two-player game. It would be good to extend the model presented here to consider the interaction between a mating pair in more detail.

2. The model assumes that mating is random. Fishelson and Hilzerman (2002) note that many pairs mate repeatedly. This has two obvious effects. Firstly, mating repeatedly with one partner affects the rate at which individuals mate. Secondly, the possibility of forming such partnerships is likely to increase the level of parental care, since individuals may play a "tit-for-tat" strategy. Also, as Ros et al. (2003) state, mating is not random, since individuals of both sexes prefer large partners. Large females lay on average more eggs and larger individuals of both sexes are able to mouthbrood more offspring.

Development of the model to take these factors into account would enable us to model and understand parental care patterns more fully.

Another interesting problem to look at would be the case of sequential brooding. *Eretmodus cyanostictus* is a mouthbrooding cichlid in which the clutch is first brooded by the female and then by the male (see Grüter and Taborsky 2005). This is a game in which the moves are obviously asynchronous. Also, this game has aspects of a war of attrition, since the male would like the female to brood the clutch for as long as possible.

In conclusion, the model explains to some degree why there is variety in the parental care strategies of mouthbrooding fish. However, the model needs to be expanded, in order to explain the pattern of parental care behaviour that is actually observed.

Acknowledgements The author is grateful for the support of Science Foundation Ireland under the BIO-SI project (no. 07MI012).

Appendix 1: Stability Conditions in Game with Simultaneous Moves

The left-hand side of each inequality is the reproduction rate of males (which is the reproduction rate of the females divided by the ASR), the first entry on the right-hand side is the reproduction rate of a mutant male and the second entry is the reproduction rate of a mutant female divided by the ASR.

Only male parental care is an ESS if

$$k_m \lambda_1 q_1^{CD} p_1^{CD} > \max \left\{ \frac{\lambda_1 \lambda_m^D q_1^{CD}}{\lambda_m^D + \lambda_1 q_1^{CD}}, \frac{k_b \lambda_1 \lambda_f^C p_1^{CD}}{\lambda_f^C + \lambda_1 r p_1^{CD}} \right\}. \tag{8.44}$$

Only female parental care is an ESS if

$$k_f \lambda_1 q_1^{DC} p_1^{DC} > \max\left\{ \frac{k_b \lambda_1 \lambda_m^C q_1^{DC}}{\lambda_m^C + \lambda_1 q_1^{DC}}, \frac{\lambda_1 \lambda_f^D p_1^{DC}}{\lambda_f^D + \lambda_1 r p_1^{DC}} \right\}. \tag{8.45}$$

Parental care by both sexes is an ESS if

$$k_b \lambda_1 q_1^{CC} p_1^{CC} > \max\left\{ \frac{k_f \lambda_1 \lambda_m^D q_1^{CC}}{\lambda_m^D + \lambda_1 q_1^{CC}}, \frac{k_m \lambda_1 \lambda_f^D p_1^{CC}}{\lambda_f^D + \lambda_1 r p_1^{CC}} \right\}. \tag{8.46}$$

Appendix 2: Stability Conditions in Game with Asynchronous Moves

No parental care with unconditional desertion by males, $[(D, D), D]$, is neutrally stable when

$$\max\{k_f, \frac{k_b}{k_f}\} < \frac{\lambda_m^D(\lambda_m^C + \lambda_1 q_1^{DD})}{\lambda_m^C(\lambda_m^D + \lambda_1 q_1^{DD})} \tag{8.47}$$

$$k_f < \frac{\lambda_f^D(\lambda_f^C + \lambda_1 r p_1^{DD})}{\lambda_f^C(\lambda_f^D + \lambda_1 r p_1^{DD})}. \tag{8.48}$$

This strategy profile is strongly stable when, in addition

$$k_b < \frac{\lambda_f^D(\lambda_f^C + \lambda_1 r p_1^{DD})}{\lambda_f^C(\lambda_f^D + \lambda_1 r p_1^{DD})}.$$

Biparental care with unconditional care from males, $[(C, C), C]$, is neutrally stable when

$$\min\{k_f, \frac{k_b}{k_f}\} > \frac{\lambda_m^D(\lambda_m^C + \lambda_1 q_1^{CC})}{\lambda_m^C(\lambda_m^D + \lambda_1 q_1^{CC})} \tag{8.49}$$

$$\frac{k_b}{k_f} > \frac{\lambda_f^D(\lambda_f^C + \lambda_1 r p_1^{CC})}{\lambda_f^C(\lambda_f^D + \lambda_1 r p_1^{CC})}. \tag{8.50}$$

The condition required for strong stability, i.e.

$$k_b > \frac{\lambda_f^D(\lambda_f^C + \lambda_1 r p_1^{CC})}{\lambda_f^C(\lambda_f^D + \lambda_1 r p_1^{CC})},$$

is a weaker condition than Condition (8.50). Hence, if $[(C, C), C]$ is neutrally stable, then it is strongly stable.

Finally, it can be shown that $[(C, D), C]$, i.e. biparental care with male care being conditional on the female giving care, is neutrally stable when

$$k_f < \frac{\lambda_m^D(\lambda_m^C + \lambda_1 q_1^{CC})}{\lambda_m^C(\lambda_m^D + \lambda_1 q_1^{CC})} < \frac{k_b}{k_f} \tag{8.51}$$

$$k_b > \frac{\lambda_f^D(\lambda_f^C + \lambda_1 r p_1^{CC})}{\lambda_f^C(\lambda_f^D + \lambda_1 r p_1^{CC})}. \tag{8.52}$$

This strategy profile is strongly stable when, in addition

$$\frac{k_b}{k_f} > \frac{\lambda_f^D(\lambda_f^C + \lambda_1 p_1^{CC} r)}{\lambda_f^C(\lambda_f^C(\lambda_f^D + \lambda_1 p_1^{CC} r))}.$$

References

Balshine-Earn S (1997) The benefits of uniparental versus biparental mouth brooding in Galilee St. Peter's fish. J Fish Biol 50:371–381

Balshine-Earn S, Earn DJD (1997) An evolutionary model of parental care in St. Peter's fish. J Theor Biol 184:423–431

Broom M, Ruxton GD (1998) Evolutionarily stable stealing: Game theory applied to kleptoparatism. Behav Ecol 9(4):397–403

Clutton-Brock TH, Parker GA (1992) Potential reproductive rates and the operation of sexual selection. Q Rev Biol 67:437–456

Dawkins R, Carlisle TR (1976) Parental investment and mate desertion: A fallacy. Nature 262: 131–133

Emlen ST, Oring LW (1977) Ecology, sexual selection, and the evolution of mating systems. Science 197:215–223

Eriksson A, Lindgren K, Lundh T (2004) War of attrition with implicit time costs. J Theor Biol 203:319–332

Fishelson L, Hilzerman F (2002) Flexibility in reproductive styles of male St. Peter's tilapia *Sarotherodon Galilaeus* (Cichlidae). Environ Biol Fishes 63:173–182

Grafen A, Sibly R (1978) A model of mate desertion. Anim Behav 26:645–652

Grüter C, Taborsky B (2005) Sex ratio and the sexual conflict about brood care in a biparental mouthbrooder. Behav Ecol Sociobiol 58:44–52

Gubernick DJ, Teferi T (2000) Adaptive significance of male parental care in a monogamous mammal. Proc R Soc Lond B 267:147–150

Hayward A, Gillooly JF (2011) The cost of sex: Quantifying energetic investment in gamete production by males and females. PLoS ONE 6(1):e16,557

Houston AI, McNamara JM (2002) A self-consistent approach to paternity and parental effort. Phil Trans R Soc Lond B 357:351–362

Houston AI, McNamara JM (2005) John Maynard Smith and the importance of consistency in evolutionary game theory. Biol Philos 20:933–950

Jennions M, Polakow D (2001) The effect of partial brood loss on male desertion in a cichlid fish: An experimental test. Behav Ecol 12(1):84–92

Jennions MD, Kokko H (2010) Sexual selection. In: Westneat DF, Fox CW (eds) Evolutionary behavioral ecology. Oxford University Press, Oxford, pp 343–364

Kokko H, Jennions M (2003) It takes two to tango. Trends Ecol Evol 18:103–104

Kokko H, Jennions MD (2008) Parental investment, sexual selection and sex ratios. J Evol Biol 21:919–948

Kokko H, Rankin DJ (2006) Lonely hearts or sex in the city? density-dependent effects in mating systems. Phil Trans R Soc B 361:319–334

Maynard Smith J (1977) Parental investment: A prospective analysis. Anim Behav 25:1–9

Maynard Smith J (1982) Evolution and the theory of games. Cambridge University Press, Cambridge

McNamara JM, Székely T, Webb JN, Houston AI (2005) A dynamic game-theoretic model of parental care. J Theor Biol 205:605–623

Ramsey DM (2010) A large population parental care game: Polymorphisms and feedback between patterns of care and the operational sex ratio. J Theor Biol 266:675–690

Ros AFH, Zeilstra I, Oliveira RF (2003) Mate choice in the Galilee St. Peter's fish, *Sarotherodon Galilaeus*. Behaviour 140:1173–1188

Trivers RL (1972) Parental investment and sexual selection. In: Campbell B (ed) Sexual selection and the descent of man 1871–1971. Aldine Press, Chicago, pp 139–179

Webb JN, Houston AI, McNamara JM, Székely T (1999) Multiple patterns of parental care. Anim Behav 58:983–993

Wright HWY (2006) Paternal den attendance is the best predictor of offspring survival in the socially monogamous bat-eared fox. Anim Behav 71:503–510

Yamamura N, Tsuji N (1993) Parental care as a game. J Evol Biol 6:103–127

Yaniv O, Motro U (2004) The parental investment conflict in continuous time: St. Peter's fish as an example. J Theor Biol 228:377–388

Chapter 9
Conditions for Cooperation and Trading in Value-Cost Dynamic Games

Jürgen Scheffran

Abstract In value-cost dynamic games multiple agents adjust the flow and allocation of investments to action pathways that affect the value of other agents. This article determines conditions for cooperation among agents who invest to gain value from each other. These conditions are specified in a game-theoretic setting for agents that invest to realize cooperative benefits and value targets. The dynamic interaction of allocation priorities and the stability of equilibrium concepts is analyzed. One focus is to determine solutions concepts based on cost-exchange ratios and benefit-exchange ratios that represent trade-offs between the agents, as a function of the action and interaction effects of the respective action pathways. The general approach is applied to the trading between buyers and sellers of goods to determine conditions for mutually beneficial market exchange, the price of goods, and the specialization between consumers and producers.

Keywords Cooperative games • Conjectural variation • Economic trading • Market price • Interaction stability • Value-cost dynamic game

9.1 Introduction

One of the challenges in dynamic game theory is to understand the emergence of cooperation and the transition between conflict and cooperation among multiple players. Particularly relevant are phenomena of competitive and cooperative interactions in trading processes involving buyers and sellers (respectively, producers and consumers) who compete or cooperate with each other in changing constellations. These issues have been analyzed in different frameworks.

J. Scheffran (✉)
Institute of Geography, KlimaCampus, University of Hamburg, 20144 Hamburg, Germany
e-mail: juergen.scheffran@zmaw.de

V. Křivan and G. Zaccour (eds.), *Advances in Dynamic Games*, Annals of the International Society of Dynamic Games 13, DOI 10.1007/978-3-319-02690-9_9,

In *differential games* solution concepts are based on payoff optimization and Nash equilibria (Başar and Olsder 1982; Dockner et al. 2000). Studies analyzed cooperative equilibria (Tolwinski et al. 1986); conditions for the existence of Pareto optima (Engwerda 2010); semi-cooperative strategies (Bressan and Shen 2004); cooperation in games with incomplete information (Petrosjan 2004); and the price of anarchy, information, and cooperation (Başar and Zhu 2011).

In *oligopoly theory and market games* firms compete according to reaction functions to each other's quantities and prices (Szidarovszky and Li 2000; Zhang and Zhang 1996). Conjectural variations among players lead to an iterative hill climbing process called "*tâtonnement*" (Bresnehan 1981; Figuières et al. 2004). Well-known are the Cournot model, where each firm chooses its output as a function of the output of other firms, and the Bertrand-Edgeworth model where profit maximizing output is selected at a given price. The Nash/Bertrand conjectures result in competitive equilibria such that firms set price equal to marginal cost for lacking capacity constraints, or sellers set prices and buyers choose quantities at these prices (Bertrand competition) (Allen and Hellwig 1986; Dixon 1992; Simaan and Cruz 1976; Tuinstra 2000). For multi-user communication, Su and van der Schaar (2011) investigate the stability and Pareto boundaries of conjectural equilibria. In repeated games firms have an incentive to cooperate by colluding to charge the monopoly price and sharing the market. While conjectural variations have been mostly used to analyze competitive interactions, applications in cooperative interaction deserve more attention.

Cooperative game theory has been developed for players joining coalitions, with solution concepts such as the core or Shapley value transferred to a dynamic game framework (Petrosjan 1995; Yeung and Petrosjan 2006, 2012). One issue is to identify mechanisms that lead to cooperation in non-cooperative Nash equilibria such as the prisoners' dilemma (PD) (Gerber 2000). To overcome the PD in single-stage games, Axelrod (1984) performed an experimental tournament of repeated PD games in which tit for tat was the most successful rule-based response strategy. The success of sequential strategies depends on the payoffs of the players and the social context in which the games are played (Selten and Stoecker 1986). Some models use punishment to enforce cooperation, including peer and institutional punishment (Isakov and Rand 2012). Helbing and Johansson (2010) explore the robustness of cooperation in spatial public goods games and the effect of the mutation rate on the relaxation dynamics. Ohtsuki (2011) studies the stability of resource division in the Nash demand game of selfish agents, with a focus on mutation and diffusion of strategies leading to a fair split of resources. Gao et al. (2012) study the dynamics of investment and the effect of punishment on cooperation in continuous public goods games, finding an equilibrium between high-tolerance individuals with high investments and low-tolerance individuals with low investments. While higher cooperation benefits tend to increase the share of cooperators, for some payoff values the reverse is true (Nemeth and Takacs 2010). Assessing the trade-off between network throughput and energy efficiency, the Shapley value was used to determine a fair distribution of the total cooperative cost among players (Miao et al. 2012).

Evolutionary game theory analyzes the competition among populations via replica equations that select cooperative and non-cooperative strategies regarding their fitness (Hofbauer and Sigmund 1998). Crucial is the coevolution of individual strategies and social ties where the interplay of strategic updating and partner network adaptation by social learning can facilitate the escape from social dilemmas (Du and Fu 2011). The role of reinforcement learning and adaptive dynamics in social dilemma situations is discussed in Tanabe and Masuda (2012), for variants of tit-for-tat and the win-stay lose-shift strategy, where learning accelerates the evolution to optimality. Gomez Portillo (2012) shows conditions to build a cooperative system under unstable growth, depending on benefit-cost ratios and imitation capacity required for cooperation. In dynamically weighted networks, players update their strategies and weights of adjacent links depending on payoffs in evolutionary games (Cao et al. 2011). An adaptive weight adjustment mechanism dramatically promotes evolution of cooperation. Wang et al. (2010) analyze the role of asymmetry in interaction and show that the probability of cooperation increases with the payoff ratio between actors. While spatial structure and heterogeneity have been recognized as potent promoters of cooperation, coevolutionary rules may affect the interaction network, the reproduction capability, reputation, mobility, or age of players (Perc and Szolnoki 2012). Among specific examples are shops competing for different types of customers, leading to attracting price cycles (Hahn 2012), and price auctions showing evolutionary stability and convergence to a Nash equilibrium (Louge and Riedel 2012).

While dynamic game theory derives response mechanisms from optimization principles, *agent-based models* (ABMs) use behavioral rules in multi-agent dynamic settings and simulate complex multi-agent patterns of interaction, which is useful in situations of uncertainty and bounded rationality, taking into account the adaptive nature of human action under changing environmental conditions. Various tools have been applied from statistical physics, nonlinear dynamics, and complex systems science to analyze adaptive social phenomena, such as self-organization or micro-macro phase transitions (e.g. Epstein and Axtell 1997; Helbing and Johansson 2010; Weidlich 2000). Applications range from moving crowds and traffic systems to urban, demographic, and environmental planning. Due to the complexity an analytic treatment is difficult, and better understanding is required on how cooperation evolves in multi-agent settings. Agents repeatedly interact with the environment and other agents, using dynamic reinforcement learning in multi-agent coordination and network formation to update the probabilities of future action based on previous actions and received rewards (Chasparis and Shamma 2012). In ABMs resource limitation may modify the original structure of the interactions and allow for well-mixed populations of cooperators and defectors under limited resources (Requejo and Camacho 2012). To understand how cooperative behavior evolves in social networks, Rezaei and Kirley (2012) investigate the evolution of cooperation in the N-player prisoner's dilemma game where cooperative actions create and defects break social links. Computational simulations use varying population and group sizes; group formation and partner selection; and agent decision-making strategies under varying dilemma constraints (cost-to-benefit ratios). Simulation shows that the social network model is able to evolve and maintain cooperation.

Despite significant progress, there is still a methodological gap between more simple models with a few number of players which pursue optimizing game strategies subject to mathematical analysis, and more complex models with a large number of agents which interact according to behavioral rules subject to computer simulation. To overcome this gap integrated theoretical frameworks should be sufficiently complex to represent the diversity in social interaction but not too complex to preclude generalizable results.

Adaptive approaches explore the linkages between optimizing and rule base behavior. Different from differential games that search for optimal-control solutions over an extended period, adaptive mechanisms consider short-term action-reaction patterns among agents who in discrete time-steps act upon a system within their resource and capacity limits according to response strategies to achieve target values. When multiple agents act on the same environment, they may interact with each other in conflictive or cooperative ways. Conflicts can be diminished by compromising or win–win solution concepts. While agents may pursue optimal utility strategies in some cases, they may follow heuristic rules based on experience, learning, traditions, and social environments in other cases.

Adaptive frameworks also offer explanations why and when agents cooperate. Generally agents cooperate to achieve individual or collective goals more efficiently. In a PD situation cooperation is difficult as agents risk to lose individual benefits by switching to cooperation, unless mechanisms guarantee the benefits of cooperation despite the possibility for individual defection. Common rules and institutional mechanisms offer a framework and may determine how much output (benefit) agents receive for their input (investment). They also guarantee that agents who defect lose the benefit of cooperation and thus have no incentive to cheat. Once the benefits of cooperation are realized, a key question is how the investments are allocated and the benefits are distributed among the agents. In extreme cases one agent receives the full benefit and the other none. Here fair mechanisms of distribution and sharing are required which are shaped by the power structures and fairness principles between the agents.

A prominent example of cooperation is trading, which is a mutual transaction where one agent commits an act beneficial to another agent who replies in turn. Trading between buyers and sellers of economic goods is based on the supply (production) and demand (consumption) of these goods. Related issues occur in auctions to determine the price of goods and in communication between sender and receiver of information. A framework of cooperation investigates sequences of mutual actions and price formation in transactions, taking into account the willingness to pay and requested benefits of buyers and sellers, as a function of the impacts of the agents on each other. This goes beyond competitive mechanisms of price formation between firms.

To analyze these phenomena an integrated model of value-cost dynamic games is presented, connecting non-cooperative and cooperative dynamic games with evolutionary games and multi-agent models. Individual action and social interaction of agents are characterized by capabilities and efforts (costs) invested to change the natural and social environment, using rule-based allocation of investments to action pathways as main control variables and adaptive response mechanisms to achieve value targets (Scheffran 2001). The framework allows to model dynamic interactions, such as conflict and cooperation, group learning and adaptation, coalition formation and breakup (Scheffran 2006). The value-cost dynamic game has been used in different applications, including arms races and arms control, economic production and environmental sustainability, resource conflicts in water and fishery, as well as energy security and climate change (for an overview, see Scheffran and Hannon 2007). One focus has been on the conditions for stability of the interaction matrix (Scheffran 2001).

This article analyzes conditions for cooperation and trading among agents that invest part of their capital to cooperation with other agents, to achieve value targets and benefits compared to unilateral action. Based on conjectural variations in two-agent interaction, equilibria and their stability are determined, which serve as a basis to relate cost exchange and benefit exchange according to trade-off mechanisms. A focus is on trading between buyers and sellers of goods to explicitly determine the market price of goods as a function of the tradeoff rules.

9.2 The Framework of the Value-Cost Dynamic Game

9.2.1 *Model Outline*

Value-cost dynamic games describe the behavior of individual agents who use part of their capital to invest in different action paths to change system variables which may affect the value of each of the agents. When agents repeatedly adapt the flow of investment and its allocation to meet their target values, they interact in complex dynamic ways.

Definition 9.1. The *value-cost dynamic game* is characterized by the elements:

- **Agents** $i = 1, \ldots, n$
- **Time periods** $t = 1, \ldots, T$ during which agents act.
- **System state** $x(t) = (x^1(t), \ldots, x^m(t))$ of system variables x^k $(k = 1, \ldots, m)$.
- **Action** a_i^k of agent i regarding system variable x^k results in a system change $\Delta x_i^k(t) = g_i^k(x(t), a_i^k(t))$ induced by agent i. An action path is a particular sequence of actions $a_i^k(t)$ by agent i. $a_i(t) = (a_i^1(t), \ldots, a_i^m(t))$ is the action vector of agent i, $a(t) = (a_1(t), \ldots, a_n(t))^T$ is the action matrix of all agents.
- **Capital** $K_i(t)$ represents the capability of agent i to act.
- **Cost** $C_i(t) = \kappa_i(t)K_i(t) \leq K_i(t)$ is the invested fraction $0 \leq \kappa_i \leq 1$ of the capital K_i into action a_i of agent i. The cost allocated to each of the action variables a_i^k is $C_i^k = c_i^k \cdot a_i^k$ where c_i^k is the unit cost of action variable a_i^k such that $C_i = \sum_k c_i^k \cdot a_i^k$ are the total costs of agent i. The cost vector $C = (C_1, \ldots, C_n)$ comprises the costs of all agents.[1]

[1]Capital and investment can be expressed in terms of financial units (money) but other capital resources could also be considered, such as time, labor, energy, and natural resources. The unit chosen is specific to the respective application area.

- **Allocation priorities** $0 \leq p_i^k \leq 1$ determine the fraction of investment costs C_i allocated to action $a_i^k = C_i \cdot p_i^k / c_i^k$, with $\sum_k p_i^k = 1$. $p_i(t) = (p_i^1(t), \ldots, p_i^m(t))$ is the allocation vector of agent i, while $p(t) = (p_1(t), \ldots, p_n(t))^T$ is the allocation matrix of all agents.
- **Value functions** $V_i(t) = f_i(C(t), p(t), x(t))$ represent the evaluation of the system state and the action variables $C(t)$ and $p(t)$ of all agents in a given time period t. $V_i^*(t)$ indicates a target value set by agent i.

The *multi-agent interactive targeting problem* is to select priorities $p_i^k(t)$ to allocate cost $C_i(t)$ towards actions a_i^k to meet or approach the value target $V_i(t) = f_i(C(t), p(t), x(t)) = V_i^*(t)$ for all agents $i = 1, \ldots, n$.

Model Specifications:

1. The following analysis will be restrained to value functions that are not explicitly dependent on state $x(t)$ and add up the value impacts of the actions of all agents $j = 1, \ldots, n$, leading to a linear value function of agent i (neglecting t for the respective time period to simplify notation):

$$V_i = \sum_j V_{ji} = \sum_{k=1}^{m} \sum_{j=1}^{n} v_{ji}^k a_j^k = \sum_{j=1}^{n} f_{ji} C_j, \quad (i = 1, \ldots, n) \tag{9.1}$$

   where v_{ji}^k is the unit value of agent i induced by action $a_j^k = \Delta x_j^k = C_j \cdot p_j^k / c_j^k$ of agent j $(k = 1, \ldots, m)$ where Δx_j^k is the change in system variable x^k induced by action a_j^k. Note that in v_{ji} the first index refers to the agent causing the action and the second index to the agent affected by this action. The *interaction effects* for a pair of agents i and j

$$f_{ji} = \sum_{k=1}^{m} \frac{v_{ji}^k}{c_j^k} p_j^k \qquad (i, j = 1, \ldots, n)$$

   form the *value-cost interaction matrix*

$$F(p) = \begin{pmatrix} f_{11} & \cdots & f_{1i} & \cdots & f_{1n} \\ f_{21} & \cdots & f_{2i} & \cdots & f_{2n} \\ \cdots & \cdots & \cdots & \cdots & \cdots \\ f_{n1} & \cdots & f_{ni} & \cdots & f_{nn} \end{pmatrix}.$$

2. The vector equation $V = F(p) \cdot C = V^*$ corresponds to n linear equations $V_i = f_i(C, p) = V_i^*$ which for $f_{ii}(p_i) \neq 0, \forall i = 1, \ldots, n$ can be resolved for the *target costs* of agent i

$$\tilde{C}_i(t) = \frac{V_i^*(t) - \sum_{j \neq i} f_{ji}(p_j) C_j(t)}{f_{ii}(p_i)}. \tag{9.2}$$

This is a response function of the costs of all other agents. If in a given time period t the actual cost $C_i(t)$ diverts from the target cost $\tilde{C}_i(t)$, the difference equation

$$\Delta C_i(t) = C_i(t+1) - C_i(t) = \alpha_i(\tilde{C}_i(t) - C_i(t)) = \frac{\alpha_i}{f_{ii}}(V_i^*(t) - \sum_{j=1}^{n} f_{ji}\, C_j(t)) \tag{9.3}$$

provides an adaptation mechanism that depends on the interaction effects f_{ji} which are controlled by the allocation priorities p_j^k of agents $j = 1, \dots, n$, as well as the unit values v_{ji}^k and unit costs c_j^k. For $\alpha_i = 1$ it takes one time step to reach the target cost $\tilde{C}_i(t)$ which is moving due to the responses of other agents. Depending on stability conditions the dynamic interaction evolves until cost equilibria or boundaries are reached (Scheffran 2001).

3. While total costs $C_i(t)$ represent the intensity of action, the allocation priorities $p_i^k(t)$ affect the direction of action and can serve as control variables to meet target priorities $\tilde{p}_i^k(t)$, following their adaptation dynamics

$$\Delta p_i^k(t) = \alpha_i^k(\tilde{p}_i^k(t) - p_i^k(t)) \tag{9.4}$$

within the boundaries $0 \leq p_i^k(t) \leq 1$ and $\sum_k p_i^k(t) = 1$.

4. In real-world situations agents control the intensity $C_i(t)$ and direction $p_i^k(t)$ of their investment which follow response functions to achieve target values. They may evolve at different temporal scales, with two special cases:

   - *Fixed cost, variable allocation:* For a certain time period T the total cost $C_i(t) \leq K_i$ is kept constant within the capital constraint, while agents adapt allocation priority $p_i^k(t)$ to each other. This represents agents that seek to find the best cost allocation within given budget limits.
   - *Fixed allocation, variable cost:* For a certain time period T the allocation $p_i^k(t)$ is kept constant, while agents adapt their total flow of costs $C_i(t)$ to each other. This represents situations where agents only respond with the intensity of their efforts (speed of action) while their direction of action and thus their freedom of choice is restrained (e.g., due to rules and habits).

   In this article we focus on problems of the first type where agents adapt their action priorities within cost limits for a certain period, thus decisions on allocation occur faster than on total cost. For subsequent periods, the one-period act turns into a dynamic game in which all variables change over time.

5. It is assumed that investments and values have the same units, thus can be added and subtracted. In practice this can be realized by estimating how much cost C_i^V an agent is willing to invest to achieve a particular value V_i. This implies net value $V_i - C_i^V = 0$ and positive (negative) net value below (above) this cost threshold. Further, net value may be fully or partially converted into capital which affects the capability to invest in the following time period. Capital may further increase or decline according to its inherent growth dynamics.

6. The dynamic game includes the possibility that single players $i = 1, \ldots, n$ form collective agents (coalitions) $I = 1, \ldots, N$ that allocate joint investments C_I to coalition actions a_I^k and pursue coalition values V_I both which are distributed to the individual values V_i according to a distribution mechanism (Scheffran 2006). The following analysis applies to both individual and collective agents.
7. The action and interaction effects f_{ii} and f_{ji} are held constant in this article although in many cases may be time and state-dependent. For instance, they may decline with increasing investment, thus diminish the value of allocating more investment to this pathway. It is also possible that initially interaction effects increase with investment (indicating a self-enforcing beneficial relationship) and beyond some point begin to decline. To analyze these phenomena, specific models are required to represent concrete application areas.

9.2.2 Game-Theoretic Framework of Direct Interaction

In the following, we assess the linear relationship between costs C_i and values V_i of a group of agents $i = 1, \ldots, n$. Going beyond previous applications of the value-cost dynamic game which describe the interaction of agents that allocate their investment to individual actions we expand the analysis to agents that explicitly allocate a fraction of their investment to the interaction with other agents. The relationship between actions and system states is implicitly represented by the interaction effects f_{ij} between a pair of agents i and j.

- Agent i allocates a fraction $0 \leq p_{ii} \leq 1$ of its investment C_i to individual (unilateral) actions, resulting in a direct action value

$$V_{ii} = f_{ii} p_{ii} C_i.$$

  f_{ii} is the direct action effect of agent i, indicating the value-cost ratio of unilateral action. For $p_{ii} = 1$, agent i only acts unilaterally and has no direct interaction effect on another agent.
- Agent i allocates a fraction $0 \leq p_{ij} \leq 1$ of its investment C_i to the direct interaction with agent j, resulting in an interaction value to agent j

$$V_{ij} = f_{ij} p_{ij} C_i$$

  where f_{ij} is the direct interaction effect of i's investment on j's value.

Remark 9.1. In the general model framework there is no distinction between direct and indirect effects as any action that affects the value of another agent is treated as an interaction. The term "direct interaction" is introduced here to describe situations where agents direct part of their investment towards the interaction with other agents, assuming that this investment is no longer available for unilateral action. Although indirect side effects can be relevant in real-world cases (i.e., when the investment of one agent affects value of either agent besides the intended path), the focus will be in the following on the pure cases of direct interaction without side effects (see further discussion in the final section).

Definition 9.2. The following cases of interaction are distinguished:

- Conflicting effect $f_{ij} < 0$: Investment $C_i > 0$ causes value loss $V_{ij} < 0$.
- Cooperative effect $f_{ij} > 0$: Investment $C_i > 0$ causes value gain $V_{ij} > 0$.
- Unilateral loss $f_{ii} < 0$: Investment $C_i > 0$ causes value loss $V_{ii} < 0$.
- Unilateral gain $f_{ii} > 0$: Investment $C_i > 0$ causes value gain $V_{ii} > 0$.

Remark 9.2. Mixed cases are possible if one agent acts with a cooperative effect on another agent who responds with a conflicting effect, and vice versa. The definition is based on the assumption that value losses are associated with a conflicting relation and value gains with a cooperative relation. In the following we focus on unilateral gains ($f_{ii} > 0$) and cooperative relationships ($f_{ij} > 0$).

Definition 9.3. The *net value of direct interaction* among agents $i = 1, \ldots, n$ is

$$V_i = \sum_j V_{ji} - C_i = \sum_j f_{ji} p_{ji} C_j - C_i. \tag{9.5}$$

The following analysis focuses on the conditions of cooperation among a pair of agents i and j, neglecting the influence of other agents (for a discussion of the general multi-agent case, see the concluding section). To simplify notation we use $p_i \equiv p_{ij}$ as the allocation priority of agent i for cooperation with j and $p_{ii} = 1 - p_i$ as the allocation priority for unilateral action, without specifying particular action paths. For agent i the net value change becomes

$$V_i = V_{ii} + V_{ji} - C_i = f_{ii}(1 - p_i)C_i + f_{ji} p_j C_j - C_i \tag{9.6}$$

The case of positive action and interaction effects $f_{ii} > 0$ and $f_{ji} > 0$ for $i, j = 1, \ldots, n$ is used to develop general conditions for cooperative interaction. The extreme cases of full cooperation and full non-cooperation among agents i and j are represented by the following game matrix.

Proposition 9.1. *The following statements are evident from the game matrix:*

Agent j	No cooperation	Full cooperation
Agent i	$(p_j = 0)$	$(p_j = 1)$
No cooperation	$V_i = (f_{ii} - 1)C_i$	$V_i = (f_{ii} - 1)C_i + f_{ji}C_j$
$(p_i = 0)$	$V_j = (f_{jj} - 1)C_j$	$V_j = -C_j$
Full cooperation	$V_i = -C_i$	$V_i = f_{ji}C_j - C_i$
$(p_i = 1)$	$V_j = (f_{jj} - 1)C_j + f_{ij}C_i$	$V_j = f_{ij}C_i - C_j$

1. *The case of non-cooperative action* $(p_i, p_j) = (0, 0)$ *is a stable Nash equilibrium, i.e. neither player has an incentive to divert from it unilaterally for positive unilateral action effects* $f_{ii} > 0$, $f_{jj} > 0$.
2. *For* $f_{ii} < 1$ *unilateral action is too costly, and agent i needs cooperation with agent j to generate positive net value.*
3. *Provided that the value of mutual cooperation exceeds the value of mutual non cooperation, the cooperative case* $(p_i, p_j) = (1, 1)$ *is more beneficial for both agents but is not a stable Nash equilibrium against unilateral change.*
4. *The transition from non-cooperation to cooperation contains the risk of large asymmetry when one player has maximum value (unilateral and interaction gains) and the other minimum value (loss of investment). In this case, one player receives the full benefit of cooperation while the other pays for it.*

Problem 9.1. Under which conditions will cooperation emerge in the value-cost prisoner's dilemma game? Specifically, which investment fraction p_i will agent i allocate to cooperation in response to an investment fraction p_j of agent j?

9.3 Relative Benefits from Cooperative Interaction

To find ways out of the value-cost prisoner's dilemma, the challenge is to develop rules and mechanisms that guarantee both players the benefits of cooperation. One option is that both agents specify these benefits and offer investments to realize them. An enforced agreement would make sure that gains and losses of non-cooperation are compensated by mechanisms that favor cooperation by incentives and punishment. It is crucial to identify boundary conditions under which agents switch behavior and make the transition from non-cooperation to cooperation by starting an exchange of investment for mutual benefits.

9.3.1 *Equilibria and Boundary Conditions*

To determine the benefits of cooperation among two agents i and j, two conditions are distinguished (to avoid repetitions only the equations for one of the agents are shown if the behavior of the other is mirror-symmetric):

1. Agent i invests $p_i C_i$ in the interaction if the cooperation value V_{ji} gained from j in return exceeds the unilateral value V_{ii} from the same investment:

$$V_{ji}(p_j) = f_{ji} p_j C_j > f_{ii} p_i C_i = V_{ii}(p_i) \tag{9.7}$$

2. Agent i decides to invest $p_i C_i$ in the interaction if the cooperation value V_{ji} gained from j in return exceeds the investment:

$$V_{ji}(p_j) = f_{ji} p_j C_j > p_i C_i \tag{9.8}$$

Condition 2 is a minimum requirement for interaction, requesting that the interaction is worth the investment and no relative loss is incurred from it. For condition 1 investment in cooperation is related to the opportunity cost of non-cooperation. In the following only condition 1 is pursued while condition 2 is mathematically similar and identical to $f_{ii} = 1$ and $f_{jj} = 1$. One should note the different meaning of both conditions which lead to different outcomes, in particular when unilateral actions are inefficient $f_{ii} < 1$ and $f_{jj} < 1$ and agents can improve only through collaboration.

Definition 9.4. The *relative benefit of cooperation* for agent i is:

$$B_i(p_i, p_j) \equiv V_{ji}(p_j) - V_{ii}(p_i) = f_{ji} p_j C_j - f_{ii} p_i C_i \tag{9.9}$$

Once agents i and j set target benefits $B_i = B_i^*$ and $B_j = B_j^*$ for given investments C_i and C_j during the action period, this defines different response mechanisms for meeting these targets.

1. *Own benefit response:* Allocation priority of cooperation p_i is selected to meet the target benefit of agent i:

$$p_i = \frac{f_{ji} p_j C_j - B_i^*}{f_{ii} C_i} \equiv \tilde{p}_i(p_j)$$

2. *Partner benefit response:* Allocation priority of cooperation p_i is selected to meet the target benefit of agent j:

$$p_i = \frac{f_{jj} p_j C_j + B_j^*}{f_{ij} C_i} \equiv \bar{p}_i(p_j) \tag{9.10}$$

3. *Joint equilibrium response:* Allocation priority of cooperation p_i is selected to meet the target benefits of both agents i and j in the equilibrium allocation:

$$p_i^* \equiv \frac{B_i^*/f_{ii} + Z B_j^*/f_{ij}}{C_i(Z-1)} \tag{9.11}$$

   with

$$Z \equiv \frac{f_{ij} f_{ji}}{f_{ii} f_{jj}}. \tag{9.12}$$

Remark 9.3. For response 1 the aim of agent i is to adapt priority $p_i(t)$ at time t to obtain its target benefit B_i^* and thus move towards $\tilde{p}_i(p_j)$ which depends on p_j of agent j. Under response 2 the aim is to adapt priority $p_i(t)$ at time t to meet the stated target benefit B_j^* of agent j, thus moving towards $\bar{p}_i(p_j)$ as a function of p_j. In the equilibrium (p_i^*, p_j^*) both functions $\tilde{p}_i(p_j)$ and $\bar{p}_i(p_j)$ intersect, thus both response mechanisms lead to the same equilibrium allocation.

For each of the three strategies there are different information requirements. For strategy $\tilde{p}_i$, agent i needs information about cost C_j and allocation p_j of agent j, as well the effect f_{ji} of j's costs on agent i. For strategy $\bar{p}_i$, agent i needs to know significantly more information about j: besides cost $p_j C_j$ the action effects f_{jj} and interaction effects f_{ij} on agent j, as well as the stated target benefit of j. Equilibrium p_i^* is independent of p_j and C_j but requires complete information about the action and interaction effects of both agents. The existence of a mutually satisfying equilibrium depends on $Z > 1$ which is the case if the product of the cooperative interaction effects $f_{ij} f_{ji}$ exceeds the product of the unilateral action effects $f_{ii} f_{jj}$. The question is which combinations of B_i^* and B_j^* can be actually realized for $0 \leq p_i^* \leq 1$ and $0 \leq p_j^* \leq 1$.

Proposition 9.2. *The constraints on agreed allocations* $0 \leq p_i^* \leq 1$ *and* $0 \leq p_j^* \leq 1$ *translate into constraints for the target benefits:*

$$-\frac{f_{ji}}{f_{jj}} B_j^* \leq B_i^* \leq (Z-1) f_{ii} C_i - \frac{f_{ji}}{f_{jj}} B_j^*$$

$$-\frac{f_{ii}}{f_{ij}} B_j^* \leq B_i^* \leq (1 - 1/Z) f_{ji} C_j - \frac{f_{ii}}{f_{ij}} B_j^*.$$

The set of possible target benefits is given by the quadrangle with the coordinates for agents i:

$$B_i^*(p_i^* = 0, p_j^* = 0) = 0$$
$$B_i^*(p_i^* = 1, p_j^* = 0) = -f_{ii} C_i$$
$$B_i^*(p_i^* = 0, p_j^* = 1) = f_{ji} C_j$$
$$B_i^*(p_i^* = 1, p_j^* = 1) = f_{ji} C_j - f_{ii} C_i$$

$B_i^*(1,1) > 0$ *and* $B_j^*(1,1) > 0$ *is realized for*

$$\frac{f_{ii}}{f_{ji}} \leq \frac{C_j}{C_i} \leq \frac{f_{ij}}{f_{jj}}$$

which for $Z > 1$ *defines a set of possible total investments* (C_i, C_j) *that allow positive target benefits for both agents. The case* $p_i^* = p_j^* = 1$ *is Pareto optimal since each agent can improve benefit only by reducing benefit for the other.*

9.3.2 Dynamics and Stability of Interaction

When agents pursue one of the response strategies identified in the previous section, it is important whether these responses lead to a stable or unstable interaction. In the following, adaptive responses are considered that correspond to a linear difference equation of the type for agent i:

$$\Delta p_i(t) = p_i(t+1) - p_i(t) = \alpha_i(\hat{p}_i(t) - p_i(t)) = a_{ii} p_i(t) + a_{ji} p_j(t). \quad (9.13)$$

This equation describes agents that adapt their allocation priority $p_i(t)$ in each time step towards a time-dependent target priority $\hat{p}_i(t)$, where α_i indicates the adaptation rate from one time step to the next. For $\alpha_i = 1$, agent i adjusts $p_i(t)$ to the target $\hat{p}_i(t)$ in one time step, for $\alpha_i < 1$ allocation is adjusted to the target asymptotically at slower rate. a_{ii} and a_{ji} are the parameters in the linear difference equations, indicating the effect of each allocation on the respective allocation changes. As targets of adaptation $\hat{p}_i$, agents may use $\tilde{p}_i$, $\bar{p}_i$, or p_i^*, depending on the response strategy pursued. Whether the respective dynamics is stable or unstable is a crucial question.

Theorem 9.1. *The dynamic adaptation mechanism* $\Delta p_i(t) = \alpha_i(\hat{p}_i(p_j(t)) - p_i(t))$ *is unstable for response strategy* $\hat{p}_i = \tilde{p}_i(p_j)$ *and stable for response strategies* $\hat{p}_i = \bar{p}_i(p_j)$ *and* $\hat{p}_i = p_i^*$.

Proof. 1. The eigenvalues of the dynamical equation

$$\Delta p_i(t) = \alpha_i(\tilde{p}_i(p_j(t)) - p_i(t)) = a_{ii} p_i(t) + a_{ji} p_j(t) \quad (9.14)$$

with $a_{ji} = \alpha_i(f_{ji}C_j)/(f_{ii}C_i)$ and $a_{ii} = -\alpha_i$ are the solutions of $\det(A - \lambda I) = 0$:

$$\lambda_{1/2} = -\frac{\alpha_i + \alpha_j}{2} \pm \sqrt{\frac{(\alpha_i + \alpha_j)^2}{4} + \alpha_i \alpha_j (Z - 1)} \quad (9.15)$$

For $Z > 1$, one of the eigenvalues is positive, indicating instability.

2. The eigenvalues of the dynamical equation

$$\Delta p_i(t) = \alpha_i(\bar{p}_i(p_j(t)) - p_i(t)) = a_{ii} p_i(t) + a_{ji} p_j(t)$$

with $a_{ji} = \alpha_i(f_{jj}C_j)/(f_{ij}C_i)$ are

$$\lambda_{1/2} = -\frac{\alpha_i + \alpha_j}{2} \pm \sqrt{\frac{(\alpha_i + \alpha_j)^2}{4} + \alpha_i \alpha_j (1/Z - 1)}. \quad (9.16)$$

Accordingly both eigenvalues are negative for $Z > 1$ and the dynamics of response strategy $\bar{p}_i$ is asymptotically stable.

3. For $\hat{p}_i = p_i^*$ the negative eigenvalues $\lambda_1 = -\alpha_i < 0$ and $\lambda_2 = -\alpha_j < 0$ indicate asymptotic stability of the equilibrium.

□

Remark 9.4. The instability of strategy $\tilde{p}_i(p_j)$ is an expression of the prisoner's dilemma of individual rational action. To overcome the dilemma and achieve the benefits of cooperation, $\bar{p}_i$ offers an alternative mechanisms that leads to the stable joint equilibrium p_i^*. This way of realizing the benefits turns the logic of individual rational action around. The aim is not to adapt priority $p_i(t)$ at time t to obtain the own relative target benefit B_i^* and thus move towards $\tilde{p}_i(p_j(t))$, but to adapt priority $p_i(t)$ towards the stated relative target B_j^* of the counterpart and move towards $\bar{p}_i(p_j(t))$. The purpose of this response strategy is to convince the counterpart to cooperate by offering benefits of cooperation, in expectation that this would be followed by a cooperative move of j. This cooperative tit-for-tat strategy is represented in statements like "If you do me a favor, I do a favor to you." The dynamics of cooperation would evolve according to the following bargaining sequence:

1. Starting from the Nash equilibrium $(p_i, p_j) = (0, 0)$, agent i offers $\bar{p}_i(0) = B_j^*/f_{ij}C_i$ to realize the requested benefit B_j^* for agent j and takes the risk of a potential value loss $-f_{ii}\bar{p}_iC_i$ into account if j does not respond cooperatively.
2. In turn agent j offers

$$\bar{p}_j(\bar{p}_i) = \frac{f_{ii}\bar{p}_iC_i + B_i^*}{f_{ji}C_j} = \frac{B_j^* f_{ii}/f_{ij} + B_i^*}{f_{ji}C_j}$$

   of its investment to meet the benefit target B_i^* of agent i, accepting the risk of a potential value loss compared to the first step if cooperation does not continue.
3. The process continues until both agents are sufficiently close to the equilibrium (p_i^*, p_j^*) where both meet their target benefits.

While the cooperative response strategy leads to a stable interaction, it is potentially unstable against individual diversion from cooperation. Two major options exist to avoid or diminish this risk:

- Respond individually to non-cooperation by withdrawal from cooperation in return, either by not realizing an offer for cooperation ex-ante or ex-post switching to non-cooperation in the following time step (tit for tat) leading to a loss of the cooperation benefit for the counterpart.
- Formal contracts and institutional mechanisms that structurally ensure cooperation or punish unilateral withdrawal from cooperation.

In the evolution of cooperation timing matters and the magnitude of steps taken. Incremental steps (represented by $\alpha_i < 1$) and small expected benefits B_i^* may help initiating a cooperative relationship and reduce the risk of non-cooperation until sufficient trust has been established between the partners to take larger cooperative steps. When requesting too much benefit from cooperation, a deal might not be struck. During the bargaining process, agents could subsequently increase their target benefit B_i^* until the upper cost boundary is reached for one or both of the agents, provided the upper limit $(p_i, p_j) = (1, 1)$ is within the bargaining set.

$Z = \frac{f_{ij} f_{ji}}{f_{ii} f_{jj}} > 1$ is an indicator for the stability of the two-agent interaction which is a necessary condition for the existence of a bargaining set where cooperation is beneficial to both. With increasing target benefits the gap between the individual response function $\tilde{p}_i(p_j)$ and the joint response function $\bar{p}_i(p_j)$ is widening and increasing the potential risk of value loss if agent j diverts from the cooperative solution. The question then is which combinations of target benefits are acceptable for both agents within the bargaining space.

9.3.3 Trade-Offs of Benefit-Cost Exchange

As explained above, the equilibrium allocation of investment of agent i that realizes the target benefit of cooperation is given as:

$$p_i^* = \frac{B_i^*/f_{ii} + ZB_j^*/f_{ij}}{(Z-1)C_i} = \frac{f_{jj} B_i^* + f_{ji} B_j^*}{(f_{ij} f_{ji} - f_{ii} f_{jj})C_i}.$$

Accordingly, for $Z > 1$ equilibrium allocations increase with the target benefits of both agents and are positive if these are positive. A solution concept that is of particular interest is the cost exchange ratio as a function of the benefit exchange ratio $\beta_{ij} = B_i^*/B_j^*$ of cooperation

$$\gamma_{ij} = \frac{p_i^* C_i}{p_j^* C_j} = \frac{f_{ji} B_j^* + f_{jj} B_i^*}{f_{ij} B_i^* + f_{ii} B_j^*} = \frac{f_{ji} + \beta_{ij} f_{jj}}{f_{ii} + \beta_{ij} f_{ij}} \tag{9.17}$$

which is dependent on the action and interaction effects f_{ii}, f_{jj}, f_{ij}, and f_{ji}. For a given β_{ij}, the benefit-cost ratio of cooperation is constant for agent i, positive for $Z > 1$ and independent of costs:

$$\phi_i^p \equiv \frac{B_i^*}{p_i^* C_i} = \frac{f_{ij} f_{ji} - f_{ii} f_{jj}}{f_{ji}/\beta_{ij} + f_{jj}} = \frac{Z-1}{Z/(\beta_{ij} f_{ij}) + 1/f_{ii}}.$$

The overall benefit-cost ratio $\phi_i = B_i^*/C_i$ is increasing with cooperation level p_i and the benefit exchange ratio β_{ij}. The selection of β_{ij} determines the "sharing of the cake" generated by the collaboration which is subject to bargaining on allocation between both agents. The question is which cost exchange and benefit exchange ratios are agreed by both agents, following certain rules and principles of interaction. Examples are the quest for fairness or relative advantage and the power structure among the agents. In the following, a few cases for γ_{ij} and β_{ij} will be considered that represent relevant trade-offs and justifications.

1. *Asymmetric benefits:* For $\beta_{ij} \to 0$, agent i's benefit would be marginal compared to agent j, while for $\beta_{ij} \to \infty$, it is the opposite. For these cases of asymmetry the cost-exchange ratio and the benefit-cost ratio for agent i are

$$\gamma_{ij}(\beta_{ij} \to \infty) = \frac{f_{jj}}{f_{ij}}, \quad \gamma_{ij}(\beta_{ij} \to 0) = \frac{f_{ji}}{f_{ii}}$$

$$\phi_i^p(\beta_{ij} \to \infty) = f_{ii}(Z-1) \quad \phi_i^p(\beta_{ij} \to 0) = 0$$

2. *Allocation equity:* If both agents allocate the same fraction of their investment to cooperation ($p_i = p_j$), then

$$\beta_{ij} = \frac{f_{ji}C_j - f_{ii}C_i}{f_{ij}C_i - f_{jj}C_j}$$

$$\gamma_{ij} = \frac{C_i}{C_j}$$

$$\phi_i^p = \frac{Z-1}{Z/(\beta_{ij} f_{ij}) + 1/f_{ii}}$$

   Thus, agents increase their cooperative investments and benefits along the line $p_i^* = p_j^*$ until they reach the upper limit of full cooperation. For asymmetric cases of cost, $f_{ji}C_j - f_{ii}C_i < 0$ is possible for one of the agents, leading to the exploitation of that agent and a negative benefit exchange ratio.
3. *Benefit equity:* If both agents claim the same benefit $B_i^* = B_j^* = B^*$ ($\beta_{ij} = 1$), this results in the cost-exchange ratio and the benefit-cost ratio of agent i

$$\gamma_{ij} = \frac{f_{ji} + f_{jj}}{f_{ij} + f_{ii}}$$

$$\phi_i^p = \frac{f_{ii}(Z-1)}{1 + f_{ji}/f_{jj}}$$

   Thus agent i allocates investment inversely proportionate to its combined effects $f_{ij} + f_{ii}$, indicating that more efficient action and interaction can save investment in the collaboration.
4. *Cost equity:* If both agents invest the same in cooperation, the cost-exchange ratio is $\gamma_{ij} = 1$. Then benefit-exchange ratio and benefit-cost ratio are

$$\beta_{ij} = \frac{f_{ji} - f_{ii}}{f_{ij} - f_{jj}} = \frac{\phi_i^p}{\phi_j^p}$$

$$\phi_i^p = f_{ji} - f_{ii}$$

   Both are positive when the cooperation effects (f_{ji}, f_{ij}) on each agent exceed its unilateral action effects (f_{ii}, f_{jj}) on agent i, where the agent with higher benefit-cost ratio also receives higher benefits.
5. *Benefit exchange ratio equals cost-exchange ratio:* $\beta_{ij} = \gamma_{ij}$ represents agents who claim benefits proportionate to investments which leads to a quadratic equation for the benefit exchange ratio:

$$\beta_{ij} = \frac{f_{ji} + \beta_{ij} f_{jj}}{f_{ii} + \beta_{ij} f_{ij}}$$

with the solution $\beta_{ij} = a \pm \sqrt{a^2 + b}$, where $a = (f_{jj} - f_{ii})/(2f_{ij})$ and $b = f_{ji}/f_{ij}$. One of the two solutions is positive, the other negative. If agents have symmetric unilateral effects ($f_{jj} = f_{ii}$), the solution is $\beta_{ij} = \sqrt{f_{ji}/f_{ij}} = \gamma_{ij}$. If interaction effects are also symmetric ($f_{ij} = f_{ji}$), the solution is $\beta_{ij} = 1 = \gamma_{ij}$.

6. *Indifference between individual and joint benefits:* This solution concept compares individual benefits of cooperation with the individual share of joint benefit:[2]

$$\bar{B} = B_i + B_j = p_i C_i (f_{ij} - f_{ii}) + p_j C_j (f_{ji} - f_{jj}) = B_i^* + B_j^*.$$

Here the target benefit $B_i^* = p_i C_i (f_{ij} - f_{ii})$ is the share each agent i contributes to the joint benefit. If no additional benefit is created in joint benefit, the redistribution cannot be beneficial for both agents at the same time, thus one agent receives more, the other less compared to the individual benefits (the possibility that collective action creates additional benefits is not considered here). Both agents are indifferent between individual and joint benefit for

$$B_i = f_{ji} p_j C_j - f_{ii} p_i C_i = p_i C_i (f_{ij} - f_{ii}) = B_i^*$$

which leads to $V_{ij} = f_{ij} p_i C_i = f_{ji} p_j C_j = V_{ji}$, thus the mutually created values of cooperation are identical for both agents. In this case, the cost-exchange ratio

$$\gamma_{ij} = \frac{p_i C_i}{p_j C_j} = \frac{f_{ji}}{f_{ij}}.$$

is proportionate to the ratio of mutual interaction effects. Thus, when agent i increases its interaction effect on agent j, it can reduce its cooperative investment. Both agents strike a deal when they exchange the same value of cooperation which corresponds to a balance condition in terms of value: the value supplied by agent i is identical to the value received by agent j which can be seen as an expression of a "fair deal."

Remark 9.5. All cases of solution concepts can be considered from the perspective of fairness. Case 1 defines asymmetric boundary conditions of "unfair" deals where one agent dominates over the other who receives no relative benefit from the interaction. In case 2, allocations of both agents are equal and benefit claims are proportionate to potential relative benefits from the counterpart while the actual cost

---

[2]In the following the benefits of both agents are assumed to have the same units. In case of different units conversion factors apply.

ratio is proportionate to the capacity to invest which may be seen as unfair by those who contribute more to the cooperation. Cases 3 and 4 describe cases of formal equity in terms of costs and benefits of both agents and can be seen as dual to each other. While case 5 represents fairness by setting benefit ratios and cost ratios equal, case 6 uses equality of the value exchange between two agents as a condition of fairness. This case is particularly interesting as it also represents the indifference condition between individual and joint benefit for both agents. The cost and benefit exchange ratios of agent i generally increase with the interaction effect f_{ji} and decline with f_{ij}, in some cases compared to unilateral effects f_{ii} and f_{jj}.

9.4 Alternative Target and Solution Concepts

Rather than using relative benefits of cooperation compared to non-cooperation as targets, agents may seek absolute target values or relative growth targets of benefit.

9.4.1 Cooperative Pursuit of Target Values

The previous section has considered conditions for agents that use their investments in cooperation to achieve relative benefits compared to unilateral action. This section describes agents i and j that use their investments to achieve absolute value targets $V_i = V_i^*$ and $V_j = V_j^*$, which may be interpreted as a demand in terms of value. In particular, agents seek to avoid losses from their action and interaction, thus they want to assure $V_i^* \geq 0$. On the other hand, agents tend to aim for the best possible value outcomes $V_i^* = V_i^{max}$. For agent i the targeting problem in cooperation with agent j is expressed by the equation

$$V_i = V_{ii} + V_{ji} - C_i = f_{ii}(1 - p_i)C_i + f_{ji}p_jC_j - C_i = V_i^*.$$

If agent i is not able to meet its target value unilaterally for $p_i = 0$, i.e. $V_{ii} = (f_{ii} - 1)C_i < V_i^*$, this is an incentive to cooperate with j by increasing $p_j > 0$ to achieve its target value, provided that f_{ji} is sufficient to justify cooperation. The cooperative targeting problem can be transformed into the framework analyzed in the previous section with the relative benefit of cooperation

$$B_i = f_{ji}p_jC_j - f_{ii}p_iC_i = V_i^* + C_i(1 - f_{ii}) \equiv B_i^*.$$

For $C_i(1 - f_{ii}) = 0$ we have $B_i^* = V_i^*$ and both problems are identical. For $C_i(1 - f_{ii}) \neq 0$, the equilibrium solution becomes:

$$p_i^* = \frac{B_i^*/f_{ii} + ZB_j^*/f_{ij}}{C_i(Z-1)} = \frac{(V_i^* + C_i(1-f_{ii}))/f_{ii} + Z(V_j^* + C_j(1-f_{jj}))/f_{ij}}{C_i(Z-1)}$$

which increases with the target values of both agents. The constraints on allocation define a set of reachable target values.

Definition 9.5. The constraints $0 \leq p_i^* \leq 1$ define the *reachable set of target values* of agent i by cooperation for given investment budgets C_i and C_j:

$$-\frac{f_{ji}}{f_{jj}} V_j^* + \Theta_i \leq V_i^* \leq -\frac{f_{ji}}{f_{jj}} V_j^* + \Theta_i + (Z-1) f_{ii} C_i$$

where $\Theta_i(C_i, C_j) = (f_{ii} - 1)C_i + (1 - \frac{1}{f_{jj}}) f_{ji} C_j$.

Remark 9.6. In this value trade-off, the reachable target values obviously decline with the target values of the counterpart and generally increase with the total investments of both agents (for $Z > 1$ and $f_{ii} > 1$). The difference between the upper and lower boundaries of target values is $(Z-1) f_{ii} C_i$. As introduced in the prisoner's dilemma game of Sect. 9.2, the four boundary cases of the set of reachable target values are:

$$V_i^*(1,1) = f_{ji} C_j - C_i \quad V_i^*(0,0) = (f_{ii} - 1)C_i$$
$$V_i^*(1,0) = -C_i \quad V_i^*(0,1) = (f_{ii} - 1)C_i + f_{ji} C_j.$$

Of particular interest is the case of joint positive value $V_i^*(1,1) \geq 0$ and $V_j^*(1,1) \geq 0$ for full mutual cooperation, which leads to the cost constraint

$$\frac{1}{f_{ij}} \leq \frac{C_i}{C_j} \leq f_{ji}$$

which is feasible only for $f_{ij} f_{ji} > 1$. Thus, full mutual cooperation allows for positive target values of both agents only for this cost constraint which can be expanded for increasing interaction effects. Beyond a certain level of cost asymmetry, full mutual cooperation leads to negative values for one of the agents, defining an upper limit of cooperation. Other cases may be of particular interest, e.g. if the target value for one of the agents is $V_j^* = 0$, then the reachable target value for agent i is

$$V_i^* = (Z f_{ii} - 1)C_i + (1 - \frac{1}{f_{jj}}) f_{ji} C_j$$

which exceeds the jointly reachable target value of mutual cooperation for $C_j \leq f_{ji} C_i$. To bargain on and select combination of target values, it is possible to apply the solution concepts discussed in the previous section, using the benefit-exchange ratio

$$\beta_{ij} = \frac{B_i^*}{B_j^*} = \frac{V_i^* + C_i(1 - f_{ii})}{V_j^* + C_j(1 - f_{jj})}.$$

For instance, the principle of benefit equity $\beta_{ij} = 1$ leads to a linear relationship between target values, costs and action effects of both agents. For low value targets V_i^* and large f_{ii}, the target benefits may be negative which implies that there is no need for cooperation because target values can be better achieved unilaterally.

9.4.2 Relative Growth Targets

The equilibrium of cooperative allocations can be expressed as

$$p_i^* = \frac{B_i^*/f_{ii} + ZB_j^*/f_{ij}}{C_i(Z-1)} = \frac{\rho_i^* + \omega_{ji}\rho_j^*}{Z-1}$$

where $\rho_i^* = \frac{B_i^*}{f_{ii}C_i}$ is the ratio of target benefit and maximum unilateral value of agent i, indicating a requested *growth rate* of cooperative benefit compared to unilateral action, and $\omega_{ji} = \frac{f_{ji}C_j}{f_{ii}C_i}$ is the exchange ratio between maximum cooperative value from j and maximum unilateral value of i.

Thus the agreed level of cooperation p_i^* is driven by the target growth rates of both agents, where ρ_j^* is weigthed by ω_{ji}. The allocation boundaries $0 \leq p_i^* \leq 1$ translate into a bargaining set for feasible growth targets (ρ_i^*, ρ_j^*)

$$\rho_i^- \equiv -\omega_{ji}\rho_j^* \leq \rho_i^* \leq Z - 1 - \omega_{ji}\rho_j^* \equiv \rho_i^+.$$

where ρ_i^- and ρ_i^+ are the lower and upper constraints. Positive target growth is possible for $\rho_i^+ > 0$,

$$\rho_j^* \leq \frac{Z-1}{\omega_{ji}} = \frac{f_{ii}C_i}{f_{ji}C_j}(Z-1).$$

In case of $p_i^* = p_j^* = 1$ (full mutual cooperation), the equilibrium solution of growth targets becomes $\rho_j^* = \omega_{ij} - 1$. Accordingly, $p_i^* \leq 1$ and $p_j^* \leq 1$ lead to the cost constraint

$$(\rho_i^* + 1)\frac{f_{jj}}{f_{ij}} \leq \frac{C_i}{C_j} \leq \frac{f_{ji}}{f_{ii}}\frac{1}{(\rho_j^* + 1)}$$

This defines a bargaining set between both agents regarding their available investments and their reachable target growth which exists for

$$(1 + \rho_i^*)(1 + \rho_j^*) \leq Z.$$

Thus the upper limits for the admissible target growth of both agents are located on a hyperbola limited by Z. With $\rho_{ij}^* = \rho_i^*/\rho_j^*$ the cost-exchange ratio in the equilibrium (p_i^*, p_j^*) becomes

$$\gamma_{ij} = \frac{p_i^* C_i}{p_j^* C_j} = \frac{(\rho_{ij}^* + \omega_{ji})C_i}{(\rho_{ji}^* + \omega_{ij})C_j}.$$

With this approach, solution concepts similar to Sect. 9.3 can be applied.

9.5 Conditions for Trading and Pricing of Goods

9.5.1 Problem Description in the Model Framework

The trading of goods is an asymmetric form of cooperative interaction which fits to the model framework explained in the previous sections.

Definition 9.6. A *trading game* is characterized by several elements:

1. One agent acts as a *buyer* (agent 1) who invests into buying a good (respectively paying for an induced change of system variables x that is subject to bargaining), and a *seller* (agent 2) who sells a good and in turn receives the cooperative investment from the buyer. Both buyers and sellers invest into unilateral action and cooperative interaction and receive value in return. The relationship can be described as follows:
   - Both agents invest $(1 - p_i)C_i$ into unilateral action ($i = 1, 2$).
   - The buyer transfers the cooperative investment $p_1 C_1$ to buy a good $x_{12} = p_1 C_1 / c_{12}$ from the seller at unit cost c_{12} which corresponds to the price paid.
   - The seller invests $p_2 C_2$ to provide good $x_{21} = p_2 C_2 / c_{21}$ at unit cost c_{21}.
2. When both agents accept this deal, all goods sold are also bought (*market balance of demand and supply*):
$$x_{12} = \frac{p_1 C_1}{c_{12}} = \frac{p_2 C_2}{c_{21}} = x_{21}. \tag{9.18}$$
3. Similarly, the cooperation value of the seller can be expressed in two ways: in terms of the investment received from the buyer 1 and by the investment spent by the seller 2:
$$V_{12} = p_1 C_1 = v_{12} x_{12} = v_{12} x_{21} = \frac{v_{12}}{c_{21}} p_2 C_2$$
   where v_{12} is the unit value of the good for the seller which corresponds to the price received from the buyer.

4. The *trading price* of the good corresponds to the unit cost paid by the buyer c_{12} and the unit value v_{12} received by the seller

$$\pi_{12} \equiv v_{12} = c_{12} = \frac{p_1 C_1}{p_2 C_2} c_{21} = \gamma_{12} \cdot c_{21} \tag{9.19}$$

which is proportionate to the unit cost of the provider of the good and the cost-exchange ratio γ_{12}. p_1 and p_2 are adjusted until the balance condition is reached and the deal is accepted by both buyer and seller, fixing the mutual price $\pi \equiv \pi_{12}$ (neglecting the indices).

5. The *trading values* of the buyer and the seller are

$$V_{21} = v_{21}\, x_{21} = \frac{v_{21}}{c_{21}} p_2\, C_2 = f_{21}\, p_2\, C_2$$
$$V_{12} = p_1 C_1 = f_{12} p_1 C_1$$

and the unilateral values $V_{ii} = v_{ii}(1 - p_i)C_i/c_{ii} = f_{ii}(1 - p_i)C_i$ $(i = 1, 2)$. With the action effects $f_{ii} = v_{ii}/c_{ii}$ and the interaction effects $f_{21} = v_{21}/c_{21}$ and $f_{12} = 1$, the net values become:

$$V_1 = f_{11}(1 - p_1)C_1 + f_{21}\, p_2\, C_2 - C_1 = \frac{v_{11}}{c_{11}}(1 - p_1)C_1 + \frac{v_{21}}{c_{21}} p_2 C_2 - C_1$$
$$V_2 = f_{22}(1 - p_2)C_2 + f_{12}\, p_1\, C_1 - C_2 = \frac{v_{22}}{c_{22}}(1 - p_2)C_2 + p_1 C_1 - C_2.$$

Remark 9.7. 1. When sellers and buyers accept to trade, the price of a good exchanged between buyer and seller is equal to the unit value of the seller's investment and the unit cost of the buyer's investment. This is an expression of the dual nature of price which is the unit cost the buyer is willing to pay and the unit value the seller requests to realize the deal. The price is proportionate to the cost-exchange ratio (investment tradeoff) γ_{12} and thus increases with the investment $p_1 C_1$ the buyer is willing to invest to meet its demand of goods and inversely proportionate to the investment $p_2 C_2$ the seller is willing to invest into the supply of goods. This represents the supply-demand relationship in terms of investment. Price also depends on the unit cost of supplying the good by the seller. To avoid loss, the seller requests at least the provision cost from the buyer, thus $p_1 C_1 \geq p_2 C_2$ which defines a lower bound for the price $\pi \geq c_{21}$.

2. Accordingly, the buyer requests to receive a value that is at least worth the investment $V_{21} = p_2 C_2 v_{21}/c_{21} \geq p_1 C_1 = V_{12}$ transferred to the seller which defines a lower bound for the price $\pi \leq v_{21}$.
3. If the buyer is the consumer of a good and the seller is its producer, then c_{21} is the unit cost of production and v_{21} is the unit value of consumption. However, both agents in the trading game can be consumers and producers at the same time, each for a different good or action path in which they are more efficient than the

counterpart. Thus, agent 1 may buy one good from 2 and sell another good to agent 2, leading to an exchange ratio that determines the relative price between the two types of goods.

4. It is further possible that the seller (agent 2) has been a buyer of the product or its components from another seller (agent 3) such that the provision cost is equal to the price paid to this seller. Accordingly the buyer (agent 1) may sell the good to another buyer such that the unit value is equal to the price received from that buyer. In this sequence of buying and selling the price is set to rise if agents are inclined to increase value.

9.5.2 Relative Benefits from Trading

Agents trade goods if the relative benefits are positive, i.e. the value from trading exceeds the value from unilateral action:

$$B_1(p_1, p_2) = V_{21}(p_2) - V_{11}(p_1) = \frac{v_{21}}{c_{21}} p_2 C_2 - \frac{v_{11}}{c_{11}} p_1 C_1 > 0 \quad \text{(buyer benefit)}$$

$$B_2(p_1, p_2) = V_{12}(p_1) - V_{22}(p_2) = p_1 C_1 - \frac{v_{22}}{c_{22}} p_2 C_2 > 0 \quad \text{(seller benefit)}$$

which leads to the two trading conditions

$$p_1 < \frac{c_{11} v_{21} C_2}{v_{11} c_{21} C_1} p_2 \;,\; p_2 < \frac{c_{22} C_1}{v_{22} C_2} p_1 .$$

The constraints may fail if the buyer does not get sufficient value from trading or the seller does not get sufficient payment to justify production. A window of admissible investment allocations (p_1, p_2) exists for

$$Z = \frac{f_{12} f_{21}}{f_{11} f_{22}} = \frac{v_{21}}{c_{21}} \frac{c_{11}}{v_{11}} \frac{c_{22}}{v_{22}} > 1 \tag{9.20}$$

which implies that the unit value-cost ratio of trading exceeds the product of the unit value-cost ratios of unilateral action.

Remark 9.8. The two trading conditions provide constraints for the price:

$$\frac{v_{22}}{c_{22}} c_{21} < \pi = \frac{p_1 C_1}{p_2 C_2} c_{21} < \frac{c_{11}}{v_{11}} v_{21}$$

For $f_{11} = f_{22} = 1$ the constraints reduce to $c_{21} < \pi < v_{21}$, i.e. the unit value of the buyer exceed the unit cost of the seller, with the price in between. If buyer and seller also act as consumer and producer, the same good may serve as a basis for unilateral action (self-production and self-consumption) as well as interaction (trading). Then for $v_{21} = v_{11}$ and $c_{21} = c_{22}$, the stability index is $Z = c_{11}/v_{22}$ and the constraint becomes $v_{22} < \pi < c_{11}$, i.e. the unit cost of production of the good for the consumer exceeds the unit value of consumption of the good by the producer. In other words, self-production is too costly for the consumer compared to buying goods while self-consumption is not valuable enough for the producer compared to selling of the good. This is a basic condition that drives the specialization between consumer/buyer and producer/seller for a particular good (for another good this relationship may be just reverse).

For given target benefits $B_i = B_i^*$ and $B_j = B_j^*$, the respective response strategies $\tilde{p}_i$ and $\bar{p}_i$ and the equilibrium p_i^* can be defined as a function of the unit costs and values. For the cooperative response strategy $\bar{p}_i$, the adaptive mechanism is stable for $Z > 1$. The equilibrium allocation is

$$p_1^* \equiv \frac{B_1^* c_{11}/v_{11} + ZB_2^*}{C_1(Z-1)}$$

$$p_2^* \equiv \frac{B_2^* c_{22}/v_{22} + ZB_1^* c_{21}/v_{21}}{C_2(Z-1)}$$

At the upper allocation limits $(p_1^*, p_2^*) = (1,1)$, $B_i^*(1,1) > 0$ and $B_j^*(1,1) > 0$ is realized for the cost constraint

$$\frac{v_{11}c_{21}}{c_{11}v_{21}} \leq \frac{C_2}{C_1} \leq \frac{c_{22}}{v_{22}}$$

which for $Z > 1$ defines a set of possible total investments (C_1, C_2).

9.5.3 Trade-Offs of Benefit Cost Exchange

Using the trade-offs developed in Sect. 9.3.3, the price can be specified as

$$\pi = \gamma_{12}c_{21} = \frac{p_1 C_1}{p_2 C_2} c_{21} = \frac{v_{21}/c_{21} + \beta_{12}v_{22}/c_{22}}{v_{11}/c_{11} + \beta_{12}} c_{21}.$$

Generally the price increases with the unit value of trading for the buyer and the unit cost of the seller as well as with the value-cost ratio of unilateral action of the seller but declines with the value-cost ratio of unilateral action of the buyer. For the special case of one good ($v_{21} = v_{11}, c_{21} = c_{22}$), one obtains

$$\pi = \frac{1 + \beta_{12}v_{22}/v_{11}}{1 + \beta_{12}c_{11}/v_{11}} c_{11} < c_{11} \text{ for } Z > 1.$$

In case of $v_{11} = c_{11}$ and $v_{22} = c_{22}$ (which corresponds to the case 2 in Sect. 9.3.1), the price function is particularly simple

$$\pi = \frac{v_{21} + \beta_{12} c_{21}}{1 + \beta_{12}}$$

which for $\beta_{12} = 1$ is the average of the buyer's unit value and the seller's unit cost which shifts towards the seller's unit cost with increasing β_{12}. The price function can be specified for the respective solution concepts.

1. *Asymmetric benefits:* If buyer or seller receives the full benefit, the prices
$$\pi(\beta_{12} \to \infty) = \frac{v_{22}}{c_{22}} c_{21}, \quad \pi(\beta_{12} \to 0) = \frac{c_{11}}{v_{11}} v_{21}$$
correspond to the upper and lower limits of the trading constraints.
2. *Allocation equity:* If both agents allocate the same fraction of their investment to cooperation ($p_1 = p_2$), the price is:
$$\pi = \frac{C_1}{C_2} c_{21}.$$
Thus, the capacity of the buyer drives price up and the capacity of the seller brings it down.
3. *Benefit equity:* If buyer and seller claim the same benefit $B_1^* = B_2^* = B^*$ ($\beta_{12} = 1$), this results in a price:
$$\pi = \frac{v_{21} + c_{21} v_{22}/c_{22}}{1 + v_{11}/c_{11}}$$
which is the arithmetic mean $\pi = (v_{21} + c_{21})/2$ for $f_{11} = f_{22} = 1$.
4. *Cost equity:* If buyer and seller invest the same in cooperation ($\gamma_{ij} = 1$), price and benefit exchange ratio become:
$$\pi = c_{21} \qquad \beta_{12} = \frac{v_{21}/c_{21} - v_{11}/c_{11}}{1 - v_{22}/c_{22}} = \frac{\phi_1^p}{\phi_2^p}.$$
This implies zero buyer benefit for $f_{21} = f_{11}$ and zero seller benefit for $v_{22} = c_{22}$.
5. *Benefit exchange ratio equals cost-exchange ratio:* For $\beta_{12} = \gamma_{12}$ the price becomes:
$$\pi = \beta_{12} c_{21} = a \pm \sqrt{a^2 + b}$$
where $a = (v_{22}/c_{22} - v_{11}/c_{11})/2$ and $b = v_{21}/c_{21}$. For $f_{11} = f_{22}$, the price is the geometric mean $\pi = \sqrt{v_{21} c_{21}}$.
6. *Indifference between individual and joint benefits:* Here the buyer pays the value received from the seller, $V_{12} = p_1 C_1 = v_{21} p_2 C_2 / c_{21} = V_{21}$, and the price is equal to the buyer's unit value of the seller's good:
$$\pi = v_{21}.$$

Similarly, the general conditions in Sect. 9.4 for the cooperative pursuit of target values and relative growth targets can be applied to the interaction between buyers and sellers. For instance, the case of joint positive target values $V_1^*(1,1) \geq 0$ and $V_2^*(1,1) \geq 0$ for full mutual cooperation leads to the cost constraint

$$1 \leq \frac{C_1}{C_2} \leq \frac{v_{21}}{c_{21}}$$

which implies that the buyer's unit value exceeds the seller's unit cost ($v_{21} > c_{21}$), and total cost of the buyer C_1 exceeds total cost of the seller C_2.

Accordingly, for full mutual cooperation ($p_1^* = p_2^* = 1$) the equilibrium of growth targets becomes

$$\rho_1^* = \omega_{21} - 1 = \frac{v_{21}c_{11}C_2}{c_{21}v_{11}C_1} - 1$$

$$\rho_2^* = \omega_{12} - 1 = \frac{c_{22}C_1}{v_{22}C_2} - 1$$

The bargaining set of reachable growth targets between buyer and seller is

$$(1+\rho_i^*)(1+\rho_j^*) \leq Z = \frac{v_{21}}{c_{21}}\frac{c_{11}}{v_{11}}\frac{c_{22}}{v_{22}}(> 1)$$

which demonstrates that the cooperative bargaining set expands with increasing unit value v_{21} of the buyer and declining unit cost c_{21} of the seller and declines with more efficient unilateral action of either buyer or seller.

9.6 Conclusions and Outlook

The main focus of this article is to examine conditions for cooperation in value-cost dynamic games among two agents who invest into action pathways to achieve values. These conditions are specified in a game-theoretic setting of a prisoner's dilemma for agents that seek to realize certain benefits and value targets as a result of investment in cooperation. Considering ratios of these benefits it is possible to determine boundaries for the investment ratios in cooperation which depend on the action and interaction effects. The general approach is applied to the trading of economic goods to determine conditions for the mutually beneficial cost exchange as well as the market price of these goods which should exceed the unit cost of the seller and not exceed the unit value of the buyer. While some specifications of the model have been realized, it is possible to extend the model framework in multiple ways.

1. This article explicitly focuses on situations where agents allocate part of their investment to the interaction with other agents. An extension would include indirect side effects, i.e. when the investment $p_{ij} C_i$ of agent i in interaction with agent j causes a side effect $V_{ii}^s = f_{ii}^s p_{ij} C_i$ on i's value. An example is the impact of environmental pollution from trading between agents. Similarly, the investment $p_{ii} C_i$ in unilateral action by agent i may indirectly affect the value of other agents $V_{ij}^s = f_{ij}^s p_{ii} C_i$, e.g. when the consumption of a resource affects another agent.
2. The analysis given in this article can be fully embedded into the dynamic framework introduced in the beginning. This implies that total costs $C_i(t) = \kappa_i(t) K_i(t)$ are not fixed but change from one time period to the next within capital constraints, which depends on the degree to which capital can be converted to investment (given by $\kappa_i(t)$). Furthermore, the value obtained in one time period could be added to or subtracted from capital: $K_i(t+1) = K_i(t) + s_i(t) V_i(t)$ with a savings rate $s_i(t)$ (which depends on the possibility to convert value to capital). When multiple time steps are analyzed, agents could include long-term value effects of cooperation into consideration.
3. While the focus has been on unit costs and values, on total costs and values for a particular time period as well as exchange ratios between these variables, it is possible to perform the analysis in a differential form, using marginal costs and values instead. Accordingly, the exchange ratios can be expressed in a differential form: $\gamma_{ij} = dC_i^p / dC_j^p$ and $\beta_{ij} = dB_i^* / dB_j^*$. Then the time-discrete dynamic game could be represented by a differential game.
4. The action and interaction effects f_{ii} and f_{ij} have been treated as constant in this article but may change as a function of time and state variables. For instance, unit values v_{ij} may decline for increasing consumption due to saturation effects. Unit costs c_{ij} may decline due to learning and economies of scale or they may increase due to resource scarcity that affects production factors. If actions induce risks, these reduce net unit values, e.g. due to environmental pollution that diminishes the unit value for consumers as well as the benefit-cost ratio.
5. Costs and values of different agents are assumed to have the same units which allows to easily combine them. This assumption can be modified by conversion factors that make comparison of costs and values between agents possible. This is of particular interest in case of coalition formation where agents merge investments to achieve joint values. Then coalition action may create additional benefits of collaboration that would be distributed among the individual agents according to acceptability and fairness principles.
6. While the focus of this article is on the interaction between two agents, the extension to multiple agents is straightforward. Obviously the analysis can be applied to cooperation between any pair of agents i and j in a larger group, independent of what other agents do. Next, the impact of other agents l may be treated as a disturbance of the value function:

$$V_i = p_{ii} f_{ii} C_i + p_{ij} f_{ij} C_j + V_i^E = V_i^*$$

where $V_i^E = V_i^Y + \sum_l f_{li}^s C_l$ combines the environmental value effects of changes in the natural environment V_i^Y and of changes in the social environment induced by investments applied to actions of other agents l where f_{li}^s are the side effects of the investments C_l by agent l (including side effects by agent i and j). Provided the actions of agent l are not influenced by agents i and j, they can be treated as given and included into modified target values $\hat{V}_i^* = V_i^* + V_i^E$ for which the analysis of this article can be applied. Finally, the case of full multi-player cooperation can be treated with value functions of the type

$$V_i = \sum_{j=1}^{n} p_{ij} f_{ij} C_j = V_j^* \qquad (i = 1, \dots, n)$$

Here similar dynamic responses of allocation $\Delta p_{ij} = \alpha_{ij}(\hat{p}_{ij} - p_{ij})$ can be considered where the response functions $\hat{p}_{ij}$ as well as the equilibria p_{ij}^* of each agent depend on the cooperative allocation p_{ji} and the target benefits B_j^* of all other agents. The stability index Z corresponds to the determinant and eigenvalues of the n-dimensional interaction matrix. Accordingly, trade-offs between cost-exchange ratios and benefit-exchange ratios can be included. An in-depth analysis of the multi-agent case is left to the future.

7. For the trading between buyers and sellers the multi-agent extension is also straightforward. Whether a pair of agents i and j has an incentive to trade goods, depends on the condition $Z_{ij} > 1$ for this pair which becomes $c_{ii} > v_{jj}$ if a good is produced and sold by agent j and bought and consumed by agent i. Thus, agents can be sorted according to their unit costs and values where incentives to trade are largest between buyers and sellers with the largest Z_{ij}, respectively, the largest difference $c_{ii} - v_{jj}$ for one good. If two pairs of agents $(1, 2)$ and $(3, 4)$ have different prices for one good

$$\pi_{12} = \frac{p_1 C_1}{p_2 C_2} c_{21} < \pi_{34} = \frac{p_3 C_3}{p_4 C_4} c_{43},$$

then buyer 3 has an incentive to shift part of its investment $p_3 C_3$ to seller 2, while seller 2 has an incentive to shift part of its investment to buyer 3 until price equality is achieved. As discussed before, each agent in the trading game can be buyer and seller at the same time for different goods or action paths in which they are more efficient than the counterpart. It is further possible that the seller has been a buyer of the good or its components from another seller, and that the buyer sells the good to another buyer. Understanding these processes of price formation and trading cascades is left to future analysis.

8. The special case of multiple sellers $I = 1, \dots, N$, multiple buyers $i = 1, \dots, n$, and multiple goods x^k has been introduced in the coalition framework of the value-cost model in Scheffran (2006) and used for an application in the agricultural sector in Scheffran and Bendor (2009). The equilibrium market price for each good is given as

$$\pi^k = \frac{\sum_i p_i^k C_i}{\sum_I p_I^k C^I / c_I^k} \qquad (k = 1, \ldots, m).$$

where p_I^k and p_i^k are the allocations of investment C_I and C_i to good x^k, and c_I^k is the unit cost of the good x^k for producer I. Applying the trade-off mechanisms suggested in this article to this general price function may provide useful insights into price formation in complex market situations.

9. Finally, it is subject to research which of the mentioned trade-off mechanisms between cost exchange and benefit exchange are relevant in which contexts. This is of particular interest to determine the market price of goods, and the specialization between producers and consumers. Empirical research can help to understand the behavioral and bargaining issues involved. This is also relevant to design institutional frameworks that facilitate and ensure the benefits of cooperation.

While some specifications of the model have been realized, it is possible to extend the model framework along the lines described. In particular, it is promising to extend the individual-agent perspective to coalitions that receive investments from agents to realize more powerful actions than are possible individually, thus investments are acquired from a large number of agents to implement larger production capacities and achieve critical values. Interesting theoretical issues also emerge to analyze the transition between conflict and cooperation or the stability and instability of interaction. The general approach promises applicability in a wide range of fields where agents collaborate, including trading processes as well as resource use and environmental management. By adapting the model to specific application areas, it becomes possible to quantify the unit costs and values and develop exemplary simulations.

Acknowledgments Research for this article was funded in parts by the German Science Foundation (DFG) through the Cluster of Excellence CliSAP (EXC177). The author is grateful to Harry Dankowicz for discussion of some of the issues in an earlier version of this article.

References

Allen B, Hellwig M (1986) Bertrand-Edgeworth oligopoly in large markets. Rev Econ Stud 53:175–204

Axelrod R (1984) The evolution of cooperation. Basic Publ., New York

Başar T, Olsder G (1982) Dynamic noncooperative game theory. Academic Press, London/New York

Başar T, Zhu Q (2011) Prices of anarchy, information, and cooperation in differential games. Dyn Games Appl 1:50–73

Bresnehan T (1981) Duopoly models with consistent conjectures. Am Econ Rev 71:934–945

Bressan A, Shen W (2004) Semi-cooperative strategies for differential games. Int J Game Theory 32:561–593

Cao L, Ohtsuki H, Wang B, Aihara K (2011) Evolution of cooperation on adaptively weighted networks. J Theor Biol 272(1):8–15
Chasparis G, Shamma J (2012) Distributed dynamic reinforcement of efficient outcomes in multiagent coordination and network formation. Dyn Games Appl 2:18–50
Dixon H (1992) The perfectly competitive outcome as the equilibrium in an Edgeworthian price-quantity game. Econ J 102:301–309
Dockner E, Jorgensen S, Van Long N, Sorger G (2000) Differential games in economics and management science. Cambridge University Press, Cambridge
Du F, Fu F (2011) Partner selection shapes the strategic and topological evolution of cooperation: The power of reputation transitivity. Dyn Games Appl 1:354–369
Engwerda J (2010) Necessary and sufficient conditions for pareto optimal solutions of cooperative differential games. SIAM J Control Optim 48(6):3859–3881
Epstein J, Axtell R (1997) Growing artificial societies. Bradford Book, Cambridge
Figuières C, Jean-Marie A, Quérou N, Tidball M (2004) Theory of conjectural variations. World Scientific, Singapore
Gao J, Li Z, Cong R, Wang L (2012) Tolerance-based punishment in continuous public goods game. Physica 391(16):4111–4120
Gerber A (2000) Coalition formation in general NTU games. Rev Econ Des 5(2):149–175
Gomez Portillo I (2012) Cooperation and its evolution in growing systems with cultural reproduction. Eur Phys J B 85(12):409
Hahn M (2012) An evolutionary analysis of varianŠs model of sales. Dyn Games Appl 2:71–96
Helbing D, Johansson A (2010) Cooperation, norms, and revolutions: A unified game-theoretical approach. PLoS ONE 5(10):e12530
Hofbauer J, Sigmund K (1998) Evolutionary games and population dynamics. Cambridge University Press, Cambridge
Isakov A, Rand D (2012) The evolution of coercive institutional punishment. Dyn Games Appl 2:97–109
Louge F, Riedel F (2012) Evolutionary stability in first price auctions. Dyn Games Appl 2(1): 110–128
Miao X, Zhou X, Wu H (2012) A cooperative differential game model based on network throughput and energy efficiency in wireless networks. Optim Lett 6(1):55–68
Nemeth A, Takacs K (2010) The paradox of cooperation benefits. J Theor Biol 264(2):301–311
Ohtsuki H (2011) Evolutionary dynamics of the Nash demand game: A diffusion approach. Dyn Games Appl 1:449–461
Perc M, Szolnoki A (2012) Coevolutionary games—A mini review. Biosystems 99(2):109–125
Petrosjan L (1995) The Shapley value for differential games. In: Olsder G (ed) New trends in dynamic games and application. Birkhäuser, Boston, pp 407–417
Petrosjan L (2004) Cooperation in games with incomplete information. In: Nonlinear analysis and convex analysis, Yokohama Publ., Yokohama, pp 469–479
Requejo R, Camacho J (2012) Analytical models for well-mixed populations of cooperators and defectors under limiting resources. Phys Rev E 85(6):066112
Rezaei G, Kirley M (2012) Dynamic social networks facilitate cooperation in the N-player prisoner's dilemma. Physica 391(23):6199–6211
Scheffran J (2001) Stability and control of value-cost dynamic games. Cent Eur J Oper Res 9(7):197–225
Scheffran J (2006) The formation of adaptive coalitions. In: Haurie A, Muto S, Petrosjan L, Raghavan T (eds) Advances in dynamic games. Birkhäuser, Boston, pp 163–178
Scheffran J, Bendor T (2009) Bioenergy and land use—A spatial agent dynamic model of energy crop production in Illinois. Int J Environ Pollut 39(1/2):4–27
Scheffran J, Hannon B (2007) From complex conflicts to stable cooperation. Complexity 13(2): 78–91
Selten R, Stoecker R (1986) End behavior in sequences of finite prisoner's dilemma supergames. J Econ Behav Organ 7:47–70

Simaan M, Cruz J (1976) On the Stackelberg strategy in nonzero-sum games. In: Leitmann G (ed) Multicriteria decision making and differential games. Plenum Press, New York, pp 173–195

Su Y, van der Schaar M (2011) Linearly coupled communication games. IEEE Trans Commun 59(9):2543–2553

Szidarovszky F, Li W (2000) A note on the stability of a Cournot-Nash equilibrium: The multiproduct case with adaptive expectations. J Math Econ 33:101–107

Tanabe S, Masuda N (2012) Evolution of cooperation facilitated by reinforcement learning with adaptive aspiration levels. J Theor Biol 293:151–160

Tolwinski B, Haurie A, Leitmann G (1986) Cooperative equilibria in differential games. J Math Anal Appl 119:182–202

Tuinstra J (2000) A discrete and symmetric price adjustment process on the simplex. J Econ Dyn Contr 24(5–7):881–907

Wang R, He J, Wang Y, Shi L, Li YT 2010 (2010) Asymmetric interaction will facilitate the evolution of cooperation. Sci China Life Sci 53(8):1041–1046

Weidlich W (2000) Sociodynamics—A systematic approach to mathematical modelling in the social sciences. Harwood Academic Publishers, Amsterdam

Yeung D, Petrosjan L (2006) Cooperative stochastic differential games. Springer, New York

Yeung D, Petrosjan L (2012) Subgame consistent economic optimization: An advanced cooperative dynamic game analysis. Birkhäuser, Boston

Zhang A, Zhang Y (1996) Stability of a Cournot–Nash equilibrium: The multi-product case. J Math Econ 26:441–462

Chapter 10
Intra-seasonal Strategies Based on Energy Budgets in a Dynamic Predator–Prey Game

Kateřina Staňková, Alessandro Abate, and Maurice W. Sabelis

Abstract We propose a game-theoretical model to describe intra-seasonal predator–prey interactions between predatory mites (*Acari: Phytoseiidae*) and prey mites (also called fruit-tree red spider mites) (*Acari: Tetranychidae*) that feed on leaves of apple trees. Its parameters have been instantiated based on laboratory and field studies. The continuous-time dynamical model comprises predator and prey densities, along with corresponding energy levels, over the length of a season. It also includes time-dependent decision variables for the predator and the prey, representing the current portions of the predator and prey populations that are active, as opposed to diapausing (a state of physiological rest). Our aim is to find the optimal active/diapausing ratio during a season of interaction between predatory mites and prey mites: this is achieved by solving a dynamic game between predator and prey. We hereby extend our previous work that focused solely on the optimal strategy for the prey. Firstly, we analyze the optimal behavior of the prey. Secondly, we show that the optimal strategy for the predator is to stay active for the entire season. This result corresponds to biological observations.

Keywords Mathematical models • Predator–prey interactions • Dynamic noncooperative game theory • Diapause • Mites • Fruit orchard

---

K. Staňková (✉)
Department of Knowledge Engineering, Maastricht University, Maastricht, The Netherlands
e-mail: k.stankova@maastrichtuniversity.nl

A. Abate
Delft Center for Systems & Control, Delft University of Technology, 2600 AA Delft, The Netherlands
e-mail: a.abate@tudelft.nl

M.W. Sabelis
Institute for Biodiversity and Ecosystem Dynamics, University of Amsterdam, 1000 GG Amsterdam, The Netherlands
e-mail: M.W.Sabelis@uva.nl

V. Křivan and G. Zaccour (eds.), *Advances in Dynamic Games*, Annals of the International Society of Dynamic Games 13, DOI 10.1007/978-3-319-02690-9_10,

10.1 Introduction and Motivations

The work presented in this article is inspired by studies on the use of predatory mites (*Acari: Phytoseiidae*) for biological pest control of fruit-tree red spider mites (*Acari: Tetranychidae*) that feed on and thereby damage leaves of apple trees (Helle and Sabelis 1985a,b).

This system involves continuous interactions and overlapping generations in summer seasons, as well as discrete periods without interactions, and is therefore an example of a hybrid system, in the biological literature referred to as a semi-discrete system (Mailleret and Lemesle 2009; Patchepsky et al. 2008). Winters (covering 6–7 months) are usually harsh and as such endanger the survival of fruit-tree red spider (prey) mites (Helle and Sabelis 1985a) and (even more so) that of predatory mites (Fitzgerald and Solomon 1991; Helle and Sabelis 1985a).

Predatory mite and prey mite densities in the following summer season depend on the number of individuals in the previous year and on their survival during the winter. For the prey, this number equals to the number of prey individuals that are in a state of physiological rest (the so-called diapause state) at the end of the season, as prey that is active at the end of the summer season does not have a chance to survive. The decision to enter diapause promotes the survival of the prey individual during winter and it emerges from induction by a combination of sufficiently long night lengths and low temperatures (Veerman 1992). Focusing on a single season, in Staňková et al. (2013) we have shown that if the predator stays active the entire season the optimal strategy of the prey can be described as follows (see Fig. 10.1):

1. In the beginning of the summer season the prey can be in any state (all active, all in diapause, or anything in between), whereas at the end of the summer season all prey individuals enter diapause.
2. If all prey individuals are active in early summer, the prey will start entering diapause at a certain point in time and the proportion of diapausing individuals increases monotonically. Similarly, if only part of the prey population is active in early summer, then all prey end up being in diapause at one point in time and stay in diapause until the next year. Yet, if all prey individuals are in diapause in early summer, then they continue to stay in diapause until the next year.
3. The time (expressed in real time) of diapause onset depends on the energy of the prey, on predator population size, and on the rate of energy utilization, but it is independent of prey population size (i.e. timing of diapause does not require quorum sensing).
4. If predators are absent in the environment, all prey individuals enter diapause later than if the predators are present (see Fig. 10.2). Empirical observations on diapause of fruit tree red spider mites on apple trees in the field (Sabelis and Overmeer, unpublished data) reveal that virtually all individuals become active in early summer and starting from a certain point in time the population enters diapause, gradually. Moreover, experimental manipulation of the predator population in the field showed that the fruit-tree red spider mites enter diapause

Fig. 10.1 Scheme of possible optimal diapause ratio u^R for the prey ($u^R(t) \in [0, 1]$ for each $t \in [0, T]$). Based on the proposed dynamics and optimization problem, we have shown irreversibility and (largely) monotonicity of the strategy profile. Notice that the optimal strategies do not need to be continuous corresponding to the singular events in the outcome of the optimization problem. This figure is taken from Staňková et al. (2013)

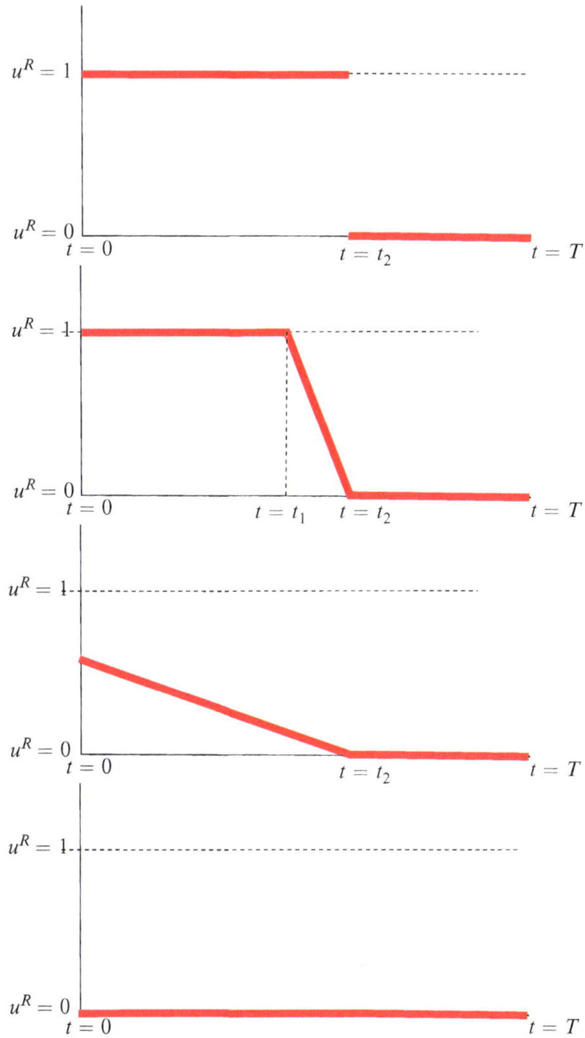

earlier in the presence of predatory mites and once in diapause they stay in diapause. However, apart from an effect of predator presence also the density of fruit-tree red spider mites had an effect on the time at which diapause was initiated, suggesting that some form of quorum sensing (possibly via spider-mite induced plant volatiles) takes place.

Using another similar spider mite species (more amenable to experimental treatment), it was experimentally shown that the decision to enter diapause also depends on predator density during summer (Kroon et al. 2004, 2005, 2008). From the point of view of the prey mite this behavior makes intuitive sense as it faces a grim future with increasing predator densities and thus an increased risk of death: it may then do better by giving up reproduction, moving away from leaves to twigs and

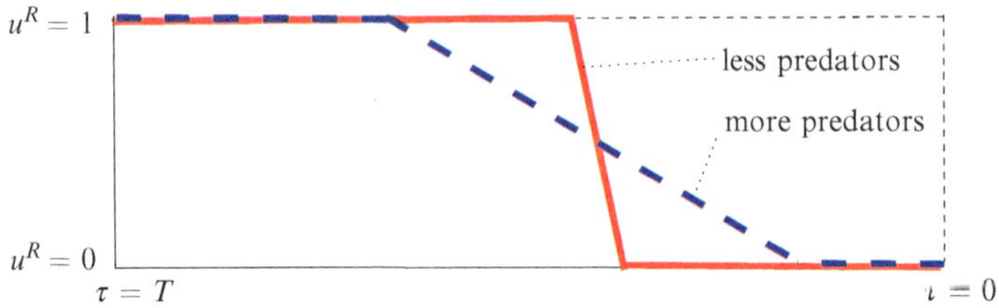

Fig. 10.2 If the number of predators increases (while all the other state variables and parameters stay the same), the prey individuals begin to enter diapause earlier, but more gradually, balancing between having enough energy to survive the diapause and escaping the predation. This figure is taken from Staňková et al. (2013)

branches (a refuge from predation, but without food) and by entering diapause earlier than indicated by the predictors of season length (night length and temperature). However, if many prey mites would make the same decision, this could create a negative feedback on the predatory mite population, which could lead them to enter diapause. Consequently, at some point in time the prey mites would profit from the decreased predation risk by terminating their diapause and returning to the leaves, which in turn could trigger the predatory mites to become active again. Another complicating factor is that an early prey diapause raises the demands on the energy level of the individual mite, which needs to cover a longer period before terminating diapause at the beginning of the next summer season—the energy level at diapause termination will determine the reproductive capacity of the prey mite (Kroon et al. 2005). Thus, the decision to enter diapause within a year will depend on the current internal energy level of the mite, as this will have far-reaching consequences for winter survival and reproduction in the summer season of the next year.

There is less information on the diapause behavior of the predatory mites. However, the predatory mites are much more flexible in entering diapause/active state and can switch multiple times during the season. Physiological decision variables depend on the predator and prey densities reached during summer, rather than only on reliable season indicators, such as night/day length and temperature (Danks 1987; Tauber et al. 1986).

This leads us to conclude that the predator's and prey's decision to enter diapause is part of a game between the two species. While we think that this game is a Stackelberg game with the population of predatory mites acting as a leader and the population of the prey mites acting as a follower (Başar and Olsder 1999; Staňková 2009), we will elaborate on this claim when analyzing the optimal behavior of both parties involved.

Notation: In the remainder of this document, unless stated otherwise, the following notation will be used:

T —length of the summer season
$R(t)$—fruit-tree red spider (prey) mite population at time $t \in [0, T]$, within the summer season

$P(t)$—predatory mite population at time $t \in [0, T]$, within the summer season
$E^R(t)$—internal energy of the prey at time $t \in [0, T]$, within the summer season
$E^P(t)$—internal energy of the predator at time $t \in [0, T]$, within the summer season
$u^R(t)$—decision variable (control) of the fruit-tree red spider mites (prey), within the summer season
$u^P(t)$—decision variable (control) of the predatory mites (predator), within the summer season
$\alpha_R(\tau)$, $\alpha_{E^R}(\tau)$, $\alpha_{E^P}(\tau)$, $\alpha_P(\tau)$, $\beta_R(\tau)$, $\beta_{E^R}(\tau)$, $\beta_{E^P}(\tau)$, $\beta_P(\tau)$—additional variables for the characteristic system in reverse time
$\mathscr{A}_P$, $\mathscr{A}_R$—singular surfaces (as used in the analysis of the game)

The article is structured as follows. Section 10.2 introduces the dynamic game between the predatory mites and the prey mites. Section 10.3 formally studies the optimal strategies of the predator and prey in this game. Section 10.4 elaborates on the biological interpretation of the results and proposes a new model to describe interactions in predator–prey systems of our interest. Section 10.5 concludes the article, discusses possible extensions, and sketches future work.

10.2 Game-Theoretical Model of the Interaction Between Predatory Mites and Fruit-Tree Red Spider Mites

Each year is divided into two parts: the *summer* and the *winter* season. The predatory mites and the fruit-tree red spider mites can consume food (prey and apple leaves, respectively) only during the summer season (which is essential for their reproduction). Furthermore, both predator and prey can enter diapause, a quiescent state that protects from the environment, from predation, or possibly lack of food. This implies a decoupling between predator and prey depending on the population fraction in diapause. During the winter season the species do not interact, and their populations independently decline at a constant rate, therefore we focus on the summer interaction only. The dynamics during winter are trivial and can be simply modeled by a reset (i.e., a decrease) of the energy and population levels.

The model that we propose describes the interactions between predatory mites (predator) and fruit-tree red spider mites (prey) within a single summer season[1] and allows characterizing the seasonal strategy of both predator and prey as a solution of a dynamic game between them.

In the remainder of the text the terms "summer season" and "winter season" are used interchangeably with the terms "summer" and "winter," respectively.

[1] Extension of our model to multiple seasons is a subject of our future research.

10.2.1 Model Formulation

The summer interactions between the predatory mites and the prey mites can be formulated as a game played with a finite horizon $[0, T]$ in which the predatory mites select a $u^{P,*}(t) \in [0, 1]$ for $t \in [0, T]$, where

$$u^{P,*} = \arg\sup_{u^P(\cdot)} \int_0^T \big(-\alpha P(t) + \beta\delta u^P(t) E^P(t) P(t)\big) \mathrm{d}t, \tag{10.1}$$

whereas the prey mites choose a $u^{R,*}(t) \in [0, 1]$ for $t \in [0, T]$, where

$$u^{R,*} = \arg\sup_{u^R(\cdot)} \int_0^T (1 - u^R(t)) E^R(t)\, R(t) \mathrm{d}t, \tag{10.2}$$

subject to the following system dynamics:

$$\frac{\mathrm{d}E^P}{\mathrm{d}t} = -ac(1 - u^P)E^P + e\, u^P u^R R - au^P E^P, \tag{10.3}$$

$$\frac{\mathrm{d}E^R}{\mathrm{d}t} = -dh(1 - u^R)E^R + f(t)g(R)u^R - d\, u^R E^R, \tag{10.4}$$

$$\frac{\mathrm{d}P}{\mathrm{d}t} = -\alpha P + \beta\delta u^P E^P P, \tag{10.5}$$

$$\frac{\mathrm{d}R}{\mathrm{d}t} = -\epsilon R + \delta u^R E^R R - \gamma u^P u^R PR. \tag{10.6}$$

In (10.3) $a > 0$ is the energy decrease rate for the predator when active, $a\,c$ (with $c \in [0, 1)$) is the energy decrease rate for the predator when in diapause, e is the energy increase rate for the predator when feeding (here the energy increase is proportional to the number of active fruit-tree red spider mites that are preyed upon and to the number of active predatory mites).

In (10.4), $d > 0$ is the energy decrease rate for the prey when active, $d\,h$ (with $h \in [0, 1)$) is the energy decrease rate for the prey when in diapause, $f(t)$ is a time-dependent function characterizing the presence of nutrients for the fruit-tree red spider mites in the environment ($0 < f(\cdot) \ll 1$), $g(R_n) \in [0, 1]$ is a non-increasing function of its variable, which represents competition among individual fruit-tree red spider mites—hence $f(t)g(R_n)u^R$ is a term representing the increase of energy in the prey due to its active state.

The number of predatory mites slowly decreases with rate $\alpha > 0$ and increases proportionally to their energy and number of active individuals with rate $\beta\,\delta$, where $\beta > 0, \delta > 0$.

The number of fruit-tree red spider mites decreases with death rate $\epsilon > 0$, increases proportionally to their energy and number of active individuals with rate

$\delta > 0$, and decreases proportionally to the number of active predatory mites and number of active fruit-tree red spider mites with rate $\gamma > 0$.

E^P and E^R refer to the energy levels of the predator and the prey, respectively. Since the energy of an organism is not a quantity that can be directly measured, we could normalize these variables as E^P, $E^R \in [0, 1]$, so that they become ratios. While for the steady state conditions (10.3)–(10.4) are invariant with respect to $E^P \in [0, 1]$, $E^R \in [0, 1]$, in general E^P and E^R might reach values outside the interval $[0, 1]$. For the sake of simplicity, in the remainder of the paper we will assume that the initial conditions $P(0)$, $R(0)$, $E^P(0)$, $E^R(0)$ are such that $E^P(t)$, $E^R(t) \in [0, 1]$ for all $t \in [0, T]$.

The fitness function for the predator (10.1) reflects the fact that all predator individuals being alive at the end of the summer season (independently whether they are active or in diapause) have a chance to survive the winter. The fitness function for the prey (10.2) reflects the fact that only the prey individuals that are in diapause at the end of the summer season have a chance to survive the winter, while the longer in diapause they are and the more internal energy they have, the higher chance of survival they have.

Note that $u^{P,*}$ and $u^{R,*}$ are optimal decisions on the population level. An additional analysis is needed to validate whether the optimal behavior on the population level coincides with the optimal behavior of an individual.

Based on Sabelis (1991) we set parameter h to $h = \frac{1}{250\,d}$. We assume that the increase of the energy of the prey from feeding (composition of the effects of the environment and competition among the prey $(f(t)\,g(R))$ equals to the decrease rate of energy of the prey when active (d), i.e., $d = f(t)\,g(R)$. Based on field data and (Sabelis 1991; Sabelis and Janssen 1993), we set $\delta = \frac{1}{5}$, $\alpha = \frac{1}{20}$. Additionally, observing that predator and prey are of the same size and their death rates are approximately equal (Sabelis 1991; Sabelis and Janssen 1993), the dynamics in (10.3)–(10.6) can be rewritten as follows (with β replaced by b in notation):

$$\frac{\mathrm{d}E^P}{\mathrm{d}t} = -\frac{1}{250}(1 - u^P)E^P + d\,u^P\,u^R\,R - d\,u^P\,E^P, \tag{10.7}$$

$$\frac{\mathrm{d}E^R}{\mathrm{d}t} = -\frac{1}{250}(1 - u^R)E^R + d\,u^R - d\,u^R\,E^R, \tag{10.8}$$

$$\frac{\mathrm{d}P}{\mathrm{d}t} = -\frac{1}{20}\,P + \frac{1}{5}\,b\,u^P\,E^P\,P, \tag{10.9}$$

$$\frac{\mathrm{d}R}{\mathrm{d}t} = -\frac{1}{20}R + \frac{1}{5}u^R E^R\,R - \frac{1}{5}u^P u^R P\,R, \tag{10.10}$$

where $E^R(t) \in [0, 1]$, $E^P(t) \in [0, 1]$, $P(t) > 0$, $R(t) > 0$ for each $t \in [0, T]$ with T known. Note that inequality $d > \frac{1}{250}$ has to be satisfied (otherwise the energy would decrease even if the prey is in diapause).

Within a summer, the goal of both predator and prey (the *players*) is to maximize their chances of survival (Cressman 2003; Weibull 1995), which translates to the optimization problems defined by (10.1) and (10.2), subject to the dynamical constraints (10.7)–(10.10).

10.3 Solution of the Game

Firstly, we formulate the problem of the predator and the problem of the prey via Hamilton–Jacobi–Bellman equations. Subsequently, we study the optimal strategies for both the predator and for the prey, and we discuss their biological relevance.

10.3.1 Hamilton–Jacobi–Bellman Formulation for the Predator

Let us introduce a reverse time scale $\tau = T - t$ and value functions for both the predator and the prey. The value function for the predator in reverse time reads as

$$V^P = \int_{T-t}^{T} \left(\frac{1}{5} b\, u^P\, E^P\, P - \frac{1}{20}\, P \right) \mathrm{d}\tau, \tag{10.11}$$

and the related Hamilton–Jacobi–Bellman (HJB) equation can be written as follows:

$$\begin{aligned} \mathcal{H}^P = \frac{\partial V^P}{\partial t} &+ \max_{u^P} \Big(\alpha_{E^P} \big(-\frac{1}{250} (1-u^P) E^P + d\, u^P\, u^R\, R - d\, u^P\, E^P \big) \\ &+ \alpha_{E^R} \big(-\frac{1}{250} (1-u^R) E^R + d\, u^R - d\, u^R\, E^R \big) \\ &+ \alpha_P \big(-\frac{1}{20}\, P + \frac{1}{5} b\, u^P\, E^P\, P \big) \\ &+ \alpha_R \big(-\frac{1}{20} R + \frac{1}{5} u^R E^R\, R - \frac{1}{5} u^P u^R PR \big) + \frac{1}{5} b u^P\, E^P\, P - \frac{1}{20} P \Big), \end{aligned} \tag{10.12}$$

with $\alpha_{E^P} = \frac{\partial V^P}{\partial E^P}$, $\alpha_P = \frac{\partial V^P}{\partial P}$, $\alpha_{E^R} = \frac{\partial V^P}{\partial E^R}$, and $\alpha_R = \frac{\partial V^P}{\partial R}$.

The corresponding system of characteristics in reverse time τ is then (with x' denoting $\frac{\mathrm{d}x}{\mathrm{d}\tau} = -\frac{\mathrm{d}x}{\mathrm{d}t}$ for a general state variable x)

$$(E^P)' = \frac{1}{250} (1-u^P) E^P - d\, u^P\, u^R\, R + d\, u^P\, E^P, \tag{10.13}$$

$$(E^R)' = \frac{1}{250} (1-u^R) E^R - d\, u^R - d\, u^R\, E^R, \tag{10.14}$$

$$P' = \frac{1}{20}\,P - \frac{1}{5} b\,u^P\,E^P\,P, \tag{10.15}$$

$$R' = \frac{1}{20} R - \frac{1}{5} u^R E^R\,R + \frac{1}{5} u^P u^R P\,R, \tag{10.16}$$

$$\alpha'_{E^P} = -\alpha_{E^P}\big(\frac{1}{250}(1-u^P) + d\,u^P\big) + \frac{1}{5}\,b\,u^P\,\alpha_P\,P + \frac{1}{5}\,b\,u^P\,P, \tag{10.17}$$

$$\alpha'_{E^R} = -\alpha_{E^R}\big(\frac{1}{250}(1-u^R) + d\,u^R\big) + \frac{1}{5}\alpha_R\,u^R\,R, \tag{10.18}$$

$$\alpha'_P = \alpha_P\big(-\frac{1}{20} + \frac{1}{5} b\,u^P\,E^P\big) - \frac{1}{5}\alpha_R u^P u^R\,R + \frac{1}{5}\,b\,u^P\,E^P - \frac{1}{20}, \tag{10.19}$$

$$\alpha'_R = d\,\alpha_{E^P}\,u^P\,u^R + \alpha_R\big(-\frac{1}{20} + \frac{1}{5} u^R E^R - \frac{1}{5} u^P u^R P\big), \tag{10.20}$$

with transversality conditions $\alpha_{E^P}(0) = \alpha_{E^R}(0) = \alpha_P(0) = \alpha_R(0) = 0$ and with $E^P(0) \in (0,1)$, $E^R(0) \in (0,1)$, $P(0) > 0$, $R(0) > 0$. The singular surface corresponding to the HJB equation (10.12) is

$$\mathscr{A}_P = \alpha_{E^P}\left(\frac{1}{250}E^P + d u^R R - dE^P\right) + \frac{1}{5} b\alpha_P\,E^P P - \frac{1}{5} u^R \alpha_R PR + \frac{1}{5} bE^P P. \tag{10.21}$$

Then, the optimal strategy for the predator is obtained as $u^P = \operatorname{Heav} \mathscr{A}_P$, i.e.,

$$u^P = \begin{cases} 1, & \text{if} \quad \mathscr{A}_P > 0, \\ 0, & \text{if} \quad \mathscr{A}_P < 0. \end{cases}$$

Moreover, $u^P \in (0,1)$ if $\mathscr{A}_P = 0$ (Melikyan 1994, 1998; Melikyan and Olsder 2010).

From the transversality conditions we can derive that $u^P(\tau = 0) = \operatorname{Heav} \mathscr{A}_P(\tau = 0) = \operatorname{Heav}\big(\frac{1}{5}b\,E^P(0)\,P(0)\big) = 1$. Note that $\mathscr{A}_P$ is independent of E^R and of α_{E^R}. Further, note that regardless of the strategy of the prey the predator is active at the end of the season.

10.3.2 Hamilton–Jacobi–Bellman Formulation for the Prey

Similarly as in Sect. 10.3.1, we can introduce the reverse time $\tau = T - t$ so that the value function for the prey becomes:

$$V^R = \int_{T-t}^{T} \big(1 - u^R\big)\,E^R\,R\,\mathrm{d}\tau, \tag{10.22}$$

and the corresponding Hamilton–Jacobi–Bellman equation is:

$$\begin{aligned}\mathscr{H}^R = \frac{\partial V^R}{\partial t} &+ \max_{u^R} \Big(\beta_{E^P}\big(-\frac{1}{250}(1-u^P)E^P + d\,u^P\,u^R\,R - d\,u^P\,E^P\big)\\ &+ \beta_{E^R}\big(-\frac{1}{250}(1-u^R)E^R + d\,u^R - d\,u^R\,E^R\big)\\ &+ \beta_P\big(-\frac{1}{20}\,P + \frac{1}{5}b\,u^P\,E^P\,P\big)\\ &+ \beta_R\big(-\frac{1}{20}R + \frac{1}{5}u^R E^R\,R - \frac{1}{5}u^P u^R P\,R\big) + \big(1-u^R\big)\,E^R\,R\Big),\end{aligned} \tag{10.23}$$

with $\beta_{E^P} = \frac{\partial V^R}{\partial E^P}$, $\beta_P = \frac{\partial V^R}{\partial P}$, $\beta_{E^R} = \frac{\partial V^R}{\partial E^R}$, and $\beta_R = \frac{\partial V^R}{\partial R}$.

The corresponding system of characteristics is then (again introducing the derivative in reverse time for any state variable x as $x' = \frac{\mathrm{d}x}{\mathrm{d}\tau} = -\frac{\mathrm{d}x}{\mathrm{d}t}$)

$$\big(E^P\big)' = \frac{1}{250}\big(1-u^P\big)\,E^P - d\,u^P\,u^R\,R + d\,u^P\,E^P, \tag{10.24}$$

$$\big(E^R\big)' = \frac{1}{250}(1-u^R)E^R - d\,u^R + d\,u^R\,E^R, \tag{10.25}$$

$$P' = \frac{1}{20}\,P - \frac{1}{5}\,b\,u^P\,E^P\,P, \tag{10.26}$$

$$R' = \frac{1}{20}\,R - \frac{1}{5}u^R E^R\,R + \frac{1}{5}u^P u^R P\,R, \tag{10.27}$$

$$\beta'_{E^P} = -\beta_{E^P}\left(\frac{1}{250}(1-u^P) + d\,u^P\right) + \frac{1}{5}\,b\,u^P \beta_P\,P, \tag{10.28}$$

$$\beta'_{E^R} = \beta_{E^R}\left(-\frac{1}{250}(1-u^R) - d\,u^R\right) + \frac{1}{5}\beta_R\,u^R\,R + \big(1-u^R\big)\;R, \tag{10.29}$$

$$\beta'_P = \beta_P\left(-\frac{1}{20} + \frac{1}{5}\beta\,u^P\,E^P\right) - \frac{1}{5}u^P u^R\,\beta_R R, \tag{10.30}$$

$$\beta'_R = \beta_R\left(-\frac{1}{20} + \frac{1}{5}u^R E^R - \frac{1}{5}u^P u^R P\right) + d\,u^P\,u^R\,\beta_{E^P} + \big(1-u^R\big)\;E^R, \tag{10.31}$$

with transversality conditions $\beta_{E^P}(0) = 0$, $\beta_{E^R}(0) = 0$, $\beta_P(0) = 0$, $\beta_R(0) = 0$, and assuming $E^R, E^P \in [0,1]$, $E^P(0) > 0$, $E^R(0) > 0$, $P(0) > 0$, $R(0) > 0$.

The singular surface corresponding to the HJB equation (10.23) is

$$\mathscr{A}_R = d\,u^P\;R + \beta_{E^R}\left(\frac{1}{250}E^R + d - d\;E^R\right) + \beta_R\left(\frac{1}{5}E^R\,R - \frac{1}{5}u^P\;P\;R\right) - E^R\;R. \tag{10.32}$$

Similarly as before, the optimal $u^R = \text{Heav}\,\mathscr{A}_R$, i.e.,

$$u^R = \begin{cases} 0, & \text{if} \quad \mathscr{A}_R < 0, \\ 1, & \text{if} \quad \mathscr{A}_R > 0, \end{cases}$$

and $u^R \in (0, 1)$ if $\mathscr{A}_R = 0$. The value of u^R for $\tau = 0$ is equal to 0 as $\mathscr{A}_R(0) = -E^R(0)\,R(0) < 0$, i.e., regardless of the strategy of the predator the prey is in diapause at the end of the season. Moreover, note that (10.32) is independent of E^P and of β_{E^P}.

10.3.3 Optimal Strategy for the Prey

In the following analysis, we confine ourselves to a specific structure for the strategy of the prey, which turned out to be optimal if the predatory mites are active the entire season (Staňková et al. 2013). More precisely, we assume that the optimal action of the prey is as shown in Fig. 10.3:

$$u = \begin{cases} 1 & \text{if} \quad t \in [0, t_1], \\ \frac{t - t_2}{t_1 - t_2} & \text{if} \quad t \in [t_1, t_2], \\ 0 & \text{if} \quad t \in [t_2, t_T]. \end{cases} \tag{10.33}$$

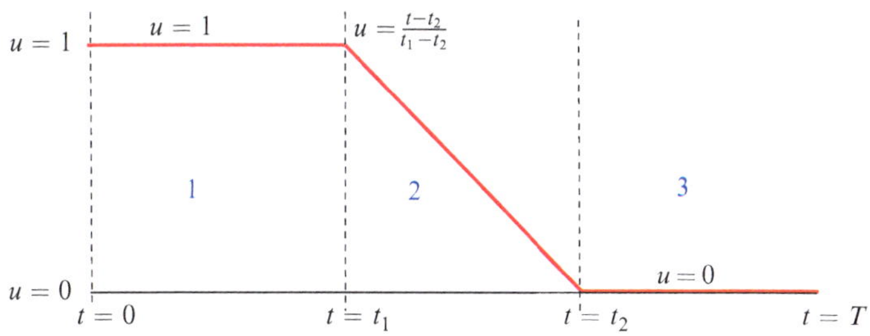

Fig. 10.3 Predicted shape for the optimal strategy of the prey mites

Then the optimization problem of the prey can be written as the solution of

$$(t_1^*, t_2^*) = \arg\sup_{t_1, t_2} \int_0^T (1 - u^R(t)) E^R(t)\, R(t) \mathrm{d}\,t. \tag{10.34}$$

subject to (10.3)–(10.6). The dynamics of the model can then be distinguished as that for $t \in [0, t_1)$ (Phase 1), for $t \in [t_1, t_2)$ (Phase 2), and for $t \in [t_2, T]$ (Phase 3), as described in the following.

Phase 1 Notice that for $t \in [0, t_1]$ that $J^R = 0$ and (10.3)–(10.6) can be rewritten as:

$$\frac{\mathrm{d}E^P}{\mathrm{d}t} = -\frac{1}{250}(1-u^P)E^P + d\,u^P\,R - d\,u^P\,E^P, \tag{10.35}$$

$$\frac{\mathrm{d}E^R}{\mathrm{d}t} = d\ - d\ E^R, \tag{10.36}$$

$$\frac{\mathrm{d}P}{\mathrm{d}t} = -\frac{1}{20}\,P + \frac{1}{5}\beta\,u^P\,E^P\,P, \tag{10.37}$$

$$\frac{\mathrm{d}R}{\mathrm{d}t} = -\frac{1}{20}R + \frac{1}{5}\,E^R\,R - \frac{1}{5}u^P\,P\,R. \tag{10.38}$$

Phase 2 Notice that for $t \in [t_1, t_2]$ we can see that $J^R = \int_{t_1}^{t_2} \frac{t_1 - t}{t_1 - t_2} E^R(t)\,r(t)\mathrm{d}t$ and (10.3)–(10.6) can be rewritten as:

$$\frac{\mathrm{d}E^P}{\mathrm{d}t} = -\frac{1}{250}(1-u^P)E^P + d\,u^P\,\frac{t-t_2}{t_1-t_2}\,R - d\,u^P\,E^P, \tag{10.39}$$

$$\frac{\mathrm{d}E^R}{\mathrm{d}t} = -\frac{1}{250}\frac{t_1-t}{t_1-t_2}E^R + d\,\frac{t-t_2}{t_1-t_2} - d\,\frac{t-t_2}{t_1-t_2}\,E^R, \tag{10.40}$$

$$\frac{\mathrm{d}P}{\mathrm{d}t} = -\frac{1}{20}\,P + \frac{1}{5}b\,u^P\,E^P\,P, \tag{10.41}$$

$$\frac{\mathrm{d}R}{\mathrm{d}t} = -\frac{1}{20}R + \frac{t-t_2}{5\,(t_1-t_2)}E^R\,R - \frac{1}{5}u^P\,\frac{t-t_2}{t_1-t_2}P\,R. \tag{10.42}$$

Phase 3 For $t \in [t_2, T]$ that $J^R = \int_{t_2}^{T} E^R(t)\,r(t)\mathrm{d}t$ and (10.3)–(10.6) can be rewritten as:

$$\frac{\mathrm{d}E^P}{\mathrm{d}t} = -\frac{1}{250}(1-u^P)E^P - d\,u^P\,E^P, \tag{10.43}$$

$$\frac{\mathrm{d}E^R}{\mathrm{d}t} = -\frac{1}{250}E^R, \tag{10.44}$$

$$\frac{\mathrm{d}P}{\mathrm{d}t} = -\frac{1}{20}\,P + \frac{1}{5}\,b\,u^P\,E^P\,P, \tag{10.45}$$

$$\frac{\mathrm{d}R}{\mathrm{d}t} = -\frac{1}{20}R. \tag{10.46}$$

With reasoning in reverse time, $\tau_1 = T - t_2$ and $\tau_2 = T - t_1$.

10.3.4 Optimal Strategy for the Predator

Since $u^P(0) = 1$ and $u^R(0) = 0$, (10.13)–(10.20) translates into

$$(E^P)' = d\,E^P, \tag{10.47}$$

$$
\begin{aligned}
(E^R)' &= \frac{1}{250} E^R, && (10.48)\\
P' &= \frac{1}{20} P - \frac{1}{5} b\, E^P\, P, && (10.49)\\
R' &= \frac{1}{20} R, && (10.50)\\
\alpha'_{E^P} &= -d\, \alpha_{E^P} + \frac{1}{5} b\, \alpha_P\, P + \frac{1}{5} b\, P, && (10.51)\\
\alpha'_{E^R} &= -\frac{1}{250} \alpha_{E^R}, && (10.52)\\
\alpha'_P &= \alpha_P \big(-\frac{1}{20} + \frac{1}{5} b\, E^P \big) + \frac{1}{5} b\, E^P - \frac{1}{20}, && (10.53)\\
\alpha'_R &= -\frac{1}{20} \alpha_R. && (10.54)
\end{aligned}
$$

The solution of (10.47)–(10.54) can be computed explicitly as follows:

$$
\begin{aligned}
E^P(\tau) &= E^P(0)\, \mathrm{e}^{d\tau}, && (10.55)\\
E^R(\tau) &= E^R(0)\, \mathrm{e}^{\frac{\tau}{250}}, && (10.56)\\
P(\tau) &= P(0)\, \mathrm{e}^{\frac{bE^P(0) bE^P(0)\, \mathrm{e}^{d\tau}}{5d} + \frac{\tau}{20}}, && (10.57)\\
R(\tau) &= R(0)\, \mathrm{e}^{\frac{\tau}{20}}, && (10.58)\\
\alpha_{E^P}(\tau) &= \big(\frac{bP(0)\, \mathrm{e}^{d\tau}}{5d} - \frac{bP(0)}{5d} \big) \mathrm{e}^{-d\tau}, && (10.59)\\
\alpha_{E^R}(\tau) &= 0, && (10.60)\\
\alpha_P(\tau) &= \big(-\mathrm{e}^{\frac{\tau}{20} - \frac{bE^P(0)\, \mathrm{e}^{d\tau}}{5d}} + \mathrm{e}^{-\frac{bE^P(0)}{5d}} \big) \mathrm{e}^{-\frac{\tau}{20} + \frac{bE^P(0)\, \mathrm{e}^{d\tau}}{5d}}, && (10.61)\\
\alpha_R(\tau) &= 0. && (10.62)
\end{aligned}
$$

Substituting (10.55)–(10.62), $u^P = 1$, and $u^R = 0$ into (10.21) yields

$$
\mathscr{A}_P = \frac{bP(0) \big(E^P(0)\, \mathrm{e}^{d\tau} - E^P(0) + 250\, d E^P(0) \big)}{1250\, d}. \tag{10.63}
$$

Note that this expression is *always positive* (because $d > 1/250$). In other words, in reverse time, the predator is initially active and remains active until all prey are in diapause.

If u^R changes from 0 to different values, the expression for the singular surface (10.21) changes. While the system of characteristics (10.13)–(10.20) with $u^P = 1$ and $u^R \in (0, 1]$ cannot be solved explicitly, we can observe (Sect. 10.3.3) that if $u^R \in (0, 1)$, then $(u^R)' = -\frac{1}{t_1 - t_2} = \frac{1}{\tau_2 - \tau_1} = \Delta$ and $(u^R)'' = 0$. If $u^P \in (0, 1)$,

Fig. 10.4 The optimal strategy for the predator is to stay active during the entire summer season

conditions $\mathscr{A}_P = 0$, $\mathscr{A}'_P = \{\mathscr{A}_P, h\} = 0$, $\mathscr{A}''_P = \{\mathscr{A}'_P, h\} = 0$, where $\{\cdot,\cdot\}$ denotes Jacobi brackets (Melikyan 1998) and h is the expression supremized in (10.12), have to be satisfied. Solving this system of three equations, with $(u^R)' = \Delta$, $(u^R)'' = 0$, and subject to (10.13)–(10.20), leads only to the degenerate solution $\alpha_{E^P} = 0$, $\alpha_P = -1$, $\alpha_R = 0$. This degenerate solution, which can be easily derived directly from (10.21), cannot be achieved when emitting characteristics (10.13)–(10.20) from their initial values. Moreover, the same degenerate solution will be found if we replace u^R in the equations $\mathscr{A}_P = 0$, $\mathscr{A}'_P = \{\mathscr{A}_P, h\} = 0$, $\mathscr{A}''_P = \{\mathscr{A}'_P, h\} = 0$ by 1. Therefore, we can conclude that the predator will not change strategy from $u^P = 1$ and will stay active the entireseason (Fig. 10.4).

Remark 10.1. In Staňková et al. (2013) a three-dimensional model, in which the energy of the predator was not included, was used to show that the optimal behavior of the prey is the one shown in Sect. 10.3.3. The underlying assumption was that the predator stays active the entire season. As this strategy turned out to be the optimal strategy of the predator in the model proposed in this article, we could use the argumentation from Staňková et al. (2013) to confirm our hypothesis regarding the structure of $u^{R,*}$. Technically, the proofs will be the same if we assume that $d \gg 0$, while for d close to $\frac{1}{250}$ the underlying analysis becomes much more complex.

One can see that as the optimal strategy of the predator can be decoupled from the optimal strategy of the prey, it does not matter whether the problem is defined as a Stackelberg game or as a Nash game because the result of these two games will coincide.

Remark 10.2. In fact, the outcome that $u^P = 1$ is optimal will hold also when u has a more general shape than the one depicted in Fig. 10.3. Namely, if $u^R = 1$ for some $t \in [0, t_1]$, nonincreasing for $t \in (t_1, t_2]$ (not necessarily linearly decreasing), and $u^R = 0$ for $t \in (t_2, T]$, the outcome for the predator will stay the same.

10.4 Discussion

In this article we have searched optimal active/diapause ratios for the predatory mites and the fruit-tree red spider mites when there are no extra energetic costs to go in or out of diapause and when their decision depends on both densities and energy levels of either species. The optimal strategy for the prey mites coincides with the results of our previous work (Staňková et al. 2013): Even if the prey mites do not encounter costs to enter diapause, their optimal strategy is to go into diapause only once per season. This implies that, once entered, the diapause is irreversible. In this article we have shown that the best response of the predatory mites to this strategy of the prey mites should be to stay active for the entire season, again assuming no energetic costs for entering or leaving the diapause state.

The outcome of our analysis regarding the prey mites is remarkably close to the empirical observations: in reality the fruit-tree red spider mites have an irreversible diapause. Additionally, the prey mites also enter a so-called deep diapause. Once the prey mites are in the deep diapause, it is not easy to bring them to a non-diapause state (e.g., they require a cold period of a certain length before they can come out of diapause). It is possible that this deep diapause evolved after the selection for an irreversible diapause predicted by our model (assuming at least initially a very flexible decision without costs for entering or leaving the diapause state). Once this choice evolved, there were probably other reasons why it was selectively advantageous to evolve a deep diapause (such reasons may be to invest more in anti-freeze chemistry at the expense of energy for other purposes such as reproduction). The deep diapause allows the spider mites to survive the winter better than, for example, predatory mites that exhibit a very flexible diapause state (crude estimates of winter survival for the prey mites are in the order of 50% whereas for predatory mites they are in the order of 5%).

The outcome of our analysis regarding the predatory mites is also rather close to real observations: while in our model the predatory mites stay active the entire season, in reality the predatory mites might enter diapause at the very end of the summer season (actually in autumn, which is part of the winter season in our model), i.e., when there is no prey. Moreover, the predatory mites have a very flexible diapause. Collecting predatory mites in the winter and bringing them to the lab to offer them prey virtually always results in the predatory mites resuming feeding within two days and reproducing within four days. This represents a great flexibility when compared to the fruit-tree red spider mites (it may take one or two months for the prey mites to become active again depending on the cold period they already experienced). This "light diapause" of the predatory mites may have as a consequence that they survive the winter less well (less than 5% of them survives) than the spider mites.

Under natural conditions the predatory mites usually keep the spider mites at very low levels, meaning that they may experience prey shortage in some periods (and possibly a motivation to enter diapause in summer). Under those conditions it is not easy to find the predatory mites on leaves as well as elsewhere on the plant. Hence, any predatory mite entering diapause will be difficult to find too. It is expected that the predatory mites respond to low prey density by entering diapause, but become active again as soon as there is prey available.

Under agricultural conditions, however, predatory mites may suffer from pesticide use (against spider mites or against pests other than spider mites) and there is much evidence that this allows the spider mites to increase in numbers and reach the status of a pest. Under those conditions, spider mites may suffer severe food competition and then they may also respond to plant food shortage by going into diapause.

Predatory mites rarely enter diapause before the end of the season under agricultural conditions and if they do they have a flexible diapause that allows them to enter and leave the diapause state, e.g. depending on temperature and prey availability. Such flexible strategies do not emerge as a solution from the model above, but they may arise as optimal strategies in different models.

Let us consider another game, in which predator and prey choose $u^{P,*}$ and $u^{R,*}$, respectively, such that

$$u^{P,*}(\cdot) = \underset{u^P(\cdot)\in[0,1]}{\arg\sup} \int_0^T \left(u^P(-P + \gamma u^R\, P\, R)\right)\mathrm{d}t, \tag{10.64}$$

$$u^{R,*}(\cdot) = \underset{u^R(\cdot)\in[0,1]}{\arg\sup} \int_0^T (1 - u^R(t))E^R(t)\, R(t)\mathrm{d}t, \tag{10.65}$$

while

$$\frac{\mathrm{d}E^R}{\mathrm{d}t} = -(1 - u^R)E^R + d\,(1 - E^R)\,u^R, \tag{10.66}$$

$$\frac{\mathrm{d}P}{\mathrm{d}t} = u^P(-P + \gamma u^R\, P\, R), \tag{10.67}$$

$$\frac{\mathrm{d}R}{\mathrm{d}t} = E^R\, R - b\,u^P\, P\, R, \tag{10.68}$$

with $\gamma \in (0, 1)$. Adopting the HJB approach again, we can show that while the optimal strategy of the prey does not change, the predator will end up in diapause unlike what was predicted by the model discussed in this article. Moreover, this new model is much simpler to solve as it is only three-dimensional and the characteristic system for both predator and prey can be solved explicitly if they adopt bang-bang actions. The comparison of different models, those including energy levels and those excluding them, is a subject of our ongoing research.

10.5 Conclusions and Future Work

In this article, a dynamical model of the predator–prey interactions between predatory mites and fruit tree red spider mites during summer has been described and analyzed. This model includes not only the dynamics of predator and prey populations but also the dynamics of their energy levels and energy decision controls for both predator and prey. We have considered the case where both predator and prey can enter diapause. We have shown that it is optimal for the predator to stay active the entire season, while the prey stay active in the beginning of the season, later enter diapause, and stay in diapause until the end of the season.

While the correspondence between theoretical predictions and empirical observations on mites is encouraging, there are also limitations (mostly analytical) that should spawn new work. Moreover, it is still to be shown that optimal summer behavior of the predator and prey populations, as derived in this study, is resistant against invasion by mutant strategies and robust against structural modifications, such as the inclusion of predator decisions to enter diapause or not. Ultimately, we hope to explain winter dynamics of predatory mites and fruit-tree red spider mites based on optimal timing of diapause induction in summer. The use of bifurcation analysis can help determining for which parameter domains the proposed optimal strategies are evolutionarily stable.

Different models of the predator–prey interactions will lead to different optimal strategies of the predator and prey. Analysis and comparison of such different models is a subject of our future research.

References

Başar T, Olsder GJ (1999) Dynamic noncooperative game theory. SIAM, Philadelphia, Pennsylvania

Cressman R (2003) Evolutionary dynamics and extensive form games. MIT Press, Cambridge

Danks H (1987) Insect dormancy: An ecological perspective. Biological Survey of Canada, Ottawa

Fitzgerald J, Solomon M (1991) Diapause induction and duration in the phytoseiid mite *Typhlodromus pyri*. Exp Appl Acarol 12:135–145

Helle W, Sabelis MW (1985a) Spider mites: their biology, natural enemies and control. World crop pests, vol 1A. Elsevier, Amsterdam

Helle W, Sabelis MW (1985b) Spider mites: their biology, natural enemies and control. World crop pests, vol 1B. Elsevier, Amsterdam, pp. 77–99

Kroon A, Veenendaal RL, Bruin J, Egas M, Sabelis MW (2004) Predation risk affects diapause induction in the spider mite *Tertanychus urticae*. Exp Appl Acarol 34:307–314

Kroon A, Veenendaal RL, Egas M, Bruin J, Sabelis MW (2005) Diapause incidence in the two-spotted spider mite increases due to predator presence, not due to selective predation. Exp Appl Acarol 35:73–81

Kroon A, Veenendaal RL, Bruin J, Egas M, Sabelis MW (2008) "Sleeping with the enemy"-predator-induced diapause in a mite. Naturwissenschaften 95:1195–1198

Mailleret L, Lemesle V (2009) A note on semi-discrete modelling in the life sciences. Philos Trans R Soc A 367(1908):4779–4799

Melikyan AA (1994) Necessary optimality conditions for a singular surface in the form of synthesis. J Optim Theory Appl 82(2):203–217

Melikyan AA (1998) Generalized characteristics of first order PDEs: Applications in optimal control and differential games. Birkhäuser, Boston

Melikyan AA, Olsder GJ (2010) Boundary singularities and characteristics of Hamilton-Jacobi equation. J Dyn Control Syst 16(1): 77–99

Patchepsky E, Nisbet RM, Murdoch WW (2008) Between discrete and continuous: Consumer-resource dynamics with synchronized reproduction. Ecology 89(1):280–288

Sabelis MW (1991) Life-history evolution of spider mites. In: Schuster R, Murphy PW (eds) The Acari: Reproduction, development and life-history strategies. Chapman & Hall, London, pp 23–50

Sabelis MW, Janssen A (1993) Evolution of life-history patterns in the phytoseiidae. In: Houck MA (ed) Mites: Ecological and evolutionary analyses of life-history patterns. Chapman & Hall, New York, pp 70–98

Staňková K (2009) On Stackelberg and inverse Stackelberg games & their applications in the optimal toll design problem, the energy market liberalization problem, and in the theory of incentives. PhD thesis, Delft University of Technology, Delft, the Netherlands

Staňková K, Abate A, Sabelis MW (2013) Irreversible prey diapause as an optimal strategy of a physiologically extended Lotka-Volterra model. J Math Biol 66(4):767–794

Tauber M, Tauber C, Masaki S (1986) Seasonal adaptations of insects. Oxford University Press, New York

Veerman A (1992) Diapause in phytoseiid mites: a review. Exp Appl Acarol 14:1–60

Weibull J (1995) Evolutionary game theory. MIT Press, MA

Chapter 11
On a Game-Theoretic Model of Environmental Pollution Problem

Marianna Troeva and Vassili Lukin

Abstract We consider a dynamic conflict model of environmental pollution in which n enterprises contaminate a water reservoir by dumping a pollutant in the production process. This process is formalized by the n-person differential game with separated dynamics and continuous payoff functions. The existence of the ε-Nash equilibrium for the class of the piecewise-programmed strategies is proved in this game. Some numerical examples are presented.

Keywords Differential games • Value function • Equilibrium point • Environmental pollution • Finite-difference scheme • Dynamic programming

11.1 Introduction

A game-theoretic approach plays a significant role in studying dynamic conflict models of environmental pollution. The dynamic games that deal with these problems were investigated in many works, see, e.g., Petrosyan and Zaharov (1997), Jørgensen and Zaccour (2001), Breton et al. (2005), Petrosyan and Zaccour (2003), Zenkevich and Kozlovskaya (2010), Mazalov and Rettieva (2010), Rettieva (2011), Krasovskiy and Tarasyev (2011), Martín-Herrán et al. (2006), Botkin et al. (2011), and references therein.

The differential games modeling joint implementation in international environmental agreements were studied in Breton et al. (2005), Petrosyan and Zaccour

M. Troeva (✉)
North-Eastern Federal University, 58, Belinskogo str., Yakutsk 677000, Russia
e-mail: troeva@mail.ru

V. Lukin
Scientific Research Institute of Mathematics, NEFU, 48, Kulakovskogo str., Yakutsk 677000, Russia
e-mail: lukinvasiliy@hotmail.com

V. Křivan and G. Zaccour (eds.), *Advances in Dynamic Games*, Annals of the International Society of Dynamic Games 13, DOI 10.1007/978-3-319-02690-9_11,

(2003). A model of territorial environmental production was studied in Zenkevich and Kozlovskaya (2010). Discrete dynamic models of bioresource management (fish catching) were developed in Mazalov and Rettieva (2010), Rettieva (2011). The noncooperative dynamic game of emission reduction trading was studied in Krasovskiy and Tarasyev (2011). In Martín-Herrán et al. (2006) the differential game dealing with deforestation was investigated.

In this paper a dynamic conflict model of an environmental pollution problem is considered. Enterprises (agents) contaminate a water reservoir by dumping a pollutant (harmful substance) of the same type in the production process.

This environmental pollution problem is formalized by the noncooperative n-person differential game. The dynamics of each agent is described by the initial boundary value problem for the parabolic equation involving Dirac measure.

The parabolic equations appear in game-theoretical models of innovation, information and technology diffusion, and diffusion processes of ecological systems (see, e.g., Cappoza 2003; Petrosyan and Zaharov 1997; Rogers 2003).

Noncooperative differential games for the different class strategies were investigated in many works, e.g., Petrosyan (1993), Kleimenov (1993), Başar and Olsder (1999), Kolokoltsov and Malafeyev (2010), Malafeyev (2000). The n-person differential game in the finite-dimensional Euclidean space was investigated in Malafeyev (2000). The existence of the ε-equilibrium for the class of the piecewise-programmed strategies was proved in this game using a similar approach such as the one developed in Varaiya and Lin (1969) for zero-sum differential games.

In Troeva (2012) the results of Malafeyev (2000) were generalized to the n-person differential game in Banach space. The dynamics of each player is described by the initial value problem for the parabolic operator-differential equation involving Dirac measure. The existence of the unique solution of this initial value problem follows from the results of Amann (2005). In Amann (2005) the existence of a unique solution of abstract parabolic evolution equations involving Banach space-valued Radon measures was proved.

The main purpose of this paper is to prove the existence of ε-Nash equilibrium in our model, and we did this using the results of Troeva (2012). Some numerical simulations are provided.

11.2 The Model

A closed water reservoir (f.e. lake) is considered. The n enterprises dump a pollutant (harmful substance) of the same type into this water reservoir in the process of production.

Furthermore it is assumed that reservoir has a water intake. The level of pollution at the water intake point (the total concentration of the pollutant released by all enterprises) must not exceed the maximum permissible value. It is assumed that if this value is exceeded, all the enterprises will pay a fine (penalty) as a percentage of their income.

It is also assumed that the company has certain expenses associated with the cleaning of the pollutant.

The spread of the harmful substances in the reservoir occurs by diffusion. Besides, a pollutant is decomposed with the rate $r > 0$.

The total income of an enterprise depends on its volume of production, which is tightly linked with its total volume of dumped polutant. Besides, the total income depends on the overall cleaning expenses and possible pollution fines.

The aim of each enterprise is to maximize the total income for finite period of time.

11.3 Differential Game

The above-mentioned problem is formalized by n-person differential game $\Gamma(c_0, T)$ with a prescribed duration $T < \infty$ and an initial position of the game $c_0 = (c_0^1, \ldots, c_0^n)$. Let $I = \{i\} = \{1, \ldots, n\}$ be a set of the agents (enterprises).

Let us denote by $u_i(t)$ the intensity of dumping the pollutant of the agent $i \in I$ at the moment t. Let us assume that the intensity of dumping the pollutant satisfies the following conditions:

$$0 \leq u_i(t) \leq G_i(t), \quad i \in I, \tag{11.1}$$

at any moment $t \in [0, T]$. Here $G_i(t) > 0$ is a given square integrable function which describes the maximal intensity of dumping the pollutant of the agent i at the moment t. Let us assume that the costs of treatment per unit volume of the pollutant of the enterprise i are constant and equal to $M_i > 0$, $i \in I$.

Let us denote by $z^i(x, y, t)$ the pollutant concentration of the agent i at the point $(x, y) \in \mathbf{R}^2$ at the moment t.

Let us consider the closed water reservoir as a bounded two-dimensional domain $\Omega \in \mathrm{R}^2$ with the boundary $S \in C^2$, $\overline{\Omega} = \Omega \cup S$, $t \in [0, T]$.

Let us denote by (x_w, y_w) the coordinates of the water intake location inside the domain Ω.

The dynamics of the agent $i \in I$ in the game $\Gamma(c_0, T)$ is described by the initial boundary value problem for the following differential equation:

$$\frac{\partial z^i}{\partial t} = \frac{\partial}{\partial x}\left(D(x, y, t)\frac{\partial z^i}{\partial x}\right) + \frac{\partial}{\partial y}\left(D(x, y, t)\frac{\partial z^i}{\partial y}\right) - \\ -rz^i + u_i \psi_i(x, y), \quad (x, y) \in \Omega, \ t > 0. \tag{11.2}$$

Here $D(x, y, t) > 0$ is the diffusion coefficient; $r > 0$ is the coefficient characterizing the pollution decomposition; $u_i \in U_i$ is a control parameter of the agent i, $U_i \subset R^{m_i}$ is a compact set in Euclidean space. The function $\psi_i(x, y) = \delta(x - x_i, y - y_i)$ gives the location of the agent i inside the domain Ω.

Let the function $z^i(x, y, t)$ satisfies the following boundary and initial conditions:

$$D(t, x, y)\frac{\partial z^i}{\partial m} = 0, \qquad (x, y) \in S, \quad t \in [0, T], \tag{11.3}$$

$$z^i(x, y, 0) = c_0^i(x, y), \quad (x, y) \in \Omega, \quad t = 0. \tag{11.4}$$

where m is an outward normal to the boundary surface $S \times [0, T]$, $c_0^i(x, y)$ is some given function describing the initial distribution of the pollutant concentration of the agent i in the water reservoir at the initial moment $t = 0$.

Definition 11.1. A measurable function $u_i = u_i(t)$, satisfying the condition (11.1) for all $t \in [0, T]$ is called the admissible control of the agent $i \in I$. Let us denote by $\overline{\mathscr{U}}_i \subset L_p(0, T)$, $i \in I$ the set of admissible controls (measurable functions) $u_i(t)$, $t \in [0, T]$.

Let the function $f_i(t) \in [0, 1]$ be the percentage of the income of the agent i, which determines the amount of the penalty for exceeding the maximum permissible value of pollution at the water intake point (x_w, y_w). The function $f_i(t)$ is defined as follows:

$$f_i(t) = \begin{cases} 0, & \text{if } \sum\limits_{j=1}^{n} z^j(x_w, y_w, t) \le C_w, \\ \dfrac{z_i(x_w, y_w, t)}{\sum\limits_{j=1}^{n} z^j(x_w, y_w, t)} \cdot \dfrac{\sum\limits_{j=1}^{n} z^j(x_w, y_w, t) - C_w}{C_w}, & \text{if } \sum\limits_{j=1}^{n} z^j(x_w, y_w, t) > C_w, \end{cases}$$

where C_w is the maximum permissible value of pollution at the water intake point.

Then the payoff of the agent i at time T is defined by the following functional:

$$H_i(z, u_i) = \int\limits_0^T h(u_i(\tau))(1 - pf_i(\tau))d\tau - \int\limits_0^T M_i u_i(\tau)d\tau. \tag{11.5}$$

where $z = (z^1, z^2, \ldots, z^n)$.

Let us represent the problem (11.2)–(11.4) as the initial-value problem for the following operator-differential equation

$$\frac{dc^i(t)}{dt} - A(t)c^i(t) = v^i(t), \ t \in [0, T] \tag{11.6}$$

$$c^i(0) = z_0^i = c_0^i, \tag{11.7}$$

where $c^i(t) = z^i(x, y, t)$, $v^i(t) = u_i(t) \cdot \delta(x - x_i, y - y_i)$.

The operator $Ac = \partial_x(D(t,x,y)c_x) + \partial_y(D(t,x,y)c_y) - rc$ allows for the boundary condition (11.3).

Equation (11.6) involves the Dirac measure. The existence of a unique solution of abstract parabolic evolution equations involving Banach space-valued Radon measures is proved in Amann (2005).

We assume that the coefficients D and r in (11.2)–(11.4) satisfy the following conditions $D(t,x,y) \in C([0,T];C^1(\overline{\Omega}))$, $r \in L_\infty(0,T;L_q(\Omega))$. According to results of Amann (2005), the unique solution $c^i \in L_p(0,T;W_p^1(\Omega))$, $c_t^i \in L_p(0,T;(W_q^1(\Omega))')$, $i \in I$ of the problem (11.6)–(11.7) exists for all $\nu^i \in L_p(0,T;(W_q^1(\Omega))')$, for all admissible control $u_i \in \overline{\mathscr{U}}_i \subset L_p(0,T)$, and for all initial condition $c_0^i \in (W_q^{2/p-1}(\Omega))'$ and $q > 2$ $(1/p + 1/q = 1)$.

Let us denote by $F_i(c_0^i,\, t_0,\, t)$ the set of the points $c^i(\cdot) \in W_p^1(\Omega)$ for which there exists an admissible control $u_i(t)$ such that the game goes from the state $c^i(t_0) = c_0^i$ to the state $c^i(t+t_0)$ for the time interval $[t_0,t]$. The set $F_i(c_0^i,\, t_0,\, t)$ is a bounded set of the space $W_p^1(\Omega)$. It is known (Ladyzhenskaya 1985) that if the boundary of the domain Ω is smooth, then a bounded set of the space $W_p^1(\Omega)$ is a compact set in $L_p(\Omega)$. This implies that $F_i(c_0^i,\, t_0,\, t)$ is a compact set for all $c_0^i \in (W_q^{2/p-1}(\Omega))'$, $t_0,\, t \in [0,T]$ as well. $F_i(c_0^i,\, t_0,\, t_0) = c_0^i$ for all $c_0^i \in (W_q^{2/p-1}(\Omega))'$, $t_0 \in [0,T]$. The set $F_i(c_0^i,\, t_0,\, t)$ has semigroup property.

One can show that the function $F_i(c_0^i,\, t_0,\, t)$ is continuous in the Hausdorff metric. So, the set $F_i(c_0^i,\, t_0,\, t)$ satisfies all axioms which define generalized dynamic systems. The set $F_i(c_0^i,\, t_0,\, t)$ is called the attainability set of the player i, $i = \overline{1,n}$ from the initial state c_0^i on the time interval $[t_0,t]$.

Let us denote by $\widehat{F}_i(c_0^i,\, t_0,\, t)$, $i \in I$ the set of trajectories $\hat{c}^i(c_0^i, t - t_0)$ of (11.6)–(11.7) which start at c_0^i at the moment t_0 and which are defined on the time interval $[t_0,t]$. The set of trajectories $\widehat{F}_i(c_0^i,\, t_0,\, t)$ is compact, e.g., in the space $L_p(0,T;W_p^{1-s}(\Omega))$ for any $s > 0$, and the function $\widehat{F}_i(c_0^i,\, t_0,\, t)$ is continuous in the corresponding Hausdorff metric.

At every moment $t \in [0,T]$ of the game $\Gamma(c_0,T)$ the agents know the realized trajectory of the game, the dynamics, and the duration T of the game.

Let $\hat{c}^i(\cdot) \in \hat{F}_i(c_0^i, 0,\, T)$ be the trajectory of (11.6)–(11.7) arising from a control u_i and $\Pi_\delta^i(\hat{c}^i)$ be the trajectory arising from the same control u_i delayed by δT. The following lemma describes the relation between these trajectories.

Lemma 11.1. *For each $\delta \in (0,1]$ there exists a map $\Pi_\delta^i : \hat{F}_i(c_0^i, 0,\, T) \to \hat{F}_i(\cdot)$ such that, if $\hat{c}^i(\tau) = \hat{c}'^i(\tau)$ for $\tau \in [0,t]$, then $\Pi_\delta^i(\hat{c}^i)(\tau) = \Pi_\delta^i(\hat{c}'^i)(\tau)$ for $\tau \in [0, t+\delta T]$ if $(t+\delta T) \le T$ and $\Pi_\delta^i(\hat{c}^i)(\tau) = \Pi_\delta^i(\hat{c}'^i)(\tau)$ for $\tau \in [0,T]$ if $(t+\delta T) > T$. Moreover,*

$$\varepsilon^i(\delta) = \sup_{\hat{c}^i \in \hat{F}_i(\cdot)} \|\hat{c}^i - \Pi_\delta^i(\hat{c}^i)\| \underset{\delta \to 0}{\longrightarrow} 0.$$

Let us fix the permutation $p = (i_1, \dots, i_k, \dots, i_n)$ and consider n-person multistep game $\Gamma_p^\delta(c_0,T)$ at every step which the agents $i_1,\dots,i_n$ choose in sequence controls $u^{i_1}, \dots, u^{i_k}, \dots, u^{i_n}$.

Definition 11.2. The strategy

$$^{\delta}\varphi^{p}_{i_k} : \hat{F}^{*}_{i_k}(\cdot) = \prod_{j \neq i_k} \hat{F}_j(\cdot) \to \hat{F}_{i_k}(\cdot),$$

of the agent i_k in the game $\Gamma^{\delta}_{p}(c_0, T)$ is a mapping such that if $\hat{c}^{j}(\tau) = \hat{c}'^{j}(\tau)$ for $j < i_k$, $\tau \in [0, l\delta T]$ and if $\hat{c}^{j}(\tau) = \hat{c}'^{j}(\tau)$ for $j > i_k$, $\tau \in [0, (l-1)\delta T]$, then $^{\delta}\varphi^{p}_{i_k}(\hat{c}^{*i_k}(\tau)) = {}^{\delta}\varphi^{p}_{i_k}(\hat{c}^{*'i_k}(\tau))$, $\tau \in [0, l\delta T]$. Here $\delta = 1/2^N$, $l = 1, 2, \ldots, 2^N$.

Let us denote by $^{\delta}\Phi^{p}_{i_k}$ the set of the strategies of the agent i_k in the game $\Gamma^{\delta}_{p}(c_0, T)$.

In the game $\Gamma^{\delta}_{p}(c_0, T)$ the players $i_1, \ldots, i_n$ choose in sequence the strategies $^{\delta}\varphi^{p}_{i_1}, \ldots, {}^{\delta}\varphi^{p}_{i_n}$. The trajectory $\chi(^{\delta}\varphi^{p})$ is uniquely defined for every n-tuple $^{\delta}\varphi^{p} = (^{\delta}\varphi^{p}_{i_1}, \ldots, {}^{\delta}\varphi^{p}_{i_n})$ stepwise on successive intervals $[0, \delta T], \ldots, [T - \delta T, T]$. The payoff function of the agent $i \in I$ in the game $\Gamma^{\delta}_{p}(c_0, T)$ is defined as follows:

$$H^{\delta}_{i}(c_0, {}^{\delta}\varphi^{p}) = H_i(\chi^{\delta}(^{\delta}\varphi^{p})), \tag{11.8}$$

here $H_i(\cdot)$ is the functional (11.5).

So, the n-person differential game $\Gamma^{\delta}_{p}(c_0, T)$ with the prescribed duration T is defined in a normal form:

$$\Gamma^{\delta}_{p}(c_0, T) = \langle I, \{^{\delta}\Phi^{p}_{i}\}^{n}_{1}, \{H^{\delta}_{i}\}^{n}_{1}\rangle.$$

With a help of the Zermelo–Neumann theorem, the existence of ε-equilibrium for any $\varepsilon > 0$ in the multistep game $\Gamma^{\delta}_{p}(c_0, T)$ can be proved.

The previous Lemma 11.1 implies the following lemma.

Lemma 11.2. *If $i_k > i_1$, $^{\delta}\varphi^{p}_{i_k} \in {}^{\delta}\Phi^{p}_{i_k}$, then $\Pi^{i_k}_{\delta} \cdot {}^{\delta}\varphi^{p}_{i_k} \in {}^{\delta}\Phi^{p_{i_k}}_{i_k}$, where $p_{i_k} = (i_k, \tilde{p})$, $\tilde{p}$ is a permutation of the set $I \setminus i_k$; moreover, for $\hat{c}^{*i_k} \in \hat{F}^{*}_{i_k}(\cdot)$*

$$\|^{\delta}\varphi^{p}_{i_k}(\hat{c}^{*i_k}) - (\Pi^{i_k}_{\delta} \cdot {}^{\delta}\varphi^{p}_{i_k})(\hat{c}^{*i_k})\| \leq \varepsilon(\delta) \underset{\delta \to 0}{\longrightarrow} 0.$$

The following lemma is valid from Malafeyev (2000).

Lemma 11.3. *Let the game $\Gamma_{H'} = \langle I, \{X'_i\}^{n}_{1}, \{H'_i\}^{n}_{1}\rangle$ be obtained from the game $\Gamma_H = \langle I, \{X_i\}^{n}_{1}, \{H_i\}^{n}_{1}\rangle$ by the epimorphic mapping $\alpha_i : X_i \to X'_i$, $i = 1, \ldots, n$, with*

$$\|H(x) - H'(\alpha x)\| \leq \varepsilon, \qquad \alpha x = (\alpha_1(x_1), \ldots, \alpha_n(x_n)).$$

Then, if x is an ε-equilibrium of the game Γ_H, then αx is the 3ε-equilibrium of the game $\Gamma_{H'}$.

Let us define the main game $\Gamma(c_0, T)$.

Definition 11.3. The pair $(\delta_i, \{{}^{\delta}\varphi_i^{p_i}\}_{\delta=1/2^N})$ is called the strategy of the agent i. Here $N \in Z$, δ_i is a range of dyadic partition of the time interval $[0, T]$ and ${}^{\delta}\varphi_i^{p_i}$ is the strategy of the agent i in the game $\Gamma_{p_i}^{\delta}(c_0, T)$ for the permutation $p_i = (i, \tilde{p})$ and $\tilde{p}$ is the permutation of the set $I \setminus i$.

For n-tuple $\varphi = (\varphi_1, \dots, \varphi_n)$ the game $\Gamma(c_0, T)$ is played as follows. The smallest $\delta_i = \delta$ is chosen and the trajectory $\chi(\cdot)$ is constructed for n-tuple ${}^{\delta}\varphi = ({}^{\delta}\varphi_1^{p_1}, \dots, {}^{\delta}\varphi_n^{p_n})$. This trajectory is unique.

The game $\Gamma(c_0, T)$ is obtained from the game $\Gamma_p^{\delta}(c_0, T)$ by the epimorphic mapping which is defined in Lemma 11.2. Since in the game $\Gamma_p^{\delta}(c_0, T)$ there exists ε-equilibrium, then the existence of the 3ε-equilibrium in the game $\Gamma(c_0, T)$ follows from Lemmas 11.2 and 11.3.

Thus, the following theorem is valid.

Theorem 11.1. *There exists ε-equilibrium in the noncooperative n-person differential game $\Gamma(c_0, T)$ for all $\varepsilon > 0$.*

11.4 Numerical Example

Let us consider the differential two-person game $\Gamma(c_0, T)$. The dynamics of the agent $i = 1, 2$ in the game $\Gamma(c_0, T)$ is described by the initial boundary value problem for the following differential equation on the domain $\overline{\Omega} = [0, l]$:

$$\frac{\partial z^i}{\partial t} = D\frac{\partial^2 z^i}{\partial x^2} - rz^i + u_i\psi_i(x), \quad x \in \Omega,\ t > 0. \tag{11.9}$$

Let the function $z^i(x, t)$ satisfies the following boundary conditions:

$$D\frac{\partial z^i}{\partial x} = 0, \quad x = 0,\ t \in [0, T], \tag{11.10}$$

$$-D\frac{\partial z^i}{\partial x} = 0, \quad x = l,\ t \in [0, T], \tag{11.11}$$

and the following initial condition:

$$z^i(x, 0) = c_0^i(x), \quad x \in \overline{\Omega},\ t = 0. \tag{11.12}$$

A measurable function $u_i = u_i(t)$, satisfying the condition $u_i = u_i(t) \in U_i = [\overline{U}_i^1, \overline{U}_i^2]$, $i = 1, 2$ for all $t \in [0, T]$ is called the admissible control of the agent $i \in I$, $\overline{U}_i^1 = \text{const}$, $\overline{U}_i^2 = \text{const}$.

Let the function $f_i(t) \in [0, 1]$ be the percentage of the income of the agent i, which determines the amount of the penalty for exceeding the maximum permissible value of pollution at the water intake point (x_w) as follows:

$$f_i(t) = \begin{cases} 0, & \text{if } \sum_{j=1}^{2} z^j(x_w,t) \le C_w, \\ \dfrac{z_i(x_w,t)}{\sum_{j=1}^{2} z^j(x_w,t)} \cdot \dfrac{\left(\sum_{j=1}^{2} z^j(x_w,t) - C_w\right)}{C_w}, & \text{if } \sum_{j=1}^{2} z^j(x_w,t) > C_w, \end{cases}$$

where C_w is the maximum permissible value of pollution at the water intake point.

Then the payoff of the agent i at time T is defined by the following functional:

$$H_i(z,u_i) = \int_0^T P_i u_i(\tau)(1 - f_i(\tau))d\tau - \int_0^T M_i u_i(\tau)d\tau, \tag{11.13}$$

where $z = (z^1, z^2, \dots, z^n)$. The goal of the agent i is to maximize $H_i(\cdot)$.

The numerical method based on the dynamic programming method (Bellman 1960) and the finite difference method (Samarsky 1989) is proposed for the numerical solving of the auxiliary multistep game $\Gamma_p^\delta(c_0, T)$.

On the domain $\overline{\Omega} = [0, l]$ we construct the uniform net with steps h on x

$$\overline{\omega}_h = \{x_k = kh,\ k = 0, \dots, N_1;\ x_0 = 0,\ x_{N_1} = l\}. \tag{11.14}$$

Here $x_i = \bar{x}_i$ is a location of the agent i, $i = 1, 2$.

On the every interval $[t_s, t_{s+1}]$, $s = \overline{0, N_\sigma - 1}$ we construct the uniform net with step τ

$$\overline{\omega}_{\tau,s} = \{\bar{t}_j = j\tau,\ j = \overline{0, N_2};\ \bar{t}_0 = t_s,\ \bar{t}_{N_2} = t_{s+1}\}.$$

Here $t_s \in \sigma$, where σ is the time interval partition

$$\sigma = \{t_0 = 0 < t_1 < \dots < t_{N_\sigma} = T\}.$$

On the admissible control parameters set $U_i = [\overline{U}_i^1, \overline{U}_i^2]$, $i = 1, 2$ we construct the following partition:

$$\Delta_i = \{u_{i,0} = \overline{U}_i^1 < u_{i,1} < \dots < u_{i,N_3} = \overline{U}_i^2\},\ i = 1, 2.$$

Let us denote by ${}^i y_k^{j,s}$ the function defined on the net $\overline{\omega}_{h\tau} = \overline{\omega}_h \times \overline{\omega}_{\tau,s}$.

We construct for the problem (11.9)–(11.12) following purely implicit difference schemes (Samarsky 1989) on the net $\overline{\omega}_{h\tau} = \overline{\omega}_h \times \overline{\omega}_{\tau,s}$ for any pair of admissible controls $(u_{1,\xi_1}, u_{2,\xi_2}) \in \Delta_1 \times \Delta_2$, $\xi_i \in \overline{0, N_3}$:

$$\frac{{}^i y_0^{j+1,s} - {}^i y_0^{j,s}}{\tau} = 2D \frac{{}^i y_1^{j+1,s} - {}^i y_0^{j+1,s}}{h^2} - {}^i y_0^{j,s} r, \quad k = 0, \tag{11.15}$$

$$\frac{{}^i y_k^{j+1,s} - {}^i y_k^{j,s}}{\tau} = D \frac{{}^i y_{k+1}^{j+1,s} - 2\,{}^i y_k^{j+1,s} + {}^i y_{k-1}^{j+1,s}}{h^2} - \\ - {}^i y_k^{j,s} r + \frac{1}{h} u_{i,\xi_i}^s \delta_{k,i}, \qquad k = \overline{1, N_1 - 1}, \tag{11.16}$$

$$\frac{{}^{i}y_{N_1}^{j+1,s} - {}^{i}y_{N_1}^{j,s}}{\tau} = -2D\frac{{}^{i}y_{N_1}^{j+1,s} - {}^{i}y_{N_1-1}^{j+1,s}}{h^2} - {}^{i}y_{N_1}^{j,s}r,\ k = N_1, \quad (11.17)$$
$$j = \overline{0, N_2 - 1},$$

$${}^{i}y_k^0 = {}^{i}y_k^{N_2,s},\ k = \overline{0, N_1},\ j = 0, \quad (11.18)$$
$$\xi_1 = \overline{0, N_3},$$
$$\xi_2 = \overline{0, N_3},$$
$$s = \overline{0, N_\sigma - 1},$$
$${}^{i}y_k^{0,0} = c_0(x_k),\ k = \overline{0, N_1},\ s = 0,\ j = 0. \quad (11.19)$$

Here $\delta_{k,i}$ is the Kronecker symbol.

The constructed absolutely stable difference scheme (11.15)–(11.19) is solved by the sweep method (Samarsky 1989).

The payoff function of the agent $i \in I = \{1, 2\}$ in the game $\Gamma_p^\delta(c_0, T)$ is approximated as follows:

$$\underline{H}_i(u_1^0, \dots, u_1^{N_\sigma-1}, u_2^0, \dots, u_2^{N_\sigma-1}) = \tau \sum_{s=0}^{N_\sigma-1}\sum_{j=0}^{N_2-1} P_i u_i^s(1 - f_i^j) -$$
$$- \tau \sum_{s=0}^{N_\sigma-1}\sum_{j=0}^{N_2-1} M_i u_i^s, \quad (11.20)$$

Let $\underline{V}_i^\delta(\cdot)$ be value of the payoff function of the agent i, $i = 1, 2$ at the equilibrium point. The following recurrence equations are valid:

$$\underline{V}_i^\delta({}^1y^{N_\sigma-1}, {}^2y^{N_\sigma-1}, t_{N_\sigma-1}) = \max_{u_{i,\xi_i}^{N_\sigma-1} \in \Delta_i} \{\underline{H}_i(u_{i,\xi_i}^{N_\sigma-1}, \bar{u}_{\{I\setminus i\},\xi_{\{I\setminus i\}}}^{N_\sigma-1})\}, \quad (11.21)$$

$$\underline{V}_i^\delta({}^1y^s, {}^2y^s, t_s) = \max_{u_{i,\xi_i}^s \in \Delta_i} \{\underline{H}_i(u_{i,\xi_i}^s, \bar{u}_{\{I\setminus i\},\xi_{\{I\setminus i\}}}^s) + \underline{V}_i^\delta({}^1y_{\xi_1}^{s+1}, {}^2y_{\xi_2}^{s+1}, t_{s+1})\},$$
$$s = \overline{N_\sigma - 2, 0}, \quad (11.22)$$
$$\underline{V}_i^\delta({}^1y^0, {}^2y^0, t_{N_\sigma}) = 0, \quad (11.23)$$

Here

$$\underline{H}_i(u_{i,\xi_i}^s, \bar{u}_{\{I\setminus i\},\xi_{\{I\setminus i\}}}^s) = \tau \sum_{j=0}^{N_2-1} P_i u_{i,\xi_i}^s(1 - f_i^j) - \tau \sum_{j=0}^{N_2-1} M_i u_{i,\xi_i}^s. \quad (11.24)$$

11.5 Numerical Results

The numerical experiment was realized for the following input datas: $D = 2.2$, $l = 20$, $h = 1$, $T = 180$, $\tau = 1$, $N_\sigma = 6$, $r = 0.02$, $M_1 = 4.5$, $M_2 = 5$, $P_{1(2)} = 15$, $U_1^i = \{0, 10, 20, 30\}$, $U_2^j = \{0, 10, 20, 30\}$, $\bar{x}_1 = 3$, $\bar{x}_2 = 17$, $\bar{x}_w = 10$, $C_w = 60$, $c_0(x) = 0$.

The results of numerical experiments are presented on the Figs. 11.1–11.8.

The dynamics of changing of the pollutant concentration at the intake point is presented in Fig. 11.4. The agent pays the penalty (Fig. 11.5) in the case of exceeding the maximum permissible value C_w.

The player with the lower costs receives significantly greater income (compare Fig. 11.3 with Fig. 11.6).

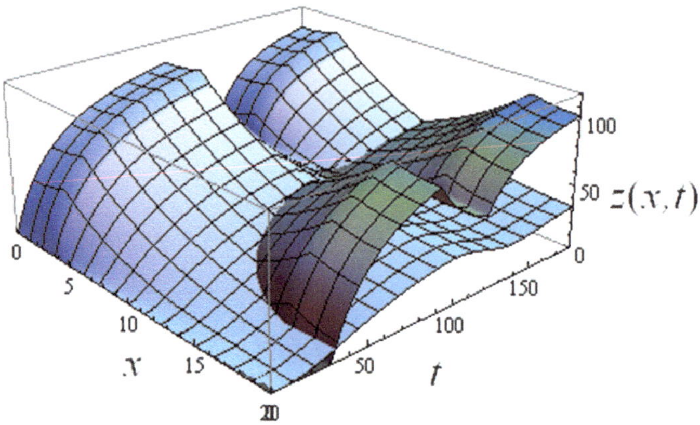

Fig. 11.1 Distribution of the pollutant concentration of agents: 1—*light color*, 2—*dark*

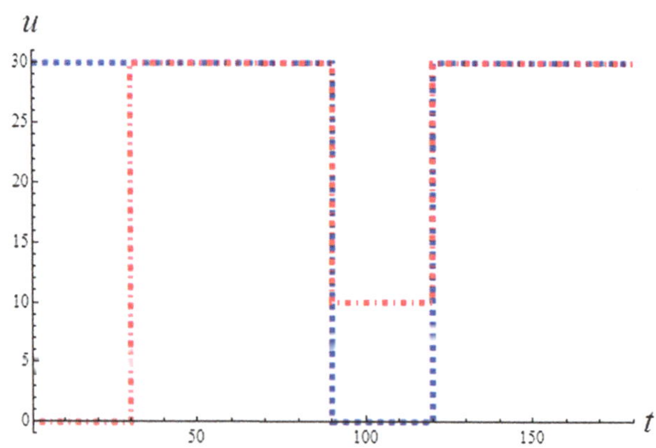

Fig. 11.2 Optimal strategies of agents: 1—*dashed line*, 2—*dot-dashed line*

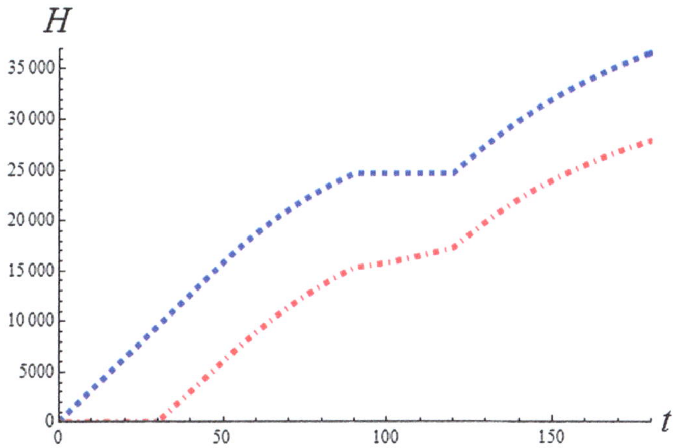

Fig. 11.3 Payoff functions of agents: 1—*dashed*, 2—*dot-dashed line*

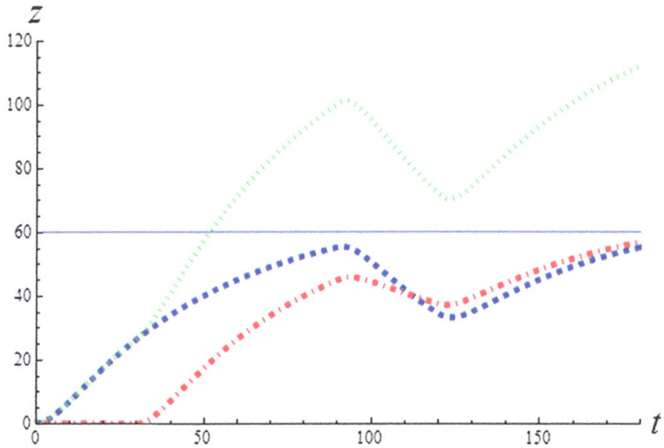

Fig. 11.4 The pollutant concentrations at the point of intake: total—*dotted*, 1—*dashed line*, 2—*dot-dashed line*

The decrease of the value $C_w = 50$ results in significant reduction of the total income of the agents (compare Fig. 11.8 with Fig. 11.3).

11.6 Conclusion

We investigated conditions for the existence of the ε-Nash equilibrium point for the class of the piecewise-programmed strategies in the noncooperative many agents differential game which describes a conflict-controlled process of the contaminating a closed water reservoir.

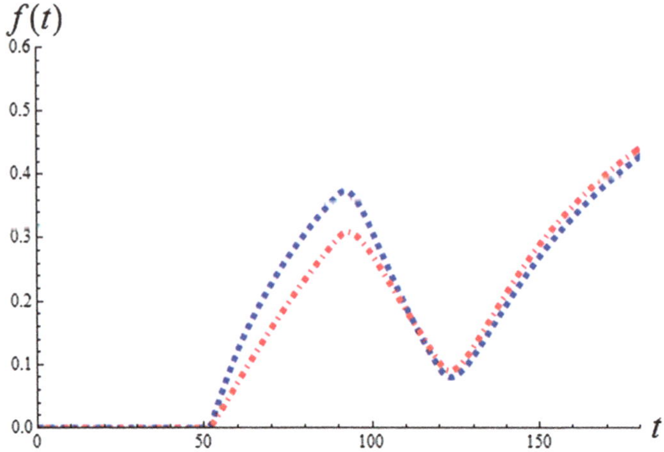

Fig. 11.5 The penalty functions $f(t)$: 1—*dashed line*, 2—*dot-dashed line*

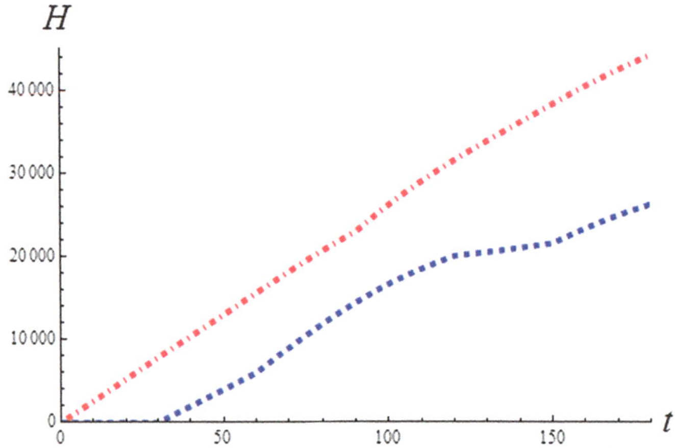

Fig. 11.6 The payoff function in case of changing $M_2 = 5 \to M_2 = 2$. 1—*dashed line*, 2—*dot-dashed*

The proposed numerical algorithm to solve the considered differential game is based on the dynamic programming method and the finite difference method and has been applied to compute the auxiliary multistep game.

Acknowledgements This work was partly supported by grant No. 4402 of the Ministry of Education and Science of the Russian Federation. The author would like to thank Prof. O. A. Malafeyev, Prof. S. G. Pyatkov, and Prof. N. D. Botkin for helpful suggestions and discussions.

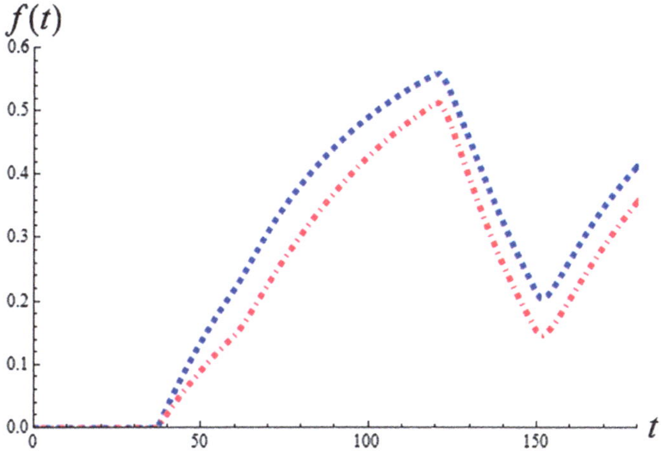

Fig. 11.7 The penalty functions $f(t)$ for increasing value $D = 5$: 1—*dashed line*, 2—*dot-dashed line*. Compare with Fig. 11.5

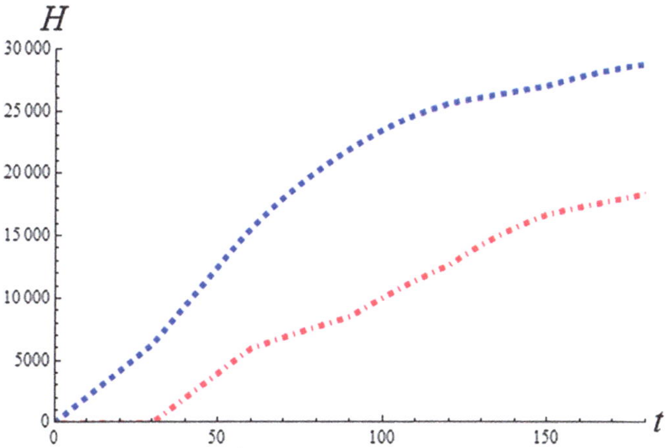

Fig. 11.8 Total income of players for $C_w = 50$

References

Amann H (2005) Nonautonomous parabolic equations involving measures. J Math Sci 130(4):4780–4802

Başar T, Olsder G (1999) Dynamic noncooperative game theory. Academic Press, New York

Bellman R (1960) Dynamic programming. Inostrannaya Literatura, Moscow

Botkin N, Hoffmann KH, Turova V (2011) Optimal control of ice formation in living cells during freezing. Appl Math Model 35(8):4044–4057

Breton M, Zaccour G, Zahaf M (2005) A differential game of joint implementation of environmental projects. Automatica 41(10):1737–1749

Cappoza I (2003) A dynamic games of techology diffusion under an emission trading regulation: A pilot experiment. series 8, Southern Europe Research in Economic Studies
Jørgensen S, Zaccour G (2001) Incentive equilibrium strategies and welfare allocation in a dynamic game of pollution control. Automatica 3(1):29–36
Kleimenov A (1993) Nonantagonistic positional differential games. Nauka, Ekaterinburg, in Russian
Kolokoltsov V, Malafeyev O (2010) Understanding game theory (Introduction to the analysis of multi agent systems with competition and cooperation). World Scientific, Singapore
Krasovskiy N, Tarasyev A (2011) Decomposition algorithm of searching equilibria in the dynamical game. Game Theory Appl 3(4):49–88
Ladyzhenskaya O (1985) The boundary-value problems of mathematical physics. Springer, Berlin, Heidelberg, New York
Malafeyev O (2000) Dynamical control systems of conflict. St. Petersburg University Press, St. Petersburg (in Russian)
Martín-Herrán G, Cartigny P, Motte E, Tidball M (2006) Deforestation and foreign transfers: A Stackelberg differential game approach. Comput Oper Res 33(2):386–400
Mazalov V, Rettieva A (2010) Fish wars and cooperation maintenance. Ecol Model 221:1545–1553
Petrosyan L (1993) Differential games of pursuit. World Scientific, Singapore
Petrosyan L, Zaccour G (2003) Time-consistent Shapley value allocation of pollution cost reduction. J Econ Dyn Control 27(3):381–398
Petrosyan L, Zaharov V (1997) Mathematical models in ecology. St.Petersburg University Press (in Russian)
Rettieva A (2011) Stable coalitional partition in discrete-time bioresourse management problem. Game Theory Appl 3(3):39–66
Rogers E (2003) Diffusion of innovations, 5th edn. Free Press, New York
Samarsky AA (1989) Difference schemes theory. Nauka, Moscow (in Russian)
Troeva M (2012) Existence of equilibrium point for noncooperative differential game in banach space. In: Simos T, Psihoyios G, Tsitouras C, Anastassi Z (eds) Numerical Analysis and Applied Mathematics (ICNAAM 2012), vol 1479, AIP Conf. Proc., American Institute of Physics, New York, pp 1234–1237
Varaiya P, Lin J (1969) Existence of saddle points on differential games. SIAM J Control Optim 7(1):141–157
Zenkevich N, Kozlovskaya N (2010) Stable Shapley value in cooperative game of territorial enviromental production. Game Theory Appl 2:67–92

Part III
Pursuit-Evasion Games

Chapter 12
Open-Loop Solvability Operator in Differential Games with Simple Motions in the Plane

Liudmila Kamneva and Valerii Patsko

Abstract The paper deals with an open-loop solvability operator in two-person zero-sum differential games with simple motions. This operator takes a given terminal set to the set defined at the initial instant whence the first player can bring the control system to the terminal set if the player is informed about the open-loop control of the second player. It is known that the operator possesses the semigroup property in the case of a convex terminal set. In the paper, sufficient conditions ensuring the semigroup property in the non-convex case are formulated and proved for problems in the plane. Examples are constructed to illustrate the relevance of the formulated conditions. The connection with the Hopf formula is analysed.

Keywords Planar differential games • Semigroup property • Simple motions • Open-loop solvability operator

12.1 Introduction

The paper concerns the simplest model description of dynamics in the differential game theory:

$$\dot{x} = p + q, \quad p \in P, \quad q \in Q.$$

The system has no state variable x at the right-hand side, and the state velocity $\dot{x}$ is defined only by controls $p \in P$ and $q \in Q$ of the first and second players, where

L. Kamneva • V. Patsko (✉)
Institute of Mathematics and Mechanics, S. Kovalevskaya str. 16, Ekaterinburg, 620990, Russia
e-mail: kamneva@imm.uran.ru; patsko@imm.uran.ru

V. Křivan and G. Zaccour (eds.), *Advances in Dynamic Games*, Annals of the International Society of Dynamic Games 13, DOI 10.1007/978-3-319-02690-9_12,

the constraints P and Q do not depend on the time. In Isaacs (1965), games with such dynamics are called games with simple motions.

In numerical methods of the differential game theory, the simple motion dynamics arises absolutely naturally under a local approximation of linear or nonlinear dynamics when the capabilities of the players are "frozen" in the time and state variables. In the framework of the simple motion dynamics, one calculates the next step of an iterative procedure to construct the value function of the game.

For example, one important class of differential games consists of the games with linear dynamics, a fixed terminal time, and a continuous terminal payoff function. For these games, the transfer to new variables is known (Krasovskii and Subbotin 1974, pp. 159–161; Krasovskii and Subbotin 1988, pp. 89–91). The new variables can be regarded as forecasting the state variables to the terminal instant by the "free" motion of the system under zero controls of the players. The transfer is performed by the Cauchy matrix of the original problem. The new dynamic system has no state variables at the right-hand side, but the controls of the players are multiplied by coefficients depending on time.

Under the numerical construction of level sets of the value function, the time interval is divided by a step, and the coefficients of the dynamics are frozen on each small time interval (Botkin 1984; Isakova et al. 1984). So, at each step we get some dynamics of simple motions. Being given a level set of the payoff function as the terminal set and going backward from the terminal set, one recalculates the level set at each time step using the dynamics of simple motions. Then one passes to the limit as the step of the partition goes to zero. If the operator of recalculation is properly chosen (at one step), then the limit set coincides with the level set (Lebesgue set) of the value function.

This is the scheme. To perform it effectively, it is very important to choose an operator to use at each step of the backward iterative procedure. It is most desirable that the operator possesses the semigroup property: if the dynamics is frozen on some time interval, then the use of any additional points of the partition does not change the result of the iterative procedure.

We investigate the operator known as the programmed absorption operator in Russian literature on the differential game theory (Krasovskii and Subbotin 1974, p. 122). It can be called the open-loop solvability operator as well. For the simple motion games, the semigroup property was established earlier (Pshenichnyy and Sagaydak 1971) in the case of the operator dealing with convex sets. In this paper, sufficient conditions providing the semigroup property in the non-convex case are formulated and proved for the problems in the plane. Examples are constructed to illustrate the relevance of the formulated conditions. In the appendix, we describe the connection between the question under investigation and the Hopf formula known in the differential game theory and the theory of partial-differential equations.

The results obtained in the paper can be useful for developments and justifications of numerical methods in the differential game theory.

12.2 Problem Statement: Open-Loop Solvability Operator

Consider a conflict-control dynamic system with simple motions (Isaacs 1965):

$$\frac{dx}{dt} = p + q, \quad p \in P, \quad q \in Q, \quad x \in \mathbb{R}^n. \tag{12.1}$$

Here, $t \in [0, \vartheta]$; p, q are controls of the first and second players; P, Q are convex compact sets in $\mathbb{R}^n$. Let M be a compact terminal set in $\mathbb{R}^n$.

For a differential game, the notion of the maximal stable bridge $W_0 \subset [0, \vartheta] \times \mathbb{R}^n$ terminating at the instant ϑ on the set M (i.e., $W_0(\vartheta) = M$) was introduced in Krasovskii and Subbotin (1974, p. 67), Krasovskii and Subbotin (1988, p. 61). Here, the notation $W_0(t)$ for a t-section of the set W_0 is used:

$$W_0(t) = \{x \in \mathbb{R}^n : (t, x) \in W_0\}, \quad t \in [0, \vartheta].$$

To guarantee the inclusion $x(\vartheta) \in M$, the positional strategy of the first player can be constructed (Krasovskii and Subbotin 1974, 1988) by the procedure of extremal aiming to the maximal stable bridge W_0. The set W_0 coincides with the solvability set in the problem of guidance over non-anticipating strategies (Bardi and Capuzzo-Dolcetta 1997; Subbotin 1995). The notion of the maximal stable bridge W_0 is very close to the notion of the viability kernel (Aubin 1991; Cardaliaguet et al. 1999), and its t-section $W_0(t)$ is well known as the alternating Pontryagin integral (Pontryagin 1967, 1981).

In the case of a convex set M, the Pshenichnyi formula describing constructively the sections $W_0(t)$, $t \in [0, \vartheta]$, is known (Pshenichnyy and Sagaydak 1971):

$$W_0(t) = (M - (\vartheta - t)P) \stackrel{*}{-} (\vartheta - t)Q. \tag{12.2}$$

Here, operations of the algebraic sum (the Minkowski sum) $A + B$ and the geometric difference (the Minkowski difference) $A \stackrel{*}{-} B$ of the sets $A, B \subset \mathbb{R}^n$ are used (see, for example, Hadwiger (1957); Polovinkin and Balashov (2004); Pontryagin (1967, 1981)):

$$A + B := \{d \in \mathbb{R}^n : \; d = a + b, \; a \in A, \; b \in B\},$$

$$A \stackrel{*}{-} B := \{d \in \mathbb{R}^n : \; d + B \subseteq A\} = \bigcap_{b \in B} (A - b).$$

The set $A + B$ is convex if the both sets A and B are convex. The set $A \stackrel{*}{-} B$ is convex in the case of a convex set A.

Define the open-loop solvability operator (the programmed absorption operator):

$$M \mapsto T_\tau(M) := (M - \tau P) \stackrel{*}{-} \tau Q, \quad \tau = \vartheta - t.$$

By (12.2), for a convex set M, we have

$$W_0(t) = T_{\vartheta-t}(M). \tag{12.3}$$

It is of interest to try to establish some conditions providing equality (12.3) in the case of a non-convex set M.

For any compact (generally speaking, non-convex) set M, the representation

$$W_0(t) = \bigcap_{\tau_1+\tau_2+\cdots+\tau_m=\vartheta-t,\ m\in\mathbb{N}} T_{\tau_1}(T_{\tau_2}(\ldots T_{\tau_m}(M)\ldots)) =: \tilde{T}_{\vartheta-t}(M)$$

is true (Pshenichnyy and Sagaydak 1971). Its right-hand side defines the operator with multiple recomputations:

$$M \mapsto \tilde{T}_\tau(M), \quad \tau = \vartheta - t.$$

Therefore, the operators T_τ and $\tilde{T}_\tau$ are equal (i.e., $T_\tau(M) = \tilde{T}_\tau(M)$ for all $\tau \in [0, \vartheta]$) if, for any $\tau_1, \tau_2 > 0$ such that $\tau_1 + \tau_2 \leq \vartheta$, the following relation holds:

$$T_{\tau_1+\tau_2}(M) = T_{\tau_1}(T_{\tau_2}(M)). \tag{12.4}$$

Equality (12.4) is known as the semigroup property of the operator T_τ. In Pshenichnyy and Sagaydak (1971), the semigroup property was proved for any convex set M. This implies (12.2).

Thus, the question on the validity of (12.3) in the case of a non-convex set M is reduced to the formulation of conditions on the sets M, P, Q, and on the range of τ_1, τ_2 to provide equality (12.4).

12.3 Auxiliary Results

Let us remark two obvious properties:

$$T_\tau(M) = \bigcap_{q\in Q}(M - \tau(P + q)); \tag{12.5}$$

$$x \in T_\tau(M) \quad \Leftrightarrow \quad \forall q \in Q \quad (x + \tau(P + q)) \cap M \neq \varnothing. \tag{12.6}$$

The following two results are known (Pshenichnyy and Sagaydak 1971).

Lemma 12.1.

$$T_{\tau_1}(T_{\tau_2}(M)) \subseteq T_{\tau_1+\tau_2}(M). \tag{12.7}$$

Proof. Fix $x \in T_{\tau_1}(T_{\tau_2}(M))$. By (12.6), for any $q \in Q$ there exists $p_1 \in P$ such that

$$x + \tau_1 q + \tau_1 p_1 \in T_{\tau_2}(M),$$

and there exists $p_2 \in P$ such that

$$z := (x + \tau_1 q + \tau_1 p_1) + \tau_2 q + \tau_2 p_2 \in M.$$

Since the set P is convex, the following inclusion holds:

$$p_* := \frac{\tau_1 p_1 + \tau_2 p_2}{\tau_1 + \tau_2} \in P.$$

We have

$$x + (\tau_1 + \tau_2)(p_* + q) = z \in M.$$

Thus, for any $q \in Q$ there exists $p_* \in P$ such that

$$x + (\tau_1 + \tau_2)(p_* + q) \in M.$$

Then by (12.6), we obtain $x \in T_{\tau_1+\tau_2}(M)$. □

Lemma 12.2. *Assume the set M is convex. Then*

$$T_{\tau_1+\tau_2}(M) = T_{\tau_1}(T_{\tau_2}(M)), \quad \tau_1, \tau_2 > 0.$$

Proof. By Lemma 12.1, it remains to prove that

$$T_{\tau_1+\tau_2}(M) \subseteq T_{\tau_1}(T_{\tau_2}(M)).$$

Let $x \in T_{\tau_1+\tau_2}(M)$. Then, because of the property (12.6), for any $q_1 \in Q$ there exists $p_1 \in P$ such that

$$z_1 := x + (\tau_1 + \tau_2)(p_1 + q_1) \in M. \tag{12.8}$$

Let us prove the inclusion

$$x + \tau_1(p_1 + q_1) \in T_{\tau_2}(M). \tag{12.9}$$

Fix $q_2 \in Q$. By the same arguments as in (12.8), we find $p_2 \in P$ such that

$$z_2 := x + (\tau_1 + \tau_2)(p_2 + q_2) \in M.$$

Considering the convexity of the set M, we have

$$x + \tau_1(p_1 + q_1) + \tau_2(p_2 + q_2) = \frac{\tau_1 z_1 + \tau_2 z_2}{\tau_1 + \tau_2} \in M.$$

Thus, inclusion (12.9) holds. Therefore, by (12.6), we get $x \in T_{\tau_1}(T_{\tau_2}(M))$. □

In addition, three lemmas formulated and proved below are required. Lemma 12.3 claims inequality (12.11), which, in particular, is necessarily true if $T_{\tau_1+\tau_2}(M) = T_{\tau_1}(T_{\tau_2}(M))$. Further, a similar condition is used in the main Theorem 12.1. In Lemma 12.4, the case of a convex set M is considered, and thus, by Lemma 12.2, the semigroup property is necessarily true. The proof is based on Lemma 12.3. The lemma is also used in the proof of Lemma 12.5, which is in its turn necessary for our proof of Theorem 12.2.

Let $\rho(\,\cdot\,, A)$ be a support function of a compact set $A \subset \mathbb{R}^n$, i.e.,

$$\rho(\eta\,, A) = \max\{\langle x, \eta\rangle : x \in A\}, \quad \eta \in \mathbb{R}^n.$$

Write

$$H(s) = \max_{q\in Q}\langle q, s\rangle + \min_{p\in P}\langle p, s\rangle, \quad s \in \mathbb{R}^n.$$

Lemma 12.3. *Fix $\tau_1, \tau_2 > 0$, and assume that the sets $T_{\tau_2}(M)$, $T_{\tau_1}(T_{\tau_2}(M))$ are nonempty, $\eta \in \mathbb{R}^n$, and*

$$\rho(\eta, T_{\tau_1+\tau_2}(M)) = \rho(\eta, T_{\tau_1}(T_{\tau_2}(M))). \tag{12.10}$$

Then

$$\rho(\eta, T_{\tau_1+\tau_2}(M)) + \tau_1 H(\eta) \leq \rho(\eta, T_{\tau_2}(M)). \tag{12.11}$$

Proof. Since $T_{\tau_1}(T_{\tau_2}(M)) \subseteq T_{\tau_1+\tau_2}(M)$, we have $T_{\tau_1+\tau_2}(M) \neq \varnothing$.

The definition of the set $T_{\tau_1}(T_{\tau_2}(M))$ implies the inclusion

$$T_{\tau_1}(T_{\tau_2}(M)) + \tau_1 Q \subseteq T_{\tau_2}(M) - \tau_1 P.$$

Then

$$\rho(\eta, T_{\tau_1}(T_{\tau_2}(M))) + \tau_1 \max_{q\in Q}\langle q, \eta\rangle \leq \rho(\eta, T_{\tau_2}(M)) + \tau_1 \max_{p\in P}\langle -p, \eta\rangle.$$

So, we get (12.11) by (12.10) and taking into account the equality

$$\max_{p\in P}\langle -p, \eta\rangle = -\min_{p\in P}\langle p, \eta\rangle.$$

□

Lemma 12.4. *Let M be a convex set. Assume that $\tau_1, \tau_2 > 0$ and the sets $T_{\tau_2}(M)$, $T_{\tau_1+\tau_2}(M)$ are nonempty. Then, for any $\eta \in \mathbb{R}^n$, inequality* (12.11) *holds.*

Proof. Since the set M is convex, Lemma 12.2 implies

$$T_{\tau_1+\tau_2}(M) = T_{\tau_1}(T_{\tau_2}(M)).$$

Then $\rho(\eta, T_{\tau_1}(T_{\tau_2}(M))) = \rho(\eta, T_{\tau_1+\tau_2}(M))$. Using Lemma 12.3, we have (12.11). □

Lemma 12.5. *Let $\eta \in \mathbb{R}^n$ and $\eta \neq 0$. Assume that there exists $z_* \in M$ such that the intersection $M \cap \Pi_*$ of the set M and the half-space*

$$\Pi_* = \{x \in \mathbb{R}^n : \langle x - z_*, \eta \rangle \leq 0\}$$

is convex and its interior is nonempty.

Then there exists $\vartheta > 0$ such that, for any $\tau \in [0, \vartheta]$, the set $T_\tau(M)$ is nonempty and the function

$$\tau \mapsto \delta_\eta(\tau) := \rho(-\eta, M) - \tau H(-\eta) - \rho(-\eta, T_\tau(M))$$

increases on $[0, \vartheta]$.

Proof.

1) Define

$$\mu_* = \langle z_*, -\eta \rangle.$$

Observe that $\mu_* < \rho(-\eta, M)$. Choose any $\mu \in (\mu_*, \rho(-\eta, M))$, and write

$$\Pi_\mu = \Pi_* - (\mu - \mu_*)\eta / \|\eta\|.$$

Since the interior of the intersection $M \cap \Pi_\mu$ is nonempty (in view of the fact that the set $M \cap \Pi_*$ is convex and its interior is nonempty), for rather small $\tau > 0$, we get

$$T_\tau(M) \cap \Pi_\mu \neq \varnothing. \tag{12.12}$$

Thus, there exists $\tau_1^* > 0$ such that, for any $\tau \in [0, \tau_1^*]$, we have (12.12).

Set

$$\alpha = \min_{p \in P} \min_{q \in Q} \langle p + q, -\eta \rangle.$$

We are able to choose a value $\tau_2^* > 0$ such that

$$\tau\alpha \geq \mu_* - \mu, \quad \tau \in (0, \tau_2^*].$$

Indeed, since $\mu_* - \mu < 0$, any $\tau_2^* > 0$ can be taken if $\alpha \geq 0$; otherwise, we choose any sufficiently small $\tau_2^* > 0$.

Set $\vartheta = \min\{\tau_1^*, \tau_2^*\}$. (Thus, the value ϑ depends on the choice of μ.)

2) Fix $\tau \in [0, \vartheta]$. Let us show that

$$T_\tau(M) \cap \Pi_\mu \subseteq T_\tau(M \cap \Pi_*). \tag{12.13}$$

Choose $x \in T_\tau(M) \cap \Pi_\mu$. By (12.6), for any $q \in Q$ there exists $p_* \in P$ such that $x + \tau q + \tau p_* \in M$. Since $x \in \Pi_\mu$, we obtain $\langle x, -\eta\rangle \geq \mu$. Therefore, considering the choice of the value μ and the definition of the value α, we deduce

$$-\langle x, -\eta\rangle + \mu_* \leq -\mu + \mu_* \leq \tau\alpha \leq \tau\langle q + p_*, -\eta\rangle.$$

Hence $\langle x + \tau q + \tau p_*, -\eta\rangle \geq \mu_*$, i.e., $x + \tau q + \tau p_* \in \Pi_*$. Thus,

$$x + \tau q + \tau p_* \in M \cap \Pi_*,$$

and, consequently, $x + \tau q \in (M \cap \Pi_*) - \tau P$. Since $q \in Q$ is chosen arbitrarily, in view of (12.5), we get

$$x \in \bigcap_{q \in Q} \big((M \cap \Pi_*) - \tau(P + q)\big) = T_\tau(M \cap \Pi_*).$$

3) Fix $\tau_2, \tau_1 + \tau_2 \in [0, \vartheta]$. By (12.12), we have

$$T_{\tau_2}(M) \neq \varnothing, \quad T_{\tau_1+\tau_2}(M) \cap \Pi_\mu \neq \varnothing.$$

Considering (12.13), the convexity of the set $M \cap \Pi_*$, Lemma 12.2, and the monotonicity of T_τ, we calculate

$$T_{\tau_1+\tau_2}(M) \cap \Pi_\mu \subseteq T_{\tau_1+\tau_2}(M \cap \Pi_*) = T_{\tau_1}(T_{\tau_2}(M \cap \Pi_*)) \subseteq T_{\tau_1}(T_{\tau_2}(M)).$$

This implies that $T_{\tau_1}(T_{\tau_2}(M)) \neq \varnothing$ and

$$\rho(-\eta, T_{\tau_1+\tau_2}(M) \cap \Pi_\mu) \leq \rho(-\eta, T_{\tau_1}(T_{\tau_2}(M))). \tag{12.14}$$

Since Π_μ is a half-space with an outward normal vector η and the intersection $T_{\tau_1+\tau_2}(M) \cap \Pi_\mu$ is nonempty, we find

$$\rho(-\eta, T_{\tau_1+\tau_2}(M)) = \rho(-\eta, T_{\tau_1+\tau_2}(M) \cap \Pi_\mu).$$

We employ this identity in (12.14) to obtain the inequality

$$\rho(-\eta, T_{\tau_1+\tau_2}(M)) \leq \rho(-\eta, T_{\tau_1}(T_{\tau_2}(M))).$$

On the other hand, Lemma 12.1 implies the opposite inequality. So, equation (12.10) holds.

Hence, considering Lemma 12.3, we get inequality (12.11), which is equivalent to the inequality $\delta_\eta(\tau_1 + \tau_2) \geq \delta_\eta(\tau_2)$. Therefore, the function $\delta_\eta(\cdot)$ increases on the segment $[0, \vartheta]$. □

12.4 The Main Theorem

Now, we deal with the case of $\mathbb{R}^2$.

A set A is called *arcwise connected* (Schwartz 1967) (in the sequel, *connected* for brevity) if any two distinct points of the set A can be joined by a simple curve (arc) which lies in the set.

A set $A \subset \mathbb{R}^2$ is called *simply connected* (Schwartz 1967) if any simple closed curve can be shrunk to a point continuously in the set, i.e., the set consists of one piece and does not have any "holes."

A polygon is defined as a plane figure that is bounded by a closed path composed of a finite sequence of straight line segments (edges of the polygon).

Let us denote by $\mathscr{V}_A$ the set of all outward normal unit vectors to the edges of the polygon A. If A is a segment, we suppose that the set $\mathscr{V}_A$ consists of two opposite directed vectors that are normal to the segment A.

Let us formulate the main theorem.

Theorem 12.1. *Assume that*

(A1) $M \subset \mathbb{R}^2$ *is a simply connected compact set;*
(A2) $P \subset \mathbb{R}^2$ *is either a non-degenerate segment, or a convex polygon ;* $Q \subset \mathbb{R}^2$ *is a convex compact set;*
(A3) *for any* $x \in \mathbb{R}^2$ *and* $\nu \in \mathscr{V}_P$, *the set*

$$\Pi_M(x, \nu) = M \cap \{z \in \mathbb{R}^2 : \langle z, \nu \rangle \leqslant \langle x, \nu \rangle\}$$

is connected;
(A4) *for any* $\tau \in [0, \vartheta]$, *the set* $T_\tau(M)$ *is nonempty and connected;*
(A5) *for any* $\nu \in \mathscr{V}_P$, *the function*

$$\tau \mapsto \delta_\nu(\tau) := \rho(-\nu, M) - \tau H(-\nu) - \rho(-\nu, T_\tau(M))$$

increases on the segment $[0, \vartheta]$.

Then the operator T_τ *possesses the semigroup property on the segment* $[0, \vartheta]$. *(And, consequently,* $W_0(t) = T_{\vartheta - t}(M)$, $t \in [0, \vartheta]$.*)*

Our proof of Theorem 12.1 is based on Lemma 12.6. To formulate the lemma, let us introduce the following notations.

For the set $T_\tau(M) \neq \varnothing$, we define "an envelope set"

$$\operatorname{env}(T_\tau(M)) = \bigcap_{\nu \in \mathscr{V}_P} \{x \in \mathbb{R}^2 : \langle x, -\nu \rangle \leq \rho(-\nu, T_\tau(M))\}.$$

Note that $T_\tau(M) \subset \operatorname{env}(T_\tau(M))$. If P is a segment, then $\operatorname{env}(T_\tau(M))$ is a closed strip; if P is a polygon, then $\operatorname{env}(T_\tau(M))$ is a convex polygon.

Let $\mathscr{P}$ be the set of vertices of the segment or polygon P. For a vertex $p \in \mathscr{P}$, define a bundle of unit vectors

$$\mathscr{N}(p) = \{(p - x)/\|p - x\| : \ x \in P \setminus \{p\}\}.$$

If P is a segment, then the set $\mathscr{P}$ consists of two vertices, and the set $\mathscr{N}(p)$ consists of a unique vector for $p \in \mathscr{P}$.

Let $l(a, \eta)$ be a ray with the initial point $a \in \mathbb{R}^2$ and the direction along the vector $\eta \in \mathbb{R}^2$:

$$l(a, \eta) = \{a + \alpha\eta : \ \alpha \geq 0\}.$$

Lemma 12.6. *Assume that* $\tau_1, \tau_2 > 0$, *the sets* $T_{\tau_2}(M)$ *and* $T_{\tau_1+\tau_2}(M)$ *are nonempty, and the following conditions hold:*

(L1) *if* $y \in \mathbb{R}^2$, $q_1 \in Q$, *and*

$$(y + \tau_1(P + q_1)) \cap T_{\tau_2}(M) = \varnothing, \quad (y + \tau_1(P + q_1)) \cap \operatorname{env}(T_{\tau_2}(M)) \neq \varnothing,$$

then there exist $p_* \in \mathscr{P}$ *and* $q_2 \in Q$ *such that*

$$\forall \eta \in \mathscr{N}(p_*) \quad l(y + \tau_1(p_* + q_1) + \tau_2(p_* + q_2), -\eta) \cap M = \varnothing;$$

(L2) *for any* $\nu \in \mathscr{V}_P$, *we have*

$$\rho(-\nu, T_{\tau_1+\tau_2}(M)) + \tau_1 H(-\nu) \leq \rho(-\nu, T_{\tau_2}(M)).$$

Then

$$T_{\tau_1+\tau_2}(M) = T_{\tau_1}(T_{\tau_2}(M)). \tag{12.15}$$

Proof. Suppose that (12.15) is false. Then, by Lemma 12.1, we can find

$$y \in T_{\tau_1+\tau_2}(M) \setminus T_{\tau_1}(T_{\tau_2}(M)) \neq \varnothing.$$

Since $y \notin T_{\tau_1}(T_{\tau_2}(M))$, by (12.6), we find $q_1 \in Q$ such that

$$G_1 \cap T_{\tau_2}(M) = \varnothing, \quad G_1 := y + \tau_1(P + q_1). \tag{12.16}$$

1) Assume

$$G_1 \cap \operatorname{env}(T_{\tau_2}(M)) \neq \varnothing. \tag{12.17}$$

a) We now assert that there exists $q_2 \in Q$ such that

$$(G_1 + \tau_2(P + q_2)) \cap M = \varnothing. \tag{12.18}$$

Indeed, using (12.16), (12.17), and condition (L1), we find $p_* \in \mathscr{P}$ and $q_2 \in Q$ such that

$$\forall \eta \in \mathscr{N}(p_*) \quad l(b, -\eta) \cap M = \varnothing, \quad b := y + \tau_1(p_* + q_1) + \tau_2(p_* + q_2). \tag{12.19}$$

For any $z \in G_1$ we have the representation

$$z = y + \tau_1(\bar{p} + q_1), \quad \bar{p} \in P.$$

In addition, for any $p \in P$, we can write

$$z + \tau_2(p + q_2) = b - \eta_*, \quad \eta_* := \tau_1(p_* - \bar{p}) + \tau_2(p_* - p).$$

We have

$$\frac{p_* - \bar{p}}{\|p_* - \bar{p}\|} \in \mathscr{N}(p_*) \quad (p_* \neq \bar{p}), \qquad \frac{p_* - p}{\|p_* - p\|} \in \mathscr{N}(p_*) \quad (p \neq p_*).$$

Therefore, if $\eta_* \neq 0$, then $\eta_*/\|\eta_*\| \in \mathscr{N}(p_*)$. Considering (12.19), we get

$$z + \tau_2(p + q_2) \notin M = \varnothing.$$

Hence (12.18) holds.

b) Set $\tilde{q} = (\tau_1 q_1 + \tau_2 q_2)/(\tau_1 + \tau_2)$. Then

$$y + (\tau_1 + \tau_2)(P + \tilde{q}) = y + \tau_1(P + q_1) + \tau_2(P + q_2) = G_1 + \tau_2(P + q_2).$$

Using (12.18), we get $(y + (\tau_1 + \tau_2)(P + \tilde{q})) \cap M = \varnothing$. By (12.6), we conclude $y \notin T_{\tau_1 + \tau_2}(M)$, that contradicts to our choice of y.

2) Assume $G_1 \cap \operatorname{env}(T_{\tau_2}(M)) = \varnothing$. Then, using the definition of the operator env , we write

$$G_1 \subseteq \bigcup_{\nu \in \mathscr{V}_P} \{x \in \mathbb{R}^2 : \langle x, -\nu \rangle > \rho(-\nu, T_{\tau_2}(M))\}.$$

Since G_1 is either a non-degenerate segment, or a convex polygon, and $\mathscr{V}_P$ is the set of outward normals to G_1, we deduce that there exists $\nu_0 \in \mathscr{V}_P$ such that

$$\forall z \in G_1 \quad \langle z, -\nu_0 \rangle > \rho(-\nu_0, T_{\tau_2}(M)). \tag{12.20}$$

Remark also that the inclusion $y \in T_{\tau_1+\tau_2}(M)$ implies the inequality

$$\langle y, -\nu_0 \rangle \leq \rho(-\nu_0, T_{\tau_1+\tau_2}(M)). \tag{12.21}$$

Suppose

$$p_0 \in \operatorname{Arg} \max_{p \in P} \langle p, \nu_0 \rangle, \quad z_0 := y + \tau_1(p_0 + q_1).$$

Since $z_0 \in G_1$, using (12.20), (12.21), and the relations

$$\langle p_0, -\nu_0 \rangle = \min_{p \in P} \langle p, -\nu_0 \rangle, \quad \langle q_1, -\nu_0 \rangle \leq \max_{q \in Q} \langle q, -\nu_0 \rangle,$$

we calculate

$$\begin{aligned} \rho(-\nu_0, T_{\tau_2}(M)) < \langle z_0, -\nu_0 \rangle &= \langle y, -\nu_0 \rangle + \tau_1 \langle p_0 + q_1, -\nu_0 \rangle \\ &\leq \rho(-\nu_0, T_{\tau_1+\tau_2}(M)) + \tau_1 H(-\nu_0), \end{aligned}$$

that contradicts to condition (L2).

So, assuming the violation of (12.15), we obtain the contradictions in the both cases 1) and 2). □

Proof (of Theorem 12.1). Choose $\tau_2, \tau_1 + \tau_2 \in (0, \vartheta]$. To prove the equality

$$T_{\tau_1+\tau_2}(M) = T_{\tau_1}(T_{\tau_2}(M)),$$

check conditions (L1) and (L2) of Lemma 12.6.

For any $\nu \in \mathscr{V}_P$, the increase of the function $\delta_\nu(\cdot)$ on the segment $[0, \vartheta]$ implies the inequality $\delta_\nu(\tau_1 + \tau_2) \geq \delta_\nu(\tau_2)$, which is equivalent to the inequality in (L2).

Let us verify condition (L1). Fix $y \in \mathbb{R}^2$ and $q_1 \in Q$. Set $G_1 = y + \tau_1(P + q_1)$, and assume the following conditions hold:

$$G_1 \cap T_{\tau_2}(M) = \varnothing, \tag{12.22}$$

$$G_1 \cap \operatorname{env}(T_{\tau_2}(M)) \neq \varnothing. \tag{12.23}$$

Consider the following two cases: P is a non-degenerate segment and P is a convex polygon.

I. Let P be a non-degenerate segment.

Note that the set $\mathscr{P}$ is two-element (vertices of the segment P), and for any $p \in \mathscr{P}$, the set $\mathscr{N}(p)$ consists of a unique vector.

Since G_1 is a segment, which is parallel to P, and the set env $(T_{\tau_2}(M))$ is a strip, which is parallel to P, inequality (12.23) implies

$$G_1 \subset \text{ env } (T_{\tau_2}(M)).$$

Let us remark that the boundary of the set env $(T_{\tau_2}(M))$ is formed by two supporting lines of the connected set $T_{\tau_2}(M)$. Consequently, in view of (12.22), we can find a vertex $a_* := y + \tau_1(p_* + q_1)$, $p_* \in \mathscr{P}$, of the segment G_1 such that

$$l(a_*, \eta_*) \cap T_{\tau_2}(M) \neq \varnothing, \quad \mathscr{N}(p_*) = \{\eta_*\}. \tag{12.24}$$

Since $a_* \notin T_{\tau_2}(M)$, remembering (12.6), we find $q_2 \in Q$ such that

$$(a_* + \tau_2(P + q_2)) \cap M = \varnothing. \tag{12.25}$$

Besides, (12.24) implies that

$$\exists \alpha > 0 : \ a_* + \alpha\eta_* \in T_{\tau_2}(M).$$

Thus, due to (12.6), we have

$$(a_* + \alpha\eta_* + \tau_2(P + q_2)) \cap M \neq \varnothing. \tag{12.26}$$

Eqs. Define $b_* := a_* + \tau_2(p_* + q_2)$. In view of (12.25), (12.26), and $b_* \in a_* + \tau_2(P + q_2)$, we find

$$l(b_*, \eta_*) \cap M \neq \varnothing.$$

To get property (L1), it remains to prove that

$$l(b_*, -\eta_*) \cap M = \varnothing.$$

Assume the converse. Then we can find

$$x^* \in l(b_*, -\eta_*) \cap M.$$

Choose also $x_* \in l(b_*, \eta_*) \cap M$. Due to

$$l(b_*, \pm\eta_*) \subset \Pi_M(b_*, \nu) \cap \Pi_M(b_*, -\nu), \quad \nu \in \mathscr{V}_P,$$

the assumption (A3) implies that there exist continuous arcs $\gamma_+ \subset \Pi_M(b_*, \nu)$ and $\gamma_- \subset \Pi_M(b_*, -\nu)$ connecting the points x_* and x^*. As a result, the complex arc

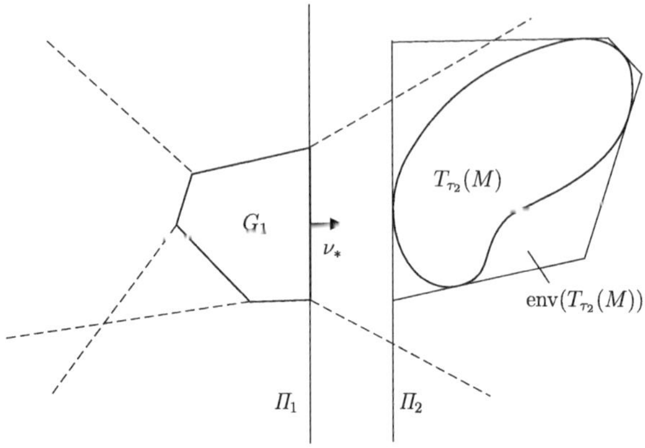

Fig. 12.1 Illustration to the proof of (12.29)

$\gamma_-\gamma_+ \subset M$ encircles the segment $a_* + \tau_2(P + q_2)$. Remembering (12.25), we get the contradiction with the simple connectedness of the set M. This contradiction completes the proof of property (L1).

II. Let P be a convex polygon.

1) We next show that there exists $p_* \in \mathscr{P}$ such that

$$\forall \eta \in \mathscr{N}(p_*) \quad l(y + \tau_1(p_* + q_1), \eta) \cap T_{\tau_2}(M) \neq \varnothing. \tag{12.27}$$

Assume the converse, i.e.,

$$\forall\, p \in \mathscr{P} \quad \exists \eta \in \mathscr{N}(p): \quad l(y + \tau_1(p + q_1), \eta) \cap T_{\tau_2}(M) = \varnothing. \tag{12.28}$$

Since G_1 is a convex polygon, we write

$$G_1 = \bigcap_{\nu \in \mathscr{V}_P} \{z \in \mathbb{R}^2 \colon\ \langle z, \nu \rangle \leq \rho(\nu, G_1)\}.$$

In view of (12.22), we have

$$T_{\tau_2}(M) \subset \mathbb{R}^2 \setminus G_1 = \bigcup_{\nu \in \mathscr{V}_P} \{z \in \mathbb{R}^2 \colon\ \langle z, \nu \rangle > \rho(\nu, G_1)\}.$$

Using (12.28) and the connectedness of the set $T_{\tau_2}(M)$, we find $\nu_* \in \mathscr{V}_P$ (Fig. 12.1) such that

$$T_{\tau_2}(M) \subset \{z \in \mathbb{R}^2 \colon\ \langle z, \nu_* \rangle > \rho(\nu_*, G_1)\}. \tag{12.29}$$

Define

$$\Pi_1 := \{z \in \mathbb{R}^2 \colon \ \langle z, \nu_* \rangle \leq \rho(\nu_*, G_1)\}.$$

Therefore

$$\mathbb{R}^2 \setminus \Pi_1 = \{z \in \mathbb{R}^2 \colon \ \langle z, -\nu_* \rangle < -\rho(\nu_*, G_1)\}.$$

Because of (12.29), we have

$$\Pi_2 := \{z \in \mathbb{R}^2 \colon \ \langle z, -\nu_* \rangle \leq \rho(-\nu_*, T_{\tau_2}(M))\} \subset \mathbb{R}^2 \setminus \Pi_1.$$

The last formula and the definition of the operator env imply

$$\mathrm{env}\,(T_{\tau_2}(M)) \subset \Pi_2 \subset \mathbb{R}^2 \setminus \Pi_1.$$

Remembering that $G_1 \subset \Pi_1$, we have

$$\mathrm{env}\,(T_{\tau_2}(M)) \cap G_1 = \varnothing,$$

that contradicts to (12.23). Thus, (12.27) holds.

2) Set $a_* = y + \tau_1(p_* + q_1)$. In view of (12.22), we get $a_* \notin T_{\tau_2}(M)$. Using (12.6), we find $q_2 \in Q$ such that

$$G_2 \cap M = \varnothing, \quad G_2 := a_* + \tau_2(P + q_2). \tag{12.30}$$

Besides, (12.27) implies that

$$\forall \eta \in \mathcal{N}(p_*) \quad \exists \alpha_\eta > 0 : \quad a_* + \alpha_\eta \eta \in T_{\tau_2}(M).$$

So, considering (12.6), we have

$$\forall \eta \in \mathcal{N}(p_*) \quad \exists \alpha_\eta > 0 : \quad (G_2 + \alpha_\eta \eta) \cap M \neq \varnothing. \tag{12.31}$$

For the chosen p_* and q_2, let us prove that

$$\forall \eta \in \mathcal{N}(p_*) \quad l(a_* + \tau_2(p_* + q_2), -\eta) \cap M = \varnothing. \tag{12.32}$$

Let p^+ and p^- be the vertices of the polygon P adjoining the vertex p_* such that going around the vertices p^-, p_*, p^+ is counterclockwise (Fig. 12.2).

We have $p^+ \neq p^-$. Set

$$b = a_* + \tau_2(p_* + q_2), \quad \eta^{\pm} = p_* - p^{\pm}.$$

By ν^+ (ν^-) denote the outward normal vector to the edge of P that has its vertices at the points p_* and p^+ (respectively, p_* and p^-).

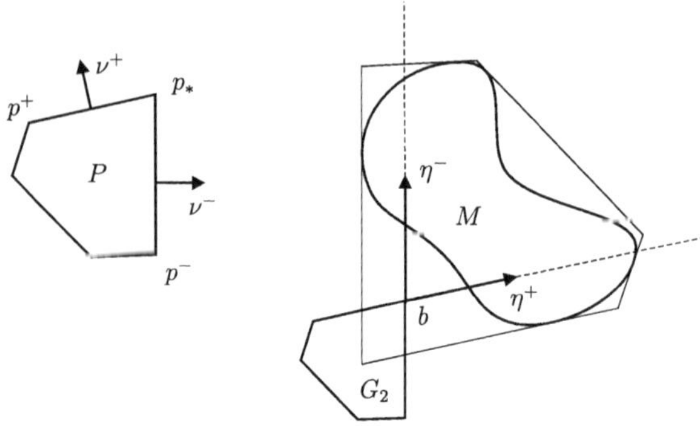

Fig. 12.2 Illustration to the verification of condition (L1)

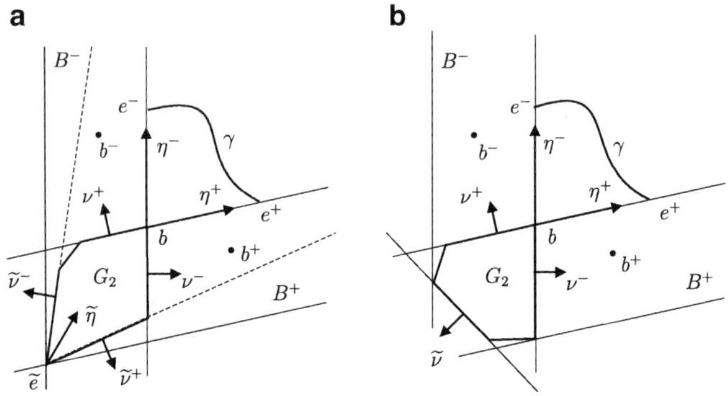

Fig. 12.3 Step 1 of the proof of (12.32)

The proof of (12.32) is divided into two steps. In the first step, we assert an auxiliary statement. At the second step, assuming that (12.32) does not hold, we obtain a contradiction with the simple connectedness of the set M.

Step 1. We claim that there exists a continuous arc γ (Fig. 12.3), which connects some points $e^+ \in l(b, \eta^+)$ and $e^- \in l(b, \eta^-)$, and the inclusion

$$\gamma \subset (b + K) \cap M \tag{12.33}$$

holds, where $K := \{\alpha\eta\colon\ \eta \in \mathcal{N}(p_*),\ \alpha \geq 0\}$.

Define

$$B^{\pm} = \{z + \alpha\eta^{\pm} : \ \alpha > 0,\ z \in G_2\} \setminus G_2.$$

Since $G_2 \cap M = \varnothing$ and (12.31) holds for $\eta = \eta^{\pm}/\|\eta^{\pm}\|$, we have $B^{\pm} \cap M \neq \varnothing$. Fix any $b^{\pm} \in B^{\pm} \cap M$. As $B^+ \cap B^- = \varnothing$, we conclude that $b^+ \neq b^-$.

a) Suppose

$$\operatorname{Arg}\min_{z \in G_2} \langle z, \nu^+\rangle = \operatorname{Arg}\min_{z \in G_2} \langle z, \nu^-\rangle =: E.$$

In this case, the convexity of G_2 implies that the set E consists of a unique vector, i.e., $E = \{\tilde{e}\}$, and $\tilde{e}$ is a vertex of the polygon G_2 (Fig. 12.3a).

By $\tilde{\nu}^+$ and $\tilde{\nu}^-$ denote the outward normals to the edges adjoining the vertex $\tilde{e}$. Assume that the normals are chosen in such a way that the counterclockwise angle from $\tilde{\nu}^-$ to $\tilde{\nu}^+$ is less than π. Set

$$\tilde{K} := \{\alpha(z - \tilde{e}) \colon \ \alpha \geq 0, \ z \in G_2\}.$$

Note that $\tilde{K} \subset K$, $\partial\tilde{K} \cap \partial K = \{0\}$, and

$$b^{\pm} \in B^{\pm} \subset \Pi_M(\tilde{e}, \tilde{\nu}^{\mp}). \tag{12.34}$$

Besides, assumption (A3) implies that the set $\Pi_M(\tilde{e}, \tilde{\nu}^{\pm})$ is connected.

Fix $\tilde{\eta} \in \tilde{K}$. Applying (12.31) for $\eta = \tilde{\eta}/\|\tilde{\eta}\|$, we can find $\tilde{\alpha} > 0$ such that

$$(G_2 + \tilde{\alpha}\tilde{\eta}) \cap M \neq \varnothing.$$

Choose $c \in (G_2 + \tilde{\alpha}\tilde{\eta}) \cap M$. Note that

$$c \in \Pi_M(\tilde{e}, \tilde{\nu}^{\pm}). \tag{12.35}$$

Consider the two possible cases: (i) $c \in (B^+ \cup B^-)$; (ii) $c \notin (B^+ \cup B^-)$.

(i) Suppose $c \in B^{\pm}$. Using (12.34), (12.35), and the connectedness of the set $\Pi_M(\tilde{e}, \tilde{\nu}^{\pm})$, we conclude that there exists a continuous arc $\gamma_1 \subset \Pi_M(\tilde{e}, \tilde{\nu}^{\pm})$ connecting the points c and $b^{\mp}$. In the set $\Pi_M(\tilde{e}, \tilde{\nu}^{\pm})$, the points c and $b^{\mp}$ are separated by the set $G_2 \cup (b + K)$. Applying (12.30), from the arc γ_1 we can single out the required continuous arc γ without self-intersections which lies in the set $b + K$ and connects some points $e^+ \in l(b, \eta^+)$ and $e^- \in l(b, \eta^-)$.

(ii) Suppose $c \notin (B^+ \cup B^-)$. In this case, the point c belongs to the interior of the set $b + K$. We have

$$c, \ b^+ \in \Pi_M(\tilde{e}, \tilde{\nu}^-), \quad c, \ b^- \in \Pi_M(\tilde{e}, \tilde{\nu}^+).$$

Using the connectedness of the sets $\Pi_M(\tilde{e}, \tilde{\nu}^+)$ and $\Pi_M(\tilde{e}, \tilde{\nu}^-)$, we get that there exists a continuous arc $\gamma_1 \subset \Pi_M(\tilde{e}, \tilde{\nu}^+)$ connecting the points c and b^+, and there exists a continuous arc $\gamma_2 \subset \Pi_M(\tilde{e}, \tilde{\nu}^-)$ connecting the points b^- and c. By (12.30), from the complex arc $\gamma_1\gamma_2$ we can single out the required continuous arc γ without self-intersections which lies in the set $b + K$ and connects some points $e^+ \in l(b, \eta^+)$ and $e^- \in l(b, \eta^-)$.

b) It remains to consider the case (Fig. 12.3b)

$$\operatorname{Arg}\min_{z \in G_2}\langle z, \nu^+\rangle \neq \operatorname{Arg}\min_{z \in G_2}\langle z, \nu^-\rangle.$$

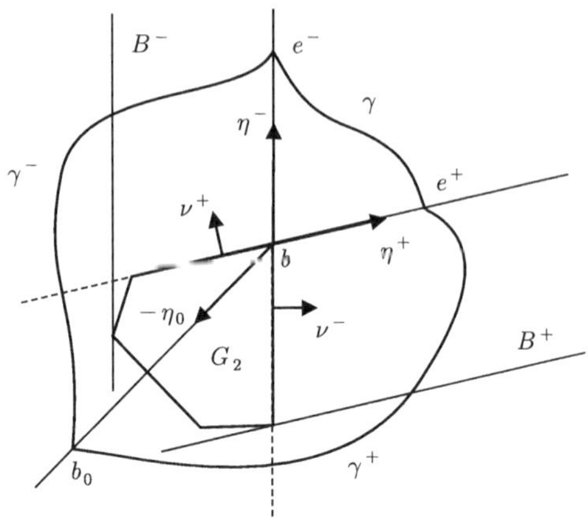

Fig. 12.4 Step 2 of the proof of (12.32)

In this case, we can find $\tilde{\nu} \in \mathscr{V}_P$ such that

$$B^{\pm} \subset \Pi_3 := \{z \in \mathbb{R}^2 : \langle z, \tilde{\nu} \rangle \leq \rho(\tilde{\nu}, G_2)\}.$$

Since $b^+, b^- \in \Pi_3$, assumption (A3) implies that there exists a continuous arc $\gamma_1 \subset \Pi_3 \cap M$ connecting the points b^+ and b^-. In the half-plain Π_3, the points b^+ and b^- are separated by the set $G_2 \cup (b + K)$. By (12.30), from the arc γ_1 we can single out the required continuous arc γ without self-intersections which lies in the set $b + K$ and connects some points $e^+ \in l(b, \eta^+)$ and $e^- \in l(b, \eta^-)$.

Thus, there exists the arc γ with the required properties.

Step 2. Suppose that (12.32) is false, i.e.,

$$\exists\, \eta_0 \in \mathscr{N}(p_*) : \quad l(b, -\eta_0) \cap M \neq \varnothing.$$

Choose $b_0 \in l(b, -\eta_0) \cap M$ (Fig. 12.4).

Let us construct a continuous closed arc without self-intersections which lies in the set M and encircles the set G_2. We have

$$e^+, b_0 \in \Pi_M(b, \nu^+), \quad e^-, b_0 \in \Pi_M(b, \nu^-).$$

By assumption (A3), there exists a continuous arc $\gamma^+ \subset \Pi_M(b, \nu^+)$ connecting the points b_0 and e^+, and there exists a continuous arc $\gamma^- \subset \Pi_M(b, \nu^-)$ connecting the points e^- and b_0. Without loss of generality, we can suppose that the arcs γ^+ and γ^- have no self-intersections.

The complex arc $\gamma^+\gamma\gamma^- \subset M$ is continuous and closed; it has no self-intersections and encircles the set G_2. Using (12.30), we obtain a contradiction with the connectedness of the set M. Thus, relation (12.32) is proved, i.e., condition (L1) holds. □

12.5 The Case of Polygon M

Assumptions (A1)–(A3) of Theorem 12.1 concern only the sets M and P; they are geometric and easy to verify. Assumptions (A4)–(A5) deal with the segment $[0, \vartheta]$.

Theorem 12.2 formulated below (based on Lemma 12.5) asserts that if M is a polygon with some geometric property with respect to P, then there exists some interval of τ where assumptions (A4)–(A5) hold.

Theorem 12.2. *Assume that*

(A1)* $M \subset \mathbb{R}^2$ *is a polygon;*

(A2) $P \subset \mathbb{R}^2$ *is either a non-degenerate segment or a convex polygon;* $Q \subset \mathbb{R}^2$ *is a convex compact set;*

(A3) *for any* $x \in \mathbb{R}^2$ *and* $\nu \in \mathscr{V}_P$, *the set*

$$\Pi_M(x, \nu) = M \cap \{z \in \mathbb{R}^2 : \langle z, \nu \rangle \leqslant \langle x, \nu \rangle\}$$

is connected.

Then there exists $\vartheta > 0$ *such that the operator* T_τ *possesses the semigroup property on the segment* $[0, \vartheta]$. *(And, consequently, we get* $W_0(t) = T_{\vartheta - t}(M)$, $t \in [0, \vartheta]$.*)*

Proof. The assumptions of the theorem contain assumptions (A1)–(A3) of Theorem 12.1.

Note that the set $T_\tau(M)$ is connected for rather small $\tau > 0$, i.e., assumption (A4) holds for rather small ϑ.

Let $\nu \in \mathscr{V}_P$. By assumption (A3) of the theorem, for any $x \in \mathbb{R}^2$ the set $\Pi_M(x, \nu)$ is connected. Since M is a polygon, we choose $z_* \in M$ such that the set $\Pi_M(z_*, \nu)$ is either a triangle or a trapezium. We obtain that the assumptions of Lemma 12.5 hold for $\eta = \nu$. Consequently, there exists $\vartheta > 0$ such that assumption (A5) of Theorem 12.1 is true.

By assumptions (A1)–(A5) of Theorem 12.1, we get that the operator T_τ possesses the semigroup property on the segment $[0, \vartheta]$. □

12.6 Examples on Violation of Assumptions (A3)–(A5)

Let us show that no one assumption from (A3)–(A5) of Theorem 12.1 is excessive, i.e., violation of only one assumption of (A3)–(A5) allows one to find sets M, P, and Q, and instants τ_1 and τ_2 such that equality (12.4) is false.

Below, we consider that P and Q are the segments and $\tau_1 = \tau_2$ for all three examples. The sets M, $\tau_1 P$, $\tau_1 Q$ (thick solid line), $T_{\tau_2}(M)$ (dash line), $T_{\tau_1}(T_{\tau_2}(M))$ (hair line), and $T_{\tau_1+\tau_2}(M)$ (dotted line) are represented in Figs. 12.5–12.7.

Figure 12.5 shows an example such that only the geometric assumption (A3) of Theorem 12.1 is violated: as the set M has a triangle excision at the right, the set

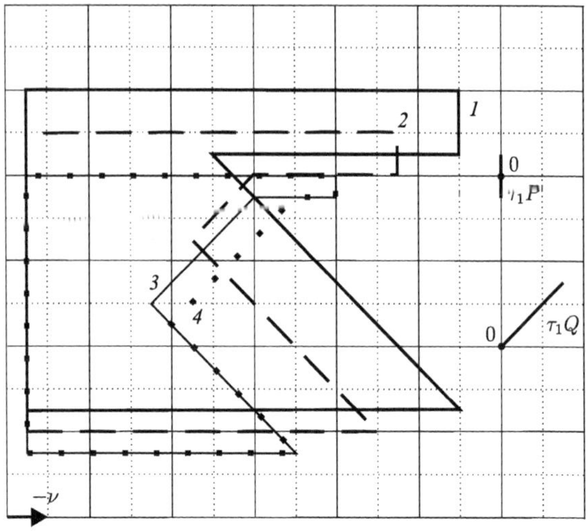

Fig. 12.5 Example 1. Assumption (A3) of Theorem 12.1 is violated; *1* is the boundary of the set M, *2* is the boundary of the set $T_{\tau_2}(M)$, *3* is the boundary of the set $T_{\tau_1}(T_{\tau_2}(M))$, *4* is the boundary of the set $T_{\tau_1+\tau_2}(M)$

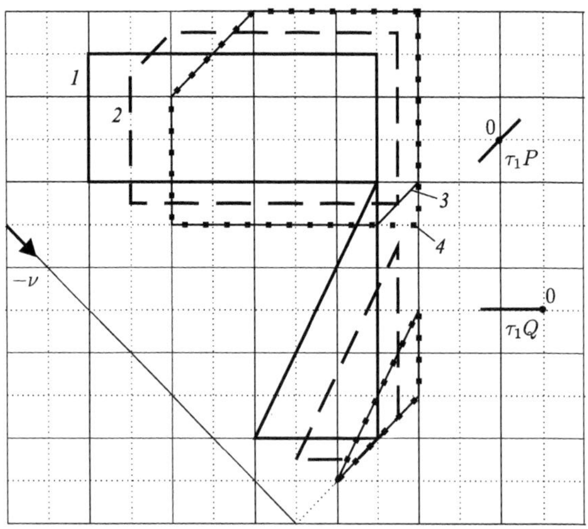

Fig. 12.6 Example 2. Assumption (A4) of Theorem 12.1 is violated; *1* is the boundary of the set M, *2* is the boundary of the set $T_{\tau_2}(M)$, *3* is the boundary of the set $T_{\tau_1}(T_{\tau_2}(M))$, *4* is the boundary of the set $T_{\tau_1+\tau_2}(M)$

$\Pi_M(x,\nu)$ is not connected for some points x (here, $\nu = (-1,0)^{\mathrm{T}}$). The difference between the boundaries of the sets $T_{\tau_1}(T_{\tau_2}(M))$ and $T_{\tau_1+\tau_2}(M)$ takes place in its middle part at the right.

The example in Fig. 12.6 is found in such a way that assumption (A4) concerning the connectedness of the set $T_{\tau_2}(M)$ is violated. The set P is a segment with the slope 45°. Each of the sets $T_{\tau_1}(T_{\tau_2}(M))$ and $T_{\tau_1+\tau_2}(M)$ consists of two disjoint parts. The underparts coincide (the triangle); the upsides are different.

Figure 12.7 gives an example such that the inequality $\delta_\nu(\tau_1+\tau_2) < \delta_\nu(\tau_2)$ holds for $\nu = (-1,0)^{\mathrm{T}}$, i.e., assumption (A5) is violated. The sets $T_{\tau_1}(T_{\tau_2}(M))$ and $T_{\tau_1+\tau_2}(M)$ are different by small triangle in its upper part at the right.

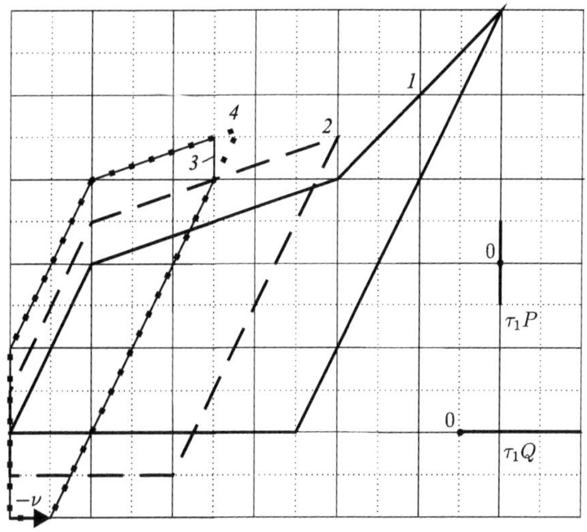

Fig. 12.7 Example 3. Assumption (A5) of Theorem 12.1 is violated: $\delta_\nu(\tau_1 + \tau_2) < \delta_\nu(\tau_2)$; *1* is the boundary of the set M, *2* is the boundary of the set $T_{\tau_2}(M)$, *3* is the boundary of the set $T_{\tau_1}(T_{\tau_2}(M))$, *4* is the boundary of the set $T_{\tau_1+\tau_2}(M)$

Acknowledgments We are grateful to the unknown reviewer for the valuable remarks. This research was carried out in the framework of the Program by Presidium of RAS "Fundamental problems of non-linear dynamics in mathematical and physical sciences" with financial support of Ural Branch of RAS (project No. 12-П-1-1012) and was also supported by RFBR, grant No. 12-01-00537.

Appendix

Let us consider a connection between the question investigated in the present work and the well-known Hopf formula (Alvarez et al. 1999; Bardi and Evans 1984; Hopf 1965).

1) For an arbitrary proper (i.e., not identically equal to $+\infty$) function $g : \mathbb{R}^n \to (-\infty, +\infty]$, we define the Legendre transform

$$g^*(s) = \sup_{x \in \mathbb{R}^n} [\langle x, s \rangle - g(x)], \quad s \in \mathbb{R}^n.$$

By $\operatorname{co} g$ denote the convex hull of the function g. Properties of the function g^* (Rockafellar 1970, Chap. 3, §16; Polovinkin and Balashov 2004) imply that if the proper function $\operatorname{co} g$ is continuous in $\mathbb{R}^n$, then

$$(\operatorname{co} g)^* = g^*. \tag{12.36}$$

The support function $\rho(\cdot, A)$ of a compact set $A \subset \mathbb{R}^n$ is connected with the indicator function

$$\sigma_A(x) = \begin{cases} 0, & x \in A, \\ +\infty, & x \notin A \end{cases}$$

of the set A by the relation

$$\rho(\cdot, A) = \sigma_A^*(\cdot). \tag{12.37}$$

2) The Hopf formula

$$w(t,x) := \sup_{s\in\mathbb{R}^n} [\langle x,s\rangle - \varphi^*(s) + (\vartheta - t)H(s)], \quad s \in R^n, \ x \in R^n, \ t \le \vartheta \tag{12.38}$$

represents the generalized (viscosity (Bardi and Capuzzo-Dolcetta 1997) or minimax (Subbotin 1991, 1995)) continuous solution of the Cauchy problem for the Hamilton–Jacobi equation

$$\begin{gathered} w_t(t,x) + H(w_x(t,x)) = 0, \, t \in (0,\vartheta), \quad x \in \mathbb{R}^n; \\ w(\vartheta, x) = \varphi(x), \, x \in \mathbb{R}^n, \end{gathered} \tag{12.39}$$

with convex continuous terminal function φ (Bardi and Evans 1984).

3) Set

$$H(s) = \max_{q\in Q}\langle q,s\rangle + \min_{p\in P}\langle p,s\rangle, \quad \varphi(s) = \sigma_M(s),$$

where M is a convex compact set. Consider the function $\overline{w}$ defined formally by the Hopf formula (12.38) for these data.

Let us show that

$$T_{\vartheta-t}(M) = \{x \in \mathbb{R}^n : \overline{w}(t,x) \le 0\}. \tag{12.40}$$

As a preliminary, for the convex sets A and B, we establish the relation

$$\rho(\,\cdot\,,\ A \overset{*}{-} B) = \operatorname{co}\big(\rho(\,\cdot\,,\ A) - \rho(\,\cdot\,,\ B)\big). \tag{12.41}$$

We use the equality

$$A \overset{*}{-} B = \bigcap_{b\in B}(A - b).$$

It is known that the support function for the intersection of an arbitrary collection of compact convex sets coincides (Rockafellar 1970) with the convex hull of the function that is the minimum of support functions for the sets used in the intersection. In our case,

$$\min_{b \in B} \rho(s,\ A - b) = \rho(s,\ A) + \min_{b \in B} \langle -b,\ s \rangle = \rho(s,\ A) - \rho(s,\ B).$$

Consequently, (12.41) holds.

Applying (12.41), for the convex set M and $\tau = \vartheta - t$, we get

$$\rho(s, T_\tau(M)) = \text{co } h_\tau(s), \tag{12.42}$$

where

$$h_\tau(s) := \rho(s, M) + \rho(s, -\tau P) - \rho(s, \tau Q) = \rho(s, M) - \tau H(s). \tag{12.43}$$

The compactness of the sets P and Q implies that the function co h_τ is continuous. Using (12.36) and (12.42), we get

$$h_\tau^*(x) = (\text{co } h_\tau)^*(x) = \sup_{s \in \mathbb{R}^n} [\langle x, s \rangle - (\text{co } h_\tau)(s)] = \sup_{s \in \mathbb{R}^n} [\langle x, s \rangle - \rho(s, T_\tau(M))]. \tag{12.44}$$

On the other hand, by (12.43) and (12.37), we calculate

$$h_\tau^*(x) = \sup_{s \in \mathbb{R}^n} [\langle x, s \rangle - h_\tau(s)] = \sup_{s \in \mathbb{R}^n} [\langle x, s \rangle - \sigma_M^*(s) + \tau H(s)]. \tag{12.45}$$

Since

$$T_{\vartheta - t}(M) = \{x \in \mathbb{R}^n :\ \sup_{s \in \mathbb{R}^n} [\langle x, s \rangle - \rho(s, T_{\vartheta - t}(M))] \leq 0\},$$

applying (12.45), we write

$$T_{\vartheta - t}(M) = \{x \in \mathbb{R}^n :\ h_\tau^*(x) \leq 0\}.$$

Comparing (12.38), (12.44), and (12.45), we get (12.40).

4) Thus, if the compact set M is convex, then

$$W_0(t) = T_{\vartheta - t}(M) = \{x \in \mathbb{R}^n : \overline{w}(t, x) \leq 0\}.$$

Now, in the case of a convex set M, we have two variants of useful description of the set $W_0(t)$, whence the guidance problem of the first player to the set M at the fixed instant ϑ is solvable, namely, by the Pshenichnyi formula and by the Hopf formula. The Pshenichnyi formula deals with sets, while the Hopf formula uses functions.

In the paper, for the problems in the plane, we obtain sufficient conditions to describe the set $W_0(t)$ by the Pshenichnyi formula for a non-convex set M. The Hopf formula does not work in the non-convex case.

References

Alvarez O, Barron E, Ishii H (1999) Hopf formulas for semicontinuous data. Indiana U Math J 48(3):993–1035

Aubin J (1991) Viability theory. Birkhäuser, Boston

Bardi M, Capuzzo-Dolcetta I (1997) Optimal control and viscosity solutions of Hamilton – Jacobi – Bellman Equations. Birkhäuser, Boston

Bardi M, Evans L (1984) On Hopf's formulas for solutions of HJ equations. Nonlinear Anal 8:1373–1381

Botkin N (1984) Numerical construction of t-sections of a program absorption set in a linear differential game. In: Subbotin A, Patsko V (eds) Algorithms and programms of solution of linear differential games. Ural Scientific Center of the Academy of Sciences of the USSR, Sverdlovsk, pp 5–38 (in Russian)

Cardaliaguet P, Quincampoix M, Saint-Pierre P (1999) Set-valued numerical analysis for optimal control and differential games. In: Bardi M, Raghavan T, Parthasarathy T (eds) Stochastic and differential games: theory and numerical methods. Annals of the International Society of Dynamic Games, vol 4. Birkhäuser, Boston, pp 177–247

Hadwiger H (1957) Vorlesungen uber inhalt, oberflache und isoperimetrie. Springer, Berlin

Hopf E (1965) Generalized solutions of nonlinear equations of first order. J Math Mech 14(6): 951–973

Isaacs R (1965) Differential games. Wiley, New York

Isakova E, Logunova G, Patsko V (1984) Construction of stable bridges in a linear differential game with fixed terminal time. In: Subbotin A, Patsko V (eds) Algorithms and programms of solution of linear differential games. Ural Scientific Center of the Academy of sciences of the USSR, Sverdlovsk, pp 127–158 (in Russian)

Krasovskii N, Subbotin A (1974) Positional differential games. Nauka, Moscow (in Russian)

Krasovskii N, Subbotin A (1988) Game-theoretical control problems. Springer, New York

Polovinkin E, Balashov M (2004) Elements of convex and strongly convex analysis. Fizmatlit, Moscow (in Russian)

Pontryagin L (1967) Linear differential games, II. Sov Math Dokl 8:910–912

Pontryagin L (1981) Linear differential games of pursuit. Math USSR Sb 40(3):285–303

Pshenichnyy B, Sagaydak M (1971) Differential games with fixed time. J Cybern 1:117–135

Rockafellar R (1970) Convex analysis. Princeton University Press, Princeton

Schwartz L (1967) Cours d'analyse, 1, 2. Hermann, Paris

Subbotin A (1991) Minimax and viscosity solutions of Hamilton – Jacobi equations. Nauka, Moscow (in Russian)

Subbotin A (1995) Generalized solutions of first-order PDEs: the dynamical optimization perspective. Birkhäuser, Boston

Chapter 13
Game with Two Pursuers and One Evader: Case of Weak Pursuers

Sergey Kumkov, Valerii Patsko, and Stéphane Le Ménec

Abstract This paper deals with a zero-sum differential game, in which the first player controls two pursuing objects, whose aim is to minimize the minimum of the misses between each of them and the evader at some given instant. The case is studied when the pursuers have equal dynamic capabilities, but are less powerful than the evader. The first player's control based on its switching lines is analyzed. Results of numeric application of this control are given.

Keywords Pursuit differential games • Fixed termination instant • Positional control • Switching lines

13.1 Introduction

In papers (Ganebny et al. 2012a,b; Le Ménec 2011), a model pursuit problem with two pursuers and one evader is considered. Three inertial objects move in a straight line. Control of each object is scalar and has a bounded value. At some prescribed instant T_1, the distance between the first pursuer and the evader is measured; also, at some instant T_2, the distance between the second pursuer and the evader is checked. The pursuers act together, and their aim is to minimize the payoff, which is the minimum of these two distances. The pursuers can be joined into one player, which will be called *the first player*. The evader is treated as *the second player*, who maximizes the payoff. The obtained problem can be considered as a pursuit

S. Kumkov (✉) • V. Patsko
Institute of Mathematics and Mechanics, S. Kovalevskaya str. 16, Ekaterinburg, 620990, Russia
e-mail: sskumk@gmail.com; patsko@imm.uran.ru

S. Le Ménec
EADS / MBDA France, 1 avenue Réaumur, 92358 Le Plessis-Robinson Cedex, France
e-mail: stephane.le-menec@mbda-systems.com

V. Křivan and G. Zaccour (eds.), *Advances in Dynamic Games*, Annals of the International Society of Dynamic Games 13, DOI 10.1007/978-3-319-02690-9_13,

game, because its practical source is a spacial pursuit, where the instant T_1 (T_2) is the instant of the rendezvous of the first (second) pursuing object with the evading object along the nominal trajectories.

From the point of view of the differential game theory, the model problem described above is interesting because the level sets of the payoff function are non-convex, and, therefore, the time sections of the level sets of the Value function are non-convex too. The authors in works (Ganebny et al. 2012a,b; Le Ménec 2011) distinguish variants of the problem parameters giving qualitatively different solutions of the problem and studied numerically corresponding level sets of the Value function.

The simplest case is the situation of "strong" pursuers when both of them have dynamic advantage over the evader. The most difficult case is when the dynamic advantage passes from a pursuer to the evader or back during the pursuit process. In particular, in this case, level sets appear, whose time sections lose connectedness during the process, and further get it back.

The main problem is to construct efficiently optimal (or quasioptimal) feedback controls of the player. The standard approaches from the differential game theory need either storing entire Value function, or fast computing its value in the neighborhood of the current point. With that, the optimal control is built using some variant of generalized gradient of the Value function (Bardi and Capuzzo-Dolcetta 1997; Isaacs 1965; Krasovskii 1985; Krasovskii and Subbotin 1974, 1988; Tarasyev et al. 1995).

The authors have experience (Botkin and Patsko 1983; Botkin et al. 1984; Patsko 2006; Patsko et al. 1994) of constructing optimal control in linear differential games with convex payoff function on the basis of switching lines and surfaces. Mentioning the switching lines, we mean some separation of the phase space at each instant into some domains, in which each component of the control keeps some of its extreme values. With that, we store the lines only without values of the Value function. In the problem with two pursuers and one evader, the payoff function is not convex, but the authors tried to extend (Ganebny et al. 2012a,b) their algorithms for constructing feedback control on the basis of switching lines for this situation too. For the case of "strong" pursuers, statements proving the optimality of corresponding controls are set forth in work (Ganebny et al. 2012a). For other variants of the game parameters, the switching lines are built also in papers (Ganebny et al. 2012a,b). But there was no strict proof of optimality of the corresponding feedback control methods.

In this paper, such a study is made for the case of equal "weak" pursuers. We assume that $T_1 = T_2$. Under these conditions, we formulate and prove statements about quasioptimality of the first player's control based on the switching lines. Also, we consider the question of stability of this control with respect to inaccuracies of numeric constructions and errors of measurements of the current position of the system.

Sections 13.2 and 13.3 of this paper deal with the formulation of the problem and passage to a two-dimensional equivalent differential game in coordinates of forecasted misses. These sections mostly repeat the corresponding text from paper (Ganebny et al. 2012a). The authors have not reduced this text to keep the readability. The remaining part of the paper is new. In Sect. 13.4, we introduce the concept of an approximating differential game, which is used to construct the switching lines. In the problem under consideration, the first player's control consists of two scalar components u_i, $i = 1, 2$, which are bounded by the constraints $|u_i| \leq \mu_i$. Each component has its own switching line depending on time. The algorithm for constructing the switching lines is described in detail in Sect. 13.5. On one side of the switching line, the corresponding control u_i takes value $u_i = +\mu_i$, on the other side, its value is $u_i = -\mu_i$. It is important that if the system has its current position just in the switching line, then the corresponding control u_i can be taken arbitrary from the interval $[-\mu_i, +\mu_i]$. Auxiliary statements concerning estimates of the Value function change along possible trajectories are proved in Sect. 13.6. Section 13.7 is devoted to the estimate of the guaranteed result, which is provided to the first player by the feedback control on the basis of the switching lines. Results of numeric simulations of the system with usage of the suggested control method are given in Sect. 13.8.

Note that there are a lot of publications dealing with group pursuit problems (multi-agent systems) (Abramyantz and Maslov 2004; Blagodatskih and Petrov 2009; Chikrii 1997; Grigorenko 1991; Hagedorn and Breakwell 1976; Levchenkov and Pashkov 1990; Petrosjan 1977; Petrosjan and Tomski 1983; Pschenichnyi 1976; Rikhsiev 1989; Stipanović et al. 2009, 2012). But these problems are difficult due to high dimension of the state vector and non-convexity of the payoff function. Therefore, usually, some strong conditions are assumed for the dynamics of the objects (for example, objects with simple motions are considered), of their initial positions, etc. In this work, where the number of objects is small, the authors obtain the solution without any essential simplifications of the problem.

13.2 Formulation of Problem

We consider a model differential game with two pursuers and one evader. All three objects move in a straight line. The dynamics descriptions for pursuers P_1 and P_2 are

$$\begin{aligned} &\ddot{z}_{P_1} = a_{P_1}, && \ddot{z}_{P_2} = a_{P_2}, \\ &\dot{a}_{P_1} = (u_1 - a_{P_1})/l_{P_1}, && \dot{a}_{P_2} = (u_2 - a_{P_2})/l_{P_2}, \\ &|u_1| \leq \mu_1, && |u_2| \leq \mu_2, \\ &a_{P_1}(t_0) = 0, && a_{P_2}(t_0) = 0. \end{aligned} \tag{13.1}$$

Here, z_{P_1} and z_{P_2} are the geometric coordinates of the pursuers; a_{P_1} and a_{P_2} are their accelerations generated by the controls u_1 and u_2. The time constants l_{P_1} and l_{P_2} define how fast the controls affect the systems.

The dynamics of the evader E is similar:

$$\ddot{z}_E = a_E, \quad \dot{a}_E = (v - a_E)/l_E, \quad |v| \le \nu, \quad a_E(t_0) = 0. \tag{13.2}$$

Unlike papers (Ganebny et al. 2012a,b; Le Ménec 2011), this work deals with the case of equal pursuers only, that is, we assume that $\mu_1 = \mu_2 = \mu$ and $l_{P_1} = l_{P_2} = l_P$.

Comparing dynamic capabilities of each pursuer P_1 and P_2 and the evader E, one can introduce the parameters (Le Ménec 2011; Shinar and Shima 2002) $\eta = \mu/\nu$, $\varepsilon = l_E/l_P$. We investigate the case of weak pursuers, that is, the situation when the inequalities

$$\eta \le 1, \quad \eta\varepsilon \le 1$$

are true and, at least, one of them is strict.

Let us fix some instant T. At this instant, the misses of the pursuers with respect to the evader are computed:

$$r_{P_1,E}(T) = \bigl|z_E(T) - z_{P_1}(T)\bigr|, \quad r_{P_2,E}(T) = \bigl|z_E(T) - z_{P_2}(T)\bigr|. \tag{13.3}$$

Assume that the pursuers act in coordination. This means that we can join them into one player P (which will be called the *first player*). This player governs the vector control $u = (u_1, u_2)$. The evader is regarded as the *second player*. The resultant miss is the following value:

$$\varphi = \min\{r_{P_1,E}(T),\ r_{P_2,E}(T)\}. \tag{13.4}$$

At any instant t, both players know exact values of all state coordinates z_{P_1}, $\dot{z}_{P_1}$, a_{P_1}, z_{P_2}, $\dot{z}_{P_2}$, a_{P_2}, z_E, $\dot{z}_E$, a_E. The vector composed of these components is denoted as z. The first player choosing its feedback control minimizes the miss φ, the second one maximizes it.

Let the game interval be $[\bar{t}, T]$, where $\bar{t} < T$.

Following Krasovskii and Subbotin (1974, 1988), feasible strategies of the first player are considered as arbitrary functions $(t, z) \mapsto U(t, z)$ with their values in the set $\bigl\{(u_1, u_2) : |u_1| \le \mu, |u_2| \le \mu\bigr\}$.

The symbol $z\bigl(\cdot; t_0, x_0, U, \Delta, v(\cdot)\bigr)$ denotes a stepwise motion of system (13.1), (13.2), which starts from the position (t_0, x_0) when the first player applies a strategy U in a discrete control scheme with the step $\Delta > 0$ and the second player uses a measurable control $v(\cdot)$ with values $\bigl|v(t)\bigr| \le \nu$. The term "discrete scheme of control" means the following. Some grid of instants t_s with the step Δ (starting at the instant t_0) is introduced. Having a position $z(t_s)$ at the instant t_s, the first player computes the vector control $u = U\bigl(t_s, z(t_s)\bigr)$. The first player's control chosen at the instant t_s is kept until the instant $t_{s+1} = t_s + \Delta$. At the position $\bigl(t_{s+1}, z(t_{s+1})\bigr)$, a new control value is chosen, etc.

Assume

$$\Gamma(t_0, z_0, U, \Delta) = \sup_{v(\cdot)} \varphi\big(z(T; t_0.z_0, U, \Delta, v(\cdot)\big).$$

Here, the supremum is taken over all measurable functions $t \mapsto v(t)$ bounded by inequality $|v(t)| \leq \nu$. The quantity $\varphi\big(z(T)\big)$ is the value of the payoff function (13.3), (13.4) at the termination instant T on the motion $z\big(\cdot; t_0, z_0, U, \Delta, v(\cdot)\big)$.

The quantity $\Gamma(t_0, z_0, U, \Delta)$ is the guarantee of the first player provided by the strategy U at the initial position (t_0, z_0) in a discrete scheme of control with the step Δ. The best guarantee of the first player for the initial position (t_0, z_0) is defined by the formula

$$\Gamma(t_0, z_0) = \min_{U} \overline{\lim_{\Delta \to 0}} \Gamma(t_0, z_0, U, \Delta),$$

where the symbol $\overline{\lim}$ denotes the upper limit. In Krasovskii and Subbotin (1974, 1988), it is shown that the minimum on feasible strategies U is reached.

It is known that the best guarantee $\Gamma(t_0, z_0)$ of the first player coincides with the best guarantee for the second player defined symmetrically. Thus, the quantity $\Gamma(t_0, z_0)$ is also called the value $V(t_0, z_0)$ of the Value function at the point (t_0, z_0).

In this paper for the case of weak equal pursuers, it is shown how to find a quasioptimal strategy of the first player (that is, a strategy close to the one optimal on the guaranteed result), which is stable with respect to inaccuracies of its numeric construction and errors of measurement of the current phase state.

13.3 Passage to Two-Dimensional Differential Game

At first, we pass to relative geometric coordinates

$$y_1 = z_E - z_{P_1}, \quad y_2 = z_E - z_{P_2} \tag{13.5}$$

in dynamics (13.1), (13.2) and payoff function (13.4). After this, we have the following notations:

$$\begin{aligned}
\ddot{y}_1 &= a_E - a_{P_1}, & \ddot{y}_2 &= a_E - a_{P_2}, \\
\dot{a}_{P_1} &= (u_1 - a_{P_1})/l_P, & \dot{a}_{P_2} &= (u_2 - a_{P_2})/l_P, \\
\dot{a}_E &= (v - a_E)/l_E, & |u_2| &\leq \mu, \\
|u_1| &\leq \mu, \ |v| \leq \nu, & \varphi &= \min\{|y_1(T)|, |y_2(T)|\}.
\end{aligned} \tag{13.6}$$

State variables of system (13.6) are y_1, $\dot{y}_1$, a_{P_1}, y_2, $\dot{y}_2$, a_{P_2}, a_E; u_1 and u_2 are controls of the first player; v is the control of the second one. The payoff function φ depends on the coordinates y_1 and y_2 at the instant T.

Let us introduce *zero effort miss coordinates* (Shima and Shinar 2002; Shinar and Shima 2002) x_1 and x_2 computed by formula

$$x_i = y_i + \dot{y}_i\tau - a_{P_i} l_P^2 h(\tau/l_P) + a_E l_E^2 h(\tau/l_E), \quad i = 1, 2. \tag{13.7}$$

Here, x_i, y_i, $\dot{y}_i$, a_{P_i}, and a_E depend on t; $\tau = T - t$. Function h is described by the relation $h(\alpha) = e^{-\alpha} + \alpha - 1$. It is very important that $x_i(T) = y_i(T)$. Let $X(t,z)$ be a two-dimensional vector composed of the variables x_1, x_2 defined by formulae (13.5), (13.7).

The dynamics in the new coordinates x_1, x_2 is the following (Le Ménec 2011):

$$\begin{aligned} \dot{x}_1 &= -l_P h(\tau/l_P)u_1 + l_E h(\tau/l_E)v, \quad && |u_1| \le \mu,\ |u_2| \le \mu, \\ \dot{x}_2 &= -l_P h(\tau/l_P)u_2 + l_E h(\tau/l_E)v, \quad && |v| \le \nu. \end{aligned} \tag{13.8}$$

The payoff function is

$$\varphi\big(x_1(T), x_2(T)\big) = \min\{|x_1(T)|,\ |x_2(T)|\}.$$

The first player governs the controls u_1, u_2 and minimizes the payoff φ; the second one has the control v and maximizes φ.

Let $x = (x_1, x_2)^{\mathrm{T}}$ and $V(t,x)$ be the value of the Value function of game (13.8) at the position (t,x). From general results of the differential game theory, it follows that $\boldsymbol{V}(t,z) = V\big(t, X(t,z)\big)$. This relation allows us to compute the Value function of the original game (13.1)–(13.4) using the Value function for game (13.8). The transformation $(t,z) \mapsto x = X(t,z)$ helps also to map the feedback controls in game (13.8) to corresponding controls in game (13.1)–(13.4).

For any $c \ge 0$, a level set (a Lebesgue set) $W_c = \big\{(t,x) : V(t,x) \le c\big\}$ of the Value function in game (13.8) can be treated as the solvability set for the considered game with the result not greater than c, that is, for a differential game with dynamics (13.8) and the terminal set

$$M_c = \big\{(T,x) : |x_1| \le c,\ |x_2| \le c\big\}.$$

Let $W_c(t) = \{x : (t,x) \in W_c\}$ be the time section (t-section) of the set W_c at the instant t.

Problem (13.8) is of the second order on the state variable and can be rewritten as follows:

$$\dot{x} = \mathscr{D}_1(t)u_1 + \mathscr{D}_2(t)u_2 + \mathscr{E}(t)v, \qquad |u_1| \le \mu,\ |u_2| \le \mu,\ |v| \le \nu. \tag{13.9}$$

Here, $x = (x_1, x_2)^{\mathrm{T}}$; vectors $\mathscr{D}_1(t)$, $\mathscr{D}_2(t)$, and $\mathscr{E}(t)$ look like

$$\begin{aligned} &\mathscr{D}_1(t) = \big(-l_P h((T-t)/l_P)^{\mathrm{T}}, 0\big), \qquad \mathscr{D}_2(t) = \big(0, -l_P h((T-t)/l_P)\big)^{\mathrm{T}}, \\ &\mathscr{E}(t) = \big(l_E h((T-t)/l_E), l_E h((T-t)/l_E)\big)^{\mathrm{T}}. \end{aligned}$$

The control of the first player has two independent components u_1 and u_2. The vector $\mathscr{D}_1(t)$ ($\mathscr{D}_2(t)$) is directed along the horizontal (vertical) axis. The second player's control v is scalar.

13.4 Approximating Differential Game

Together with system (13.9), we shall consider an approximating system

$$\dot{x} = D_1(t)u_1 + D_2(t)u_2 + E(t)v, \qquad |u_1| \leq \mu,\ |u_2| \leq \mu,\ |v| \leq \nu, \tag{13.10}$$

which will be used for numeric constructions.

As system (13.10), let us take a system with piecewise-constant functions

$$D_i(t) = \mathscr{D}_i(t_j), E(t) = \mathscr{E}(t_j), t \in [t_j, t_{j+1}), i = 1, 2,$$

which approximate the functions $\mathscr{D}_i(\cdot)$, $\mathscr{E}(\cdot)$, $i = 1, 2$, in some partition of the axis t by instants t_j.

The symbol $x^{(1)}\big(t; t_*, x_*, u(\cdot), v(\cdot)\big)$ (or, shortly, $x^{(1)}(t)$) denotes the position of system (13.9) at an instant t if its motion starts at the instant t_* from the point x_* and is generated by some feasible measurable controls $u(\cdot), v(\cdot)$. Let $x^{(2)}\big(t; t_*, x_*, u(\cdot), v(\cdot)\big)$ (or, shortly, $x^{(2)}(t)$) be the analogical denotation for system (13.10). The difference of the motions $x^{(1)}(\cdot)$ and $x^{(2)}(\cdot)$ at an instant t brought by difference of dynamics (13.9) and (13.10) can be bounded from above by the value

$$\chi(t_*, t) = \sum_{i=1}^{2} \mu \int_{t_*}^{t} \big\| D_i(s) - \mathscr{D}_i(s) \big\| ds + \nu \int_{t_*}^{t} \big\| E(s) - \mathscr{E}(s) \big\| ds.$$

The payoff function for the approximating game is the same as for game (13.9). Note that it obeys the Lipschitz condition with the constant $\lambda = 1$.

Let $V^{(2)}(t, x)$ be the value of the Value function of the approximating game at the position (t, x). Since the right-hand side of the dynamics (13.10) does not include the state variable, the Lipschitz constant of the function $x \mapsto V^{(2)}(t, x)$ for any $t \leq T$ coincides (Subbotin and Chentsov 1981, pp. 110–111) with the Lipschitz constant of the payoff function, that is, with the number $\lambda = 1$.

To find the function $V^{(2)}$, we apply a backward numeric procedure to construct t-sections

$$W_c^{(2)}(t) = \big\{x : V^{(2)}(t, x) \leq c\big\}$$

of its level sets. An algorithm developed by Ganebny uses the specificity of the plane and can be applied to problems with the dynamics piecewise-constant in time. Descriptions of the procedure can be found in works (Ganebny et al. 2012a,b).

For any $c \geq 0$ and $t \leq T$, the set $W_c^{(2)}(t)$ (if it is nonempty) is symmetric with respect to the origin of the plane x_1, x_2 because the same property is incident to dynamics (13.10) (together with the constraints onto the controls) and the payoff function. Moreover, there is the symmetry with respect to the bisectrix of the second and fourth quadrants. The latter is the consequence of the assumption about equality of the pursuers' dynamic capabilities in the game with the common terminal instant T and symmetry of the payoff function with respect to this bisectrix.

The numeric examples below are given for the game with the following parameters:

$$\eta = 0.9, \quad \varepsilon = 0.7, \quad T = 20. \tag{13.11}$$

In Fig. 13.1, the evolution in time of the set $W_c^{(2)}(t)$ for $c = 3.0$ is shown. The symbol $\tau = T - t$ in the marks denotes the backward time. The upper-left subfigure corresponds to the instant T when the game terminates. The section of the level set at this instant is a cross having infinite strips along the axes. The upper-right subfigure shows some intermediate instant when the infinite strips have not collapsed, but they become narrower. The middle-left subfigure shows the instant when the infinite strips disappear, and the t-section of the level set consists of two right triangles touching each other at the origin. Further, these triangles are compressing; also, horizontal and vertical rectilinear parts appear, which grow in the backward time (see the middle-right subfigure). At some instant, the parts parallel to the bisectrix of the first and third quadrants disappear (the lower-left subfigure), and the pentagons turn into squares. After that, the squares contract (the lower-right figure) until the t-section of the level set becomes empty.

In Fig. 13.2, one can see two three-dimensional views of the set $W_c^{(2)}$ in the space t, x_1, x_2 for $c = 3.0$. On the boundary of the set, there are contours of 30 time sections with the time step 0.55. The constructions are made with quite fine time step $\Delta t = 0.05$, so, the obtained set is a quite good approximation of the ideal level set of the Value function of game (13.9).

Figure 13.3 at the top gives the picture of sections $W_c^{(2)}(t)$ computed at the instant $t = 17.0$ ($\tau = 3.0$) for values c in the range from 0 to 50 with the step $\Delta c = 0.5$. It is important to emphasize that there are two points of minimum of the function $V^{(2)}(t, \cdot)$, which are located on the bisectrix of the second and fourth quadrants. The picture of sections at the instant $t = 8.0$ ($\tau = 12.0$) is shown in Fig. 13.2 at the bottom.

Denote by $Z(t)$, $t < T$, the set consisting of these two minimum points of the function $V^{(2)}(t, \cdot)$ in the plane x_1, x_2 at the instant t. With decreasing the direct time t, the points of the sets $Z(t)$ go away from the origin. With that, the global minimal value $c_{\min}(t)$ of the Value function grows.

An important property of dynamics (13.9) is that the directions of the vectors $\mathscr{D}_1(t)$ and $\mathscr{D}_2(t)$ do not change in time. The vectors $D_1(t)$ and $D_2(t)$ of approximating dynamics (13.10) possess the same property. Namely, the vectors $\mathscr{D}_1(t)$ and $D_1(t)$ ($\mathscr{D}_2(t)$ and $D_2(t)$) are directed horizontally (vertically) contrary to the positive direction of the axis x_1 (x_2). In particular, this property provides appearance

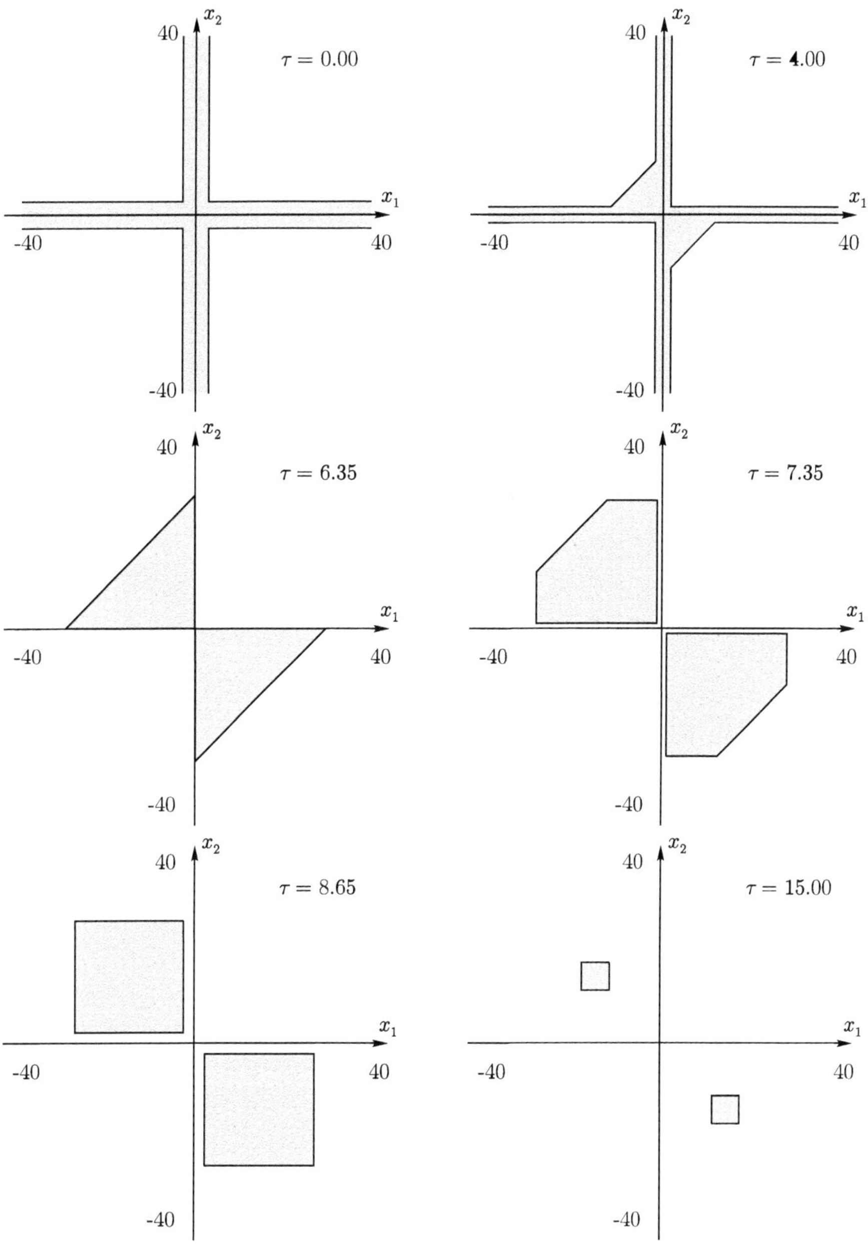

Fig. 13.1 Evolution of the set $W_{3.0}^{(2)}(t)$. The symbol $\tau = T - t$ denotes the backward time

of new horizontal and vertical parts of the boundary of the set $W_c^{(2)}(t)$ when it becomes disconnected.

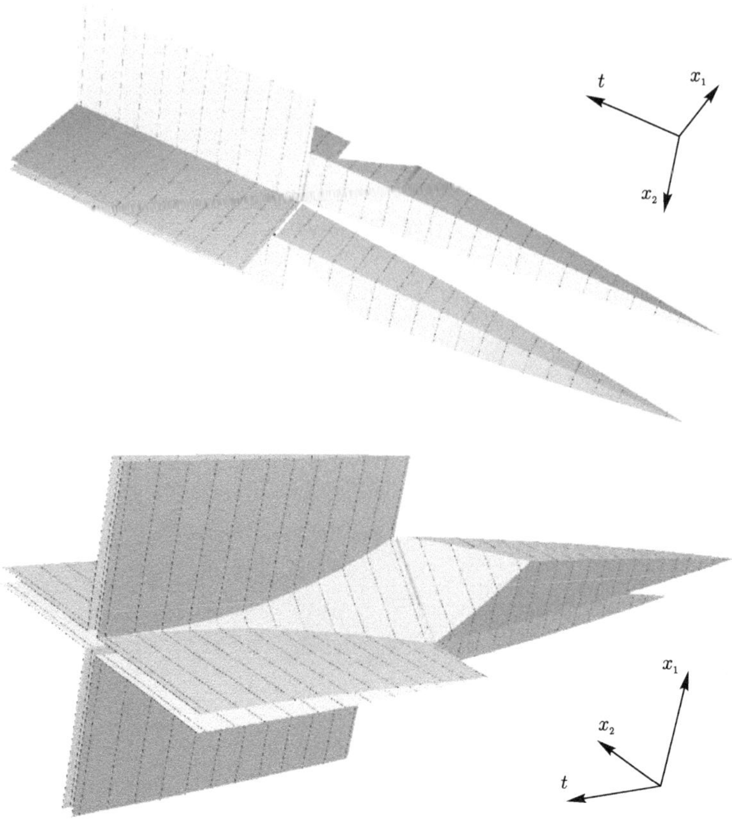

Fig. 13.2 Two three-dimensional views of the level set $W_{3.0}^{(2)}$

Denote by $\tilde{c}(t)$ the value of the function $V^{(2)}(t,\cdot)$ on the axis x_1 (which is the same for all points of the axis) at the instant t. Due to symmetry, the quantity $\tilde{c}(t)$ is also the value of the function $V^{(2)}(t,\cdot)$ on the axis x_2. The specific property of the case of weak pursuers is that the function $t \mapsto \tilde{c}(t)$ decreases with growth of t.

For $c \in [c_{\min}(t), \tilde{c}(t))$, the set $W_c^{(2)}(t)$ consists of two bounded non-intersecting parts.

13.5 Switching Lines

Taking into account that the vector $D_1(t)$ is directed horizontally, we shall study the restrictions of the function $V^{(2)}(t,\cdot)$ to horizontal lines.

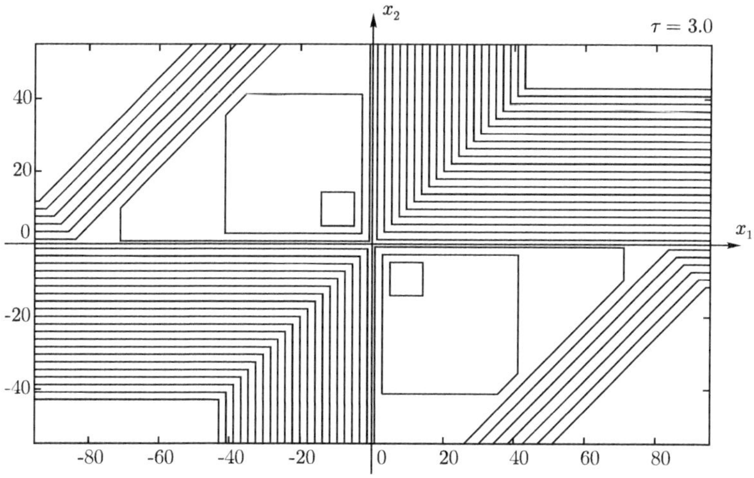

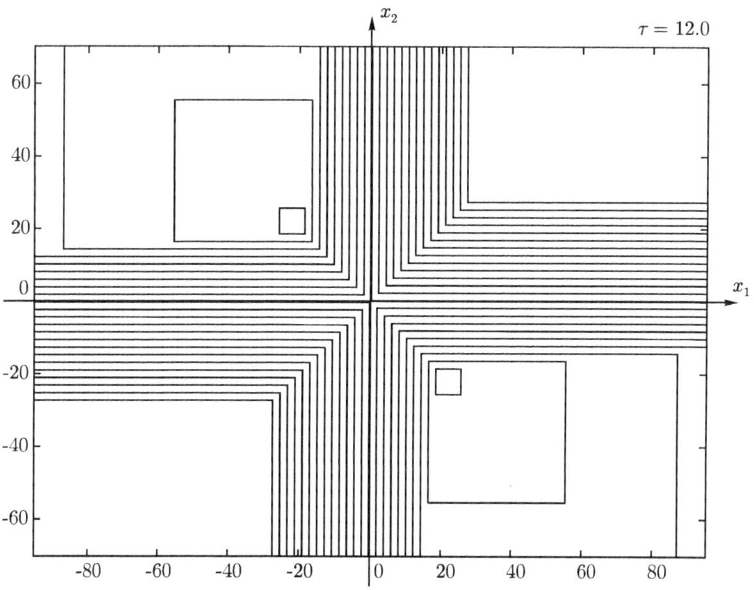

Fig. 13.3 Time sections of a system of level sets at two instants: *upper*: $t = 17.0$; *lower*: $t = 8.0$

In each horizontal line that does not cross the set $W^{(2)}_{\tilde{c}(t)}(t)$, there is only one point of minimum of the function $V^{(2)}(t,\cdot)$, and this points is located on the axis x_2. In points of the axis x_1, we have $V^{(2)}(t,x) = \tilde{c}(t)$. Therefore, the entire axis x_1 consists of points of minimum. For horizontal lines that cross the interior of the set $W^{(2)}_{\tilde{c}(t)}(t)$, there are a lot of points of minimum, and they are rectilinear part of the boundary of some set $W^{(2)}_{c}(t)$. Such a segment degenerates to a point for the line,

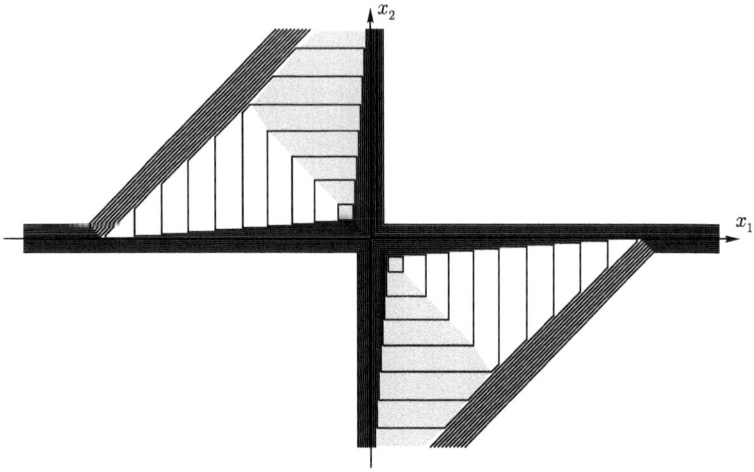

Fig. 13.4 Minima of restrictions of the Value function to horizontal lines

which passes through the point of the global minimum of the function $V^{(2)}(t,\cdot)$. In any horizontal line, the value $V^{(2)}(t,x)$ of the function $V^{(2)}(t,\cdot)$ grows if to move the point x along the line away from the segment of minima. For quite far points, the value is constant.

A set consisting of the segments of minimum is shown in Fig. 13.4 as a shadowed domain.

The same properties of the function $V^{(2)}(t,\cdot)$ take place for vertical lines too, which correspond to the vector $D_2(t)$.

Decomposing the plane x_1, x_2 into horizontal lines and considering restrictions of the function $V^{(2)}(t,\cdot)$, $t < T$, to each of them, one can take the middle point of the corresponding segment of minimum. In the horizontal axis, we take the origin as this middle point. The obtained collection of points can be seen in Fig. 13.5. Let us close this set adding two limit points at the horizontal axis. After that, let us add a horizontal segment that connects these two limit points. As a result, we get a line (Fig. 13.6), which will be denoted by $\Pi(1,t)$ and called the *switching line* for the control u_1 for system (13.10) at the instant t. On the right of this line, let $u_1 = +\mu$, on the left, $u_1 = -\mu$. On the switching line, the control u_1 can be taken arbitrary from the interval $[-\mu, +\mu]$.

The switching line $\Pi(2,t)$ (Fig. 13.7) for the control u_2 is symmetric to the line $\Pi(1,t)$ with respect to the bisectrix of the second and fourth quadrants. Above it, the control is $u_2 = +\mu$, below it, $u_2 = -\mu$. On the switching line, the control u_2 can be arbitrary from the interval $[-\mu, +\mu]$.

The lines $\Pi(1,t)$ and $\Pi(2,t)$ can be considered as exact ones for approximating system (13.10). From the further text, it will be clear that they define the optimal feedback control in system (13.10) and quasioptimal one (that is, close to the optimal one) in system (13.9). Generally speaking, we cannot compute exactly the

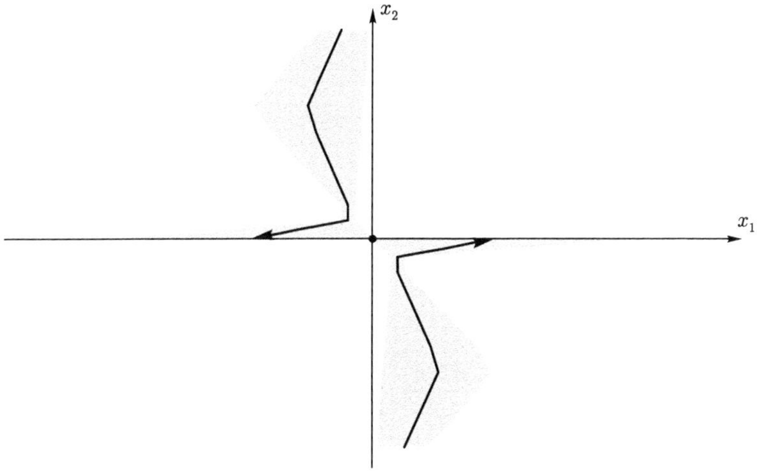

Fig. 13.5 Collection of middles of the minima intervals of the Value function restrictions to horizontal lines

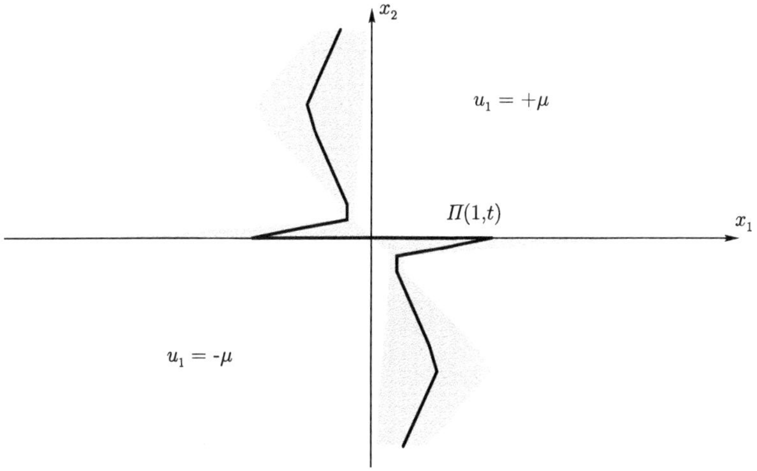

Fig. 13.6 Switching line for the control u_1

lines $\Pi(1,t)$ and $\Pi(2,t)$. For example, even if the sets $W_c^{(2)}(t)$ are constructed ideally, we work with some finite collection of them with some step on the parameter c. As a result, we get polygonal lines, which only approximate the ideal switching lines. Therefore, a very important question is what guarantee they provide for the first player.

For any $t < T$ and any horizontal (vertical) line passing through a point x, let us denote by $\mathscr{V}(1,t,x)$ (respectively, $\mathscr{V}(2,t,x)$) the minimal value of the Value function $V^{(2)}(t,\cdot)$ on this line. One has $\mathscr{V}(1,t,x) = V^{(2)}(t,x)$, when $x \in \Pi(1,t)$, and $\mathscr{V}(2,t,x) = V^{(2)}(t,x)$, when $x \in \Pi(2,t)$.

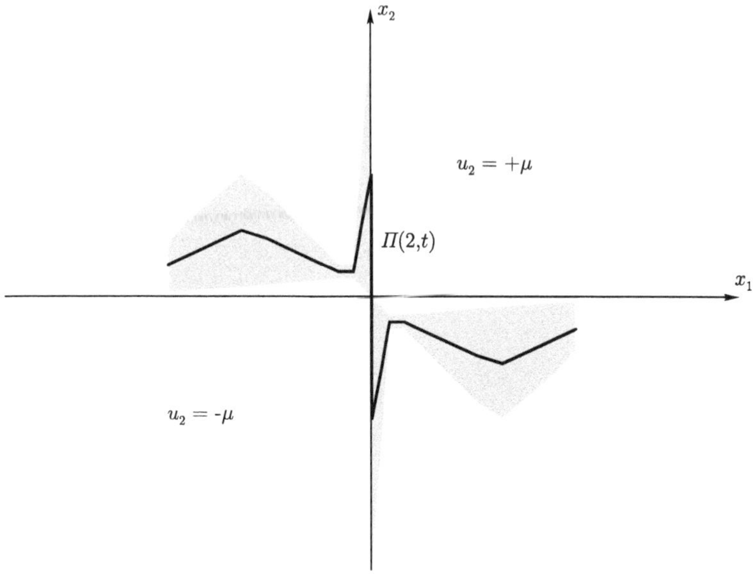

Fig. 13.7 Switching line for the control u_2

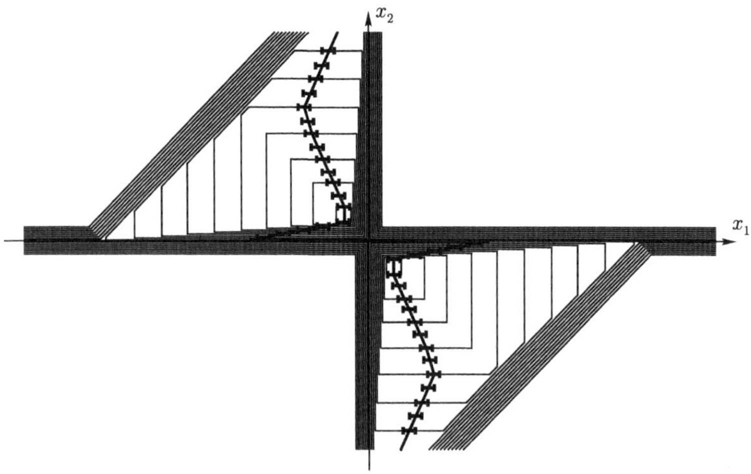

Fig. 13.8 r-extension of the switching line for the component u_1 of the first player's control

Take a number $r \geq 0$ and "expand" the line $\Pi(1,t)$ putting to it horizontal segments with the length $2r$. The obtained set (see Fig. 13.8) is denoted by $\Pi^r(1,t)$. In the same way, using vertical segments, one can construct the set $\Pi^r(2,t)$.

Geometric r-expansion of the ideal switching lines is introduced to deal with the case of inaccurate numeric construction of the switching lines. We would like to "enclose" the switching lines $\Pi(1,t)$ and $\Pi(2,t)$ by a domain, in which the

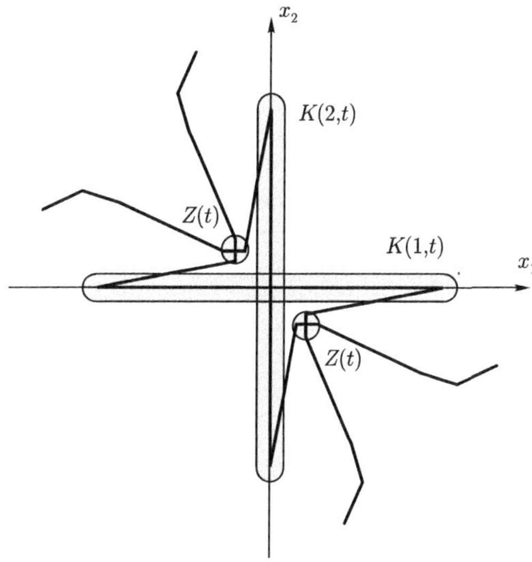

Fig. 13.9 α-neighborhoods of the set $Z(t)$ and the sets $K(1,t)$ and $K(2,t)$

inaccuracies of the construction or measurement errors can be concealed. With that, for the control u_1, it is convenient to use just the horizontal expansion, because in every horizontal line the value $\mathcal{V}(1,t,x)$ is the same for all points. Let after computations we have the following information: (1) we know the value $\mathcal{V}(1,t,x_*)$ in some point x_* at some instant t; (2) the distance from the point x_* to the switching line $\Pi(1,t)$ in the horizontal direction is not greater than r. Then we can obtain an upper estimate for $V^{(2)}(t,x_*)$:

$$V^{(2)}(t,x_*) \le \mathcal{V}(1,t,x_*) + \lambda r.$$

Due to similar reason, the vertical expansion is convenient for the control u_2.

But the expansions are inefficient at the horizontal part of the line $\Pi(1,t)$ and at the vertical part of the line $\Pi(2,t)$. Let us denote the horizontal (vertical) part of the line $\Pi(1,t)$ ($\Pi(2,t)$) by $K(1,t)$ ($K(2,t)$). Choose $\alpha > 0$ and consider closed α-neighborhoods $O\big(\alpha, K(i,t)\big)$, $i = 1, 2$, of these sets.

For further constructions, we need to "prohibit" a fast transfer from the r-expansion $\Pi^r(1,t)$ of the switching line $\Pi(1,t)$ to the r-expansion $\Pi^r(2,t)$ of the switching line $\Pi(2,t)$ and back. The lines $\Pi(1,t)$ and $\Pi(2,t)$ cross in the origin and in two points that constitute the set $Z(t)$. Let us introduce a closed α-neighborhood $O\big(\alpha, Z(t)\big)$ of the set $Z(t)$ (Fig. 13.9). Denote

$$\Pi^r_\alpha(i,t) = \mathrm{cl}\left[\Pi^r(i,t)\setminus\Big(O\big(\alpha,Z(t)\big)\bigcup O\big(\alpha,K(i,t)\big)\Big)\right], \quad i = 1,\ 2.$$

As it was said in Sect. 13.2, we assume that the initial instants in the considered games are in the interval $[\bar{t}, T]$. Let $Y = [\bar{t}, T] \times R^2$ be the space of the games.

The lines $\Pi(1,t)$ and $\Pi(2,t)$, $t < T$, depend continuously on the time. Thus, for any instant $\hat{t} \in [\bar{t}, T)$, one can find such quantities $\hat{\alpha} > 0$ and $\hat{r} > 0$ that

$$\Pi_\alpha^r(1,t) \bigcap \Pi_\alpha^r(2,t) = \varnothing, \quad t \in [\bar{t}, \hat{t}],\ \alpha \geq \hat{\alpha},\ r \in [0, \hat{r}]. \tag{13.12}$$

Moreover, for these values t, α, and r, there is an estimate $\vartheta(\hat{t}, \hat{\alpha}, \hat{r}) > 0$, which is less than the transfer time of systems (13.9) and (13.10) from one of the sets $\Pi_\alpha^r(1,\cdot)$ and $\Pi_\alpha^r(2,\cdot)$ to another.

13.6 Auxiliary Statements

Let us formulate a number of statements, which will be used during the proof of the main theorem about the guarantee when the first player applies in system (13.9) the control based on the switching lines constructed in system (13.10).

Denote by $\Pi_+(1,t)$ ($\Pi_-(1,t)$) the part of the plane, which is strictly on the right (strictly on the left) of the switching line $\Pi(1,t)$. If $x \in \Pi_+(1,t)$ ($x \in \Pi_-(1,t)$), then the control $u_1 = +\mu$ ($u_1 = -\mu$) directs the vector $D_1(t)u_1$ to the switching line, that is, to the area of less values of the Value function $V^{(2)}(t,\cdot)$. In the same way, the symbol $\Pi_+(2,t)$ ($\Pi_-(2,t)$) denotes the part of the plane above (below) the switching line $\Pi(2,t)$.

Let $\sigma = \max\{\|D_1(t)\| : t \in [\bar{t}, T]\}$. Since $\|D_1(t)\| = \|D_2(t)\|$ in the considered case of equal pursuers and equal termination instants $T_1 = T_2 = T$, the value σ is also an upper estimate for the norm of the vector $D_2(t)$ in the time interval $[\bar{t}, T]$.

Lemma 13.1. *Let us fix $i = 1, 2$. Let the position $(t_*, x_*) \in Y$ and the number $\delta > 0$, $t_* + \delta < T$, be such that $x_* \in \Pi_+(i,t_*)$ (or $x_* \in \Pi_-(i,t_*)$) and any motion of system* (13.10) *starting from the point x_* at the instant t_* stays in the set $\Pi_+(i,t)$ ($\Pi_-(i,t)$) for any instant $t \in [t_*, t_* + \delta]$. In the interval $[t_*, t_* + \delta]$, consider an arbitrary motion $x^{(1)}(\cdot)$ of system* (13.9) *starting from the point x_* at the instant t_* under some control $v(\cdot)$ of the second player and some control $u(\cdot)$ of the first player. The latter is such that $u_i \equiv +\mu$ ($u_i \equiv -\mu$) except, maybe, the interval $[t_*, t_* + \omega]$ with a length $\omega \leq \delta$.*

Then for any $t \in [t_*, t_* + \delta]$, the following estimate is true:

$$\mathscr{V}\big(\bar{i}, t, x^{(1)}(t)\big) \leq V^{(2)}(t_*, x_*) + 2\lambda\omega\sigma\mu + \lambda\chi(t_*, t_* + \delta). \tag{13.13}$$

Here, $\bar{i} = 2$ if $i = 1$, and $\bar{i} = 1$ if $i = 2$.

Remark 13.1. Let, for definiteness, $i = 2$ and from the variants + and – the sign + be taken. Then $x_* \in \Pi_+(2,t)$, and the assumption about the "correct" control u_2 type is in agreement with this. The control can differ from $u_2 \equiv +\mu$ in some interval of the length ω only. A feasible control $u_1(\cdot)$ can be arbitrary. The value ω defines the second summand in the right-hand side of estimate (13.13). The third summand is standard addition, which estimates from above the increment of the Value function $V^{(2)}(t_*, t_*)$ caused by difference of the dynamics of systems (13.9) and (13.10).

Proof. Let us assume for definiteness that $i = 2$ and the sign $+$ is chosen.

Together with the motion $x^{(1)}\big(\cdot; t_*, x_*, u(\cdot), v(\cdot)\big)$ of system (13.9), which in the formulation of the lemma is denoted as $x^{(1)}(\cdot)$, let us consider a motion $x^{(2)}\big(\cdot; t_*, x_*, u(\cdot), v(\cdot)\big)$ (or, shortly, $x^{(2)}(\cdot)$) of system (13.10), which is emanated under the same controls $u(\cdot)$ and $v(\cdot)$.

Let $c_* = V^{(2)}(t_*, x_*)$.

Fix some arbitrary instant $t \in [t_*, t_* + \delta]$.

On the basis of the open-loop control $v(\cdot)$, which is considered in the interval $[t_*, t]$, choose an open-loop control $u_{\text{st}}(\cdot)$ such that

$$x_{\text{st}}^{(2)}(t) \in W_{c*}^{(2)}(t), \tag{13.14}$$

where $x_{\text{st}}^{(2)}(\cdot) = x_{\text{st}}^{(2)}\big(\cdot; t_*, x_*, u_{\text{st}}(\cdot), v(\cdot)\big)$ is a motion of system (13.10) starting from the point x_* at the instant t_* under controls $u_{\text{st}}(\cdot)$ and $v(\cdot)$. Such a control can be chosen in any case on the basis of stability property (Krasovskii and Subbotin 1974, 1988) of the level set $W_{c*}^{(2)}$ of the Value function $V^{(2)}$. Inclusion (13.14) means that

$$V^{(2)}\big(t, x_{\text{st}}^{(2)}(t)\big) \le V^{(2)}(t_*, x_*). \tag{13.15}$$

Consider a new control $\hat{u}_{\text{st}}(\cdot)$ with components $\hat{u}_{1\text{st}}(\cdot) = u_{1\text{st}}(\cdot)$ and $\hat{u}_{2\text{st}}(\cdot) \equiv +\mu$. Let $\hat{x}_{\text{st}}^{(2)}(\cdot)$ be a motion of system (13.10) starting from the point x_* at the instant t_* under controls $\hat{u}_{\text{st}}(\cdot)$ and $v(\cdot)$.

The following component-wise relations are true:

$$\hat{x}_{1\text{st}}^{(2)}(t) = x^{(2)}{}_{1\text{st}}(t), \qquad \hat{x}_{2\text{st}}^{(2)}(t) \le x^{(2)}{}_{2\text{st}}(t). \tag{13.16}$$

Since the points $\hat{x}_{\text{st}}^{(2)}(t)$ and $x^{(2)}{}_{\text{st}}(t)$ are in the set $\Pi_+(2, t)$, it follows from (13.16) that

$$V^{(2)}\big(t, \hat{x}_{\text{st}}^{(2)}(t)\big) \le V^{(2)}\big(t, x^{(2)}{}_{\text{st}}(t)\big). \tag{13.17}$$

Due to the hypotheses of the lemma, the component $u_2(\cdot)$ of the vector control $u(\cdot)$ differs from the constant control $\hat{u}_{2\text{st}}(t) \equiv +\mu$ in some interval of the length ω only. Therefore,

$$\left| x_2^{(2)}(t) - \hat{x}_{2\text{st}}^{(2)}(t) \right| \le 2\omega\sigma\mu. \tag{13.18}$$

One has

$$\left| x_2^{(1)}(t) - x_2^{(2)}(t) \right| \le \chi(t_*, t). \tag{13.19}$$

From (13.18) and (13.19), it follows that

$$\left|x_2^{(1)}(t) - \hat{x}_{2\text{st}}^{(2)}(t)\right| \leq 2\omega\sigma\mu + \chi(t_*, t). \tag{13.20}$$

Consider a point z in the horizontal line passing through the point $x^{(1)}(t)$. The point z is the closest one to $\hat{x}_{\text{st}}^{(2)}(t)$. Due to (13.20), one gets

$$\left\|z - \hat{x}_{\text{st}}^{(2)}(t)\right\| \leq 2\omega\sigma\mu + \chi(t_*, t).$$

Consequently,

$$V^{(2)}(t, z) \leq V^{(2)}\big(t, \hat{x}_{\text{st}}^{(2)}(t)\big) + 2\lambda\omega\sigma\mu + \lambda\chi(t_*, t).$$

Taking into account (13.15) and (13.17), this gives

$$V^{(2)}(t, z) \leq V^{(2)}(t_*, x_*) + 2\lambda\omega\sigma\mu + \lambda\chi(t_*, t).$$

Thus,

$$\mathscr{V}(1, t, z) \leq V^{(2)}(t_*, x_*) + 2\lambda\omega\sigma\mu + \lambda\chi(t_*, t). \tag{13.21}$$

Since $\mathscr{V}(1, t, z) = \mathscr{V}\big(1, t, x^{(1)}(t)\big)$, relation (13.21) gives (13.13). □

Lemma 13.2. *Let $(t_*, x_*) \in Y, \delta > 0,\ t_* + \delta \leq T$. Assume that some arbitrary feasible controls $u(\cdot)$ and $v(\cdot)$ act in system (13.9) in the interval $[t_*, t_* + \delta]$. Then, along the corresponding motion $x^{(1)}(\cdot)$ starting from the point x_* at the instant t_*, the following estimate holds*

$$V^{(2)}\big(t, x^{(1)}(t)\big) \leq V^{(2)}(t_*, x_*) + 4\lambda\sigma\mu(t - t_*) + \lambda\chi(t_*, t). \tag{13.22}$$

Proof. Fix some instant $t \in [t_*, t_* + \delta]$.

Consider a motion $x^{(2)}(\cdot)$ of system (13.10) starting from the point x_* at the instant t_* under the controls $u(\cdot)$ and $v(\cdot)$ from the lemma formulation. One has

$$\left\|x^{(1)}(t) - x^{(2)}(t)\right\| \leq \chi(t_*, t). \tag{13.23}$$

Let $c_* = V^{(2)}(t_*, x_*)$. Use the stability property of the set $W_{c_*}^{(2)}$. Then on the basis of the open-loop control $v(\cdot)$, one can choose an open-loop control $u_{\text{st}}(\cdot)$ such that

$$x_{\text{st}}^{(2)}(t) = x^{(2)}\big(t; t_*, x_*, u_{\text{st}}(\cdot), v(\cdot)\big) \in W_{c_*}^{(2)}(t).$$

This means that

$$V^{(2)}\big(t, x_{\text{st}}^{(2)}(t)\big) \leq V^{(2)}(t_*, x_*). \tag{13.24}$$

Taking into account the inequality

$$\left\| x^{(2)}(t) - x_{\mathrm{st}}^{(2)}(t) \right\| \leq 4\sigma\mu(t - t_*)$$

and inequality (13.23), we obtain

$$\left\| x^{(1)}(t) - x_{\mathrm{st}}^{(2)}(t) \right\| \leq 4\sigma\mu(t - t_*) + \chi(t_*, t).$$

Therefore,

$$V^{(2)}\big(t, x^{(1)}(t)\big) \leq V^{(2)}\big(t, x_{\mathrm{st}}^{(2)}(t)\big) + 4\lambda\sigma\mu(t - t_*) + \lambda\chi(t_*, t).$$

From this due to (13.24), inequality (13.22) follows. □

Lemma 13.3. *Let* $(t_*, x_*) \in Y, \delta > 0, t_* + \delta < T$. *Assume that any motion of system* (13.10) *starting from the point* x_* *at the instant* t_* *does not reach the lines* $\Pi(i,t), i = 1, 2,$ *for* $t \in [t_*, t_* + \delta)$.

Assume that along a motion $x^{(1)}(\cdot)$ *of system* (13.9) *starting from the point* x_* *at the instant* t_* *under some feasible controls* $u(\cdot)$ *and* $v(\cdot)$, *it is true that for any* $i = 1, 2$, *in the interval* $[t_*, t_* + \delta)$

- *either* $x^{(1)}(t) \in \Pi_+(i,t)$ *and* $u_i(t) = +\mu$;
- *or* $x^{(1)}(t) \in \Pi_-(i,t)$ *and* $u_i(t) = -\mu$.

Then the following estimate is true:

$$V^{(2)}\big(t_* + \delta, x^{(1)}(t_* + \delta)\big) \leq V^{(2)}(t_*, x_*) + \lambda\chi(t_*, t_* + \delta).$$

The proof of Lemma 13.3 can be done in the same way as for Lemmas 13.1 and 13.2 using the stability property of the set $W_{c_*}^{(2)}$, where $c_* = V^{(2)}(t_*, x_*)$.

Lemma 13.4. *Let* $(t_*, x_*) \in Y, t^* \in (t_*, T)$. *Let* $0 \leq \omega \leq t^* - t_*$. *Assume that for a motion* $x^{(1)}(\cdot)$ *of system* (13.9) *starting from the point* x_* *at the instant* t_* *under some feasible controls* $u(\cdot)$ *and* $v(\cdot)$, *it be true that for any* $i = 1, 2$,

- *either* $x^{(1)}(t) \in \Pi_+(i,t)$ *in the interval* (t_*, t^*) *and* $u_i(t) = +\mu$ *in the interval* $[t_* + \omega, t^*]$;
- *or* $x^{(1)}(t) \in \Pi_-(i,t)$ *in the interval* (t_*, t^*) *and* $u_i(t) = -\mu$ *in the interval* $[t_* + \omega, t^*]$.

Then for any $t \in [t_*, t^*]$, *the following estimate is true:*

$$V^{(2)}\big(t, x^{(1)}(t)\big) \leq V^{(2)}(t_*, x_*) + 4\lambda\omega\sigma\mu + \lambda\chi(t_*, t). \tag{13.25}$$

Proof. Divide the interval $[t_*, t^*]$ by instants $\{t_s\}$, $s = 1, 2, \ldots, e$, $t_1 = t_*$, $t_e = t^*$, $t_{s+1} \leq t_s + \delta$ in such a way that for any interval $[t_s, t_{s+1}]$, $s = 2, \ldots, e - 1$ of the division, no motion of system (13.10) starting from the point $x^{(1)}(t_s)$ at the instant t_s

reaches the lines $\Pi(i,t)$, $i = 1, 2$, $t \in [t_s, t_{s+1}]$. This can be done due to continuity of the switching lines $\Pi(1,t)$ and $\Pi(2,t)$ in time and due to the assumption on the location of the points $x^{(1)}(t)$ with respect to the switching lines.

Due to Lemma 13.3 for any $s = 2, \ldots, e-1$ such that $t_s > t_* + \omega$, one has relation

$$V^{(2)}\big(t_{s+1}, x^{(1)}(t_{s+1})\big) \leq V^{(2)}\big(t_s, x^{(1)}(t_s)\big) + \lambda\chi(t_s, t_{s+1}). \tag{13.26}$$

For s such that $t_s \in [t_*, t_* + \omega]$, from Lemma 13.2, it follows that

$$V^{(2)}\big(t_{s+1}, x^{(1)}(t_{s+1})\big) \leq V^{(2)}\big(t_s, x^{(1)}(t_s)\big) + 4\lambda\delta\sigma\mu + \lambda\chi(t_s, t_{s+1}). \tag{13.27}$$

Fix $t \in [t_*, t^*]$. Using estimates (13.26) and (13.27) for $s = 1, 2, \ldots, e-1$ while $t_s < t$, we get the inequality

$$V^{(2)}\big(t, x^{(1)}(t)\big) \leq V^{(2)}(t_*, x_*) + 4\lambda(\omega + \delta)\sigma\mu + \lambda\chi(t_*, t).$$

Passing to the limit as $\delta \to 0$, one obtains estimate (13.25). □

13.7 Theorem About Guarantee

13.7.1 Estimation of Inaccuracies for Multivalued Strategy of the First Player

Take an arbitrary instant $\hat{t} \in [\bar{t}, T)$. Using $\hat{t}$, choose $\hat{\alpha} > 0$ and $\hat{r} \in (0, \hat{\alpha})$ such that in the interval $[\bar{t}, \hat{t}]$, there is an estimate $\vartheta(\hat{t}, \hat{\alpha}, \hat{r}) > 0$, which is less than the time of the transfer of systems (13.9) and (13.10) from one of the sets $\Pi_{\hat{\alpha}}^{\hat{r}}(1,\cdot)$ and $\Pi_{\hat{\alpha}}^{\hat{r}}(2,\cdot)$ to another. Then, this estimate $\vartheta(\hat{t}, \hat{\alpha}, \hat{r})$ of the transfer time is held for $\alpha \geq \hat{\alpha}$, $r \in [0, \hat{r}]$ too. Note that $r < \alpha$. Instead of $\vartheta(\hat{t}, \hat{\alpha}, \hat{r})$, we write just ϑ.

Assume

$$S(i, \alpha, r, t) = O\big(\alpha, Z(t)\big) \bigcup O\big(\alpha, K(i,t)\big) \bigcup \Pi_\alpha^r(i,t),$$
$$i = 1, 2, \quad \alpha \geq \hat{\alpha},\ r \in [0, \hat{r}],\ t \in [\bar{t}, T).$$

Let us introduce a multivalued strategy $(t,x) \mapsto \mathbf{U}(t,x)$ of the first player. Define that

$$\mathbf{U}_i(t,x) = \big\{u_i : |u_i| \leq \mu\big\}, \text{ if } x \in S(i, \alpha, r, t), \quad i = 1, 2.$$

Outside the set $S(i, \alpha, r, t)$, $t < T$, the component $\mathbf{U}_i(t,x)$, $i = 1, 2$, of the strategy $\mathbf{U}$ is one-valued. Namely, in the position (t,x), the value u_i, $i = 1, 2$, equal either to $+\mu$, or $-\mu$ is taken in such a way that the vector $D_i(t)u_i$ is directed to the switching line $\Pi(i,t)$, which is ideal for system (13.10).

Let the first player apply in system (13.9) the strategy $\mathbf{U}$ in a discrete scheme of control with a step $\Delta \leq \vartheta$. At each instant t_s of the discrete scheme, the first player computes the vector control $u \in \mathbf{U}\big(t_s, x(t_s)\big)$.

We estimate increment of the function $V^{(2)}$ along a motion $x^{(1)}(\cdot)$ starting from the point x_0 at the instant $t_0 \in [\bar{t}, T]$ under the first player's strategy $\mathbf{U}$ in a discrete scheme with a step Δ and some feasible control $v(\cdot)$ of the second player.

Assume

$$\Pi_\alpha^r(t) = \Pi_\alpha^r(1,t) \bigcup \Pi_\alpha^r(2,t), \quad K(t) = K(1,t) \bigcup K(2,t).$$

A. Let us define the following time intervals.

1. The interval $\mathcal{T}_z = [t_z, t^z]$ from the instant t_z of the first entry of the point $x^{(1)}(t)$ in the set $O\big(\alpha, Z(t)\big)$ to the instant t^z of the last leaving the set. That is

$$t_z = \min\big\{t : x^{(1)}(t) \in O\big(\alpha, Z(t)\big)\big\}, \quad t^z = \max\big\{t : x^{(1)}(t) \in O\big(\alpha, Z(t)\big)\big\}.$$

   If $\mathcal{T}_z = \varnothing$, then assume $t^z = t_0$.
2. The interval $\mathcal{T}_k = [t_k, t^k]$ from the instant t_k of the first entry of the point $x^{(1)}(t)$ in the set $O\big(\alpha, K(t)\big)$ to the instant t^k of the last leaving the set. This interval is considered only if $t_k \in [t^z, \hat{t})$.
3. The interval $\mathcal{T}_{\hat{c}} = [t_{\hat{c}}, t^{\hat{c}}]$ from the instant $t_{\hat{c}}$ of the first entry of the point $x^{(1)}(t)$ in the set $O\big(\alpha, W_{\hat{c}}^{(2)}(t)\big)$, where $\hat{c} = \tilde{c}(\hat{t})$, to the instant $t^{\hat{c}}$ of the last leaving the set. This interval is considered only
4. The interval $\mathcal{T}_\flat = [t_\flat, t^\flat]$ for $t^\flat \leq \hat{t}$. Suppose that

$$x^{(1)}(t_\flat) \in \Pi_\alpha^r(t_\flat), \quad x^{(2)}(t^\flat) \in \Pi_\alpha^r(t^\flat).$$

   Assume that the interval $\mathcal{T}_\flat$ is on the right of the instant t^z and beyond the interval $\mathcal{T}_k$. Moreover, let us agree that the interval $\mathcal{T}_\flat$ has the maximal possible length under these conditions.

From the properties conditioning the interval $\mathcal{T}_\flat$, it follows that only two cases of its location are possible: inside the interval $[t^z, t_k]$ or inside the interval $[t^k, \hat{t}]$. If the interval $\mathcal{T}_k$ is absent, then assume $t_k = \hat{t}$.

B. Compute estimates of changing the function $V^{(2)}$ along a motion $x^{(1)}(\cdot)$. The symbol $\mathrm{Var}\big(V^{(2)}, [t_*, t^*]\big)$ denotes the increment of the function $V^{(2)}$ on the interval $[t_*, t^*]$. At first, consider the intervals $\mathcal{T}_z$, $\mathcal{T}_k$, and $\mathcal{T}_{\hat{c}}$.

At the instant t^z, one has

$$V^{(2)}\big(t^z, x^{(1)}(t^z)\big) \leq c_{\min}(t^z) + \lambda\alpha \leq V^{(2)}(t_0, x_0) + \lambda\alpha. \tag{13.28}$$

At the instant t^k, the following estimate is true:

$$V^{(2)}\big(t^k, x^{(1)}(t^k)\big) \le \tilde{c}(t^k) + \lambda\alpha \le \tilde{c}(t_k) + \lambda\alpha.$$

Since

$$\tilde{c}(t_k) \le V^{(2)}\big(t_k, x^{(1)}(t_k)\big) + \lambda\alpha,$$

it holds

$$\operatorname{Var}\big(V^{(2)}, [t_k, t^k]\big) \le 2\lambda\alpha. \tag{13.29}$$

At the instant $t^{\hat{c}}$, it is true that

$$V^{(2)}\big(t^{\hat{c}}, x^{(1)}(t^{\hat{c}})\big) \le \hat{c} + \lambda\alpha. \tag{13.30}$$

C. The estimate for increment of $V^{(2)}$ along a motion $x^{(1)}(\cdot)$ on the interval $\mathscr{T}_\flat$ is not so easy. Assume for definiteness that $x^{(1)}(t_\flat) \in \Pi^r_\alpha(1, t_\flat)$.

Suppose $t_1 = t_\flat$. The symbol t_{1+} denotes the maximal instant belonging to the interval $[t_1, t_1 + \vartheta] \cap [t_1, t^b]$ such that $x^{(1)}(t) \in \Pi^r_\alpha(t)$. Since during a period of the length ϑ, the transfer from the set $\Pi^r_\alpha(1,\cdot)$ to the set $\Pi^r_\alpha(2,\cdot)$ is impossible, one has $x^{(1)}(t_{1+}) \in \Pi^r_\alpha(1, t_{1+})$. It can happen that $t_{1+} = t_1$. To estimate $V^{(2)}\big(t_{1+}, x^{(1)}(t_{1+})\big)$, we can involve Lemma 13.1.

Assume $t_{1+} < t^\flat$. Let t_2 be the minimal instant from the interval $[t_1 + \vartheta, t^\flat]$ such that $x^{(1)}(t) \in \Pi^r_\alpha(t)$. Both cases $x^{(1)}(t_2) \in \Pi^r_\alpha(1, t_2)$ and $x^{(1)}(t_2) \in \Pi^r_\alpha(2, t_2)$ are possible. In any case, the point $x^{(1)}(t)$ in the interval (t_{1+}, t_2) is outside the set $S(1, \alpha, r, t) \bigcup S(2, \alpha, r, t)$, and to estimate the quantity $\operatorname{Var}\big(V^{(2)}, [t_{1+}, t_2]\big)$ one can use Lemma 13.4. Note that $t_2 - t_1 \ge \vartheta$.

If $t_2 < t^\flat$, then introduce an instant t_{2+} defining it as the maximal one in the interval $[t_2, t_2 + \vartheta] \cap [t_2, t^\flat]$ such that $x^{(1)}(t) \in \Pi^r_\alpha(t)$. If $x^{(1)}(t_2) \in \Pi^r_\alpha(1, t_2)$, one has $x^{(1)}(t_{2+}) \in \Pi^r_\alpha(1, t_{2+})$. In the case $x^{(1)}(t_2) \in \Pi^r_\alpha(2, t_2)$, so one gets $x^{(1)}(t_{2+}) \in \Pi^r_\alpha(2, t_{2+})$. Assume that $t_{2+} < t^\flat$. Then introduce an instant t_3 defining it as the maximal one in the interval $[t_2 + \vartheta, t^\flat]$ such that $x^{(1)}(t) \in \Pi^r_\alpha(t)$, etc.

In the interval of the type $[t_j, t_{j+}]$ due to Lemma 13.1, one obtains

$$\mathscr{V}\big(i, t_{j+}, x^{(1)}(t_{j+})\big) \le V^{(2)}\big(t_j, x^{(1)}(t_j)\big) + 2\lambda\Delta\sigma\mu + \lambda\chi(t_j, t_{j+}). \tag{13.31}$$

Here, $i = 1$ if $x^{(1)}(t_j) \in \Pi^r_\alpha(1, t_j)$. The realization of the control u_2 under the strategy **U** can be "wrong" only in the interval $[t_j, t_j + \omega]$, where $\omega \le \Delta$. If $x^{(1)}(t_j) \in \Pi^r_\alpha(2, t_j)$, then assume $i = 2$ in the left-hand side of inequality (13.31).

Passing from $\mathscr{V}\big(i, t_{j+}, x^{(1)}(t_{j+})\big)$ to $V^{(2)}\big(t_{j+}, x^{(1)}(t_{j+})\big)$, we get

$$V^{(2)}\big(t_{j+}, x^{(1)}(t_{j+})\big) \le \mathscr{V}\big(i, t_{j+}, x^{(1)}(t_{j+})\big) + \lambda r.$$

Thus,

$$\mathrm{Var}\big(V^{(2)}, [t_j, t_{j+}]\big) \leq 2\lambda \Delta\sigma\mu + \lambda r + \lambda\chi(t_j, t_{j+}). \tag{13.32}$$

For intervals of the type $[t_{j+}, t_{j+1}]$, it follows that

$$\mathrm{Var}\big(V^{(2)}, [t_{j+}, t_{j+1}]\big) \leq 4\lambda \Delta\sigma\mu + \lambda\chi(t_{j+}, t_{j+1}). \tag{13.33}$$

Due to relations (13.32) and (13.33),

$$\mathrm{Var}\big(V^{(2)}, [t_j, t_{j+1}]\big) \leq 6\lambda \Delta\sigma\mu + \lambda r + \lambda\chi(t_j, t_{j+1}).$$

In the interval $[t_\flat, t^\flat]$, there are not more than $[(t^\flat - t_\flat)/\vartheta]$ intervals of the type $[t_j, t_{j+1}]$. (Here and below, $[\cdot]$ denotes the Entier operation.) The last interval terminating at the instant $t^\flat$ can be an interval of the type $[t_j, t_{j+}]$, where $t_{j+} - t_j \leq \vartheta$. Gathering estimates for all intervals, we get

$$\mathrm{Var}\big(V^{(2)}, [t_\flat, t^\flat]\big) \leq \left(\left[\frac{t^\flat - t_\flat}{\vartheta}\right] + 1\right) \cdot (6\lambda \Delta\sigma\mu + \lambda r) + \lambda\chi(t_\flat, t^\flat). \tag{13.34}$$

D. As it was mentioned above, in the interval $[t^z, \hat{t}]$, not more than two intervals of the type $\mathcal{T}_\flat$ can be located. If there are two of them, then they are separated by an interval of the type $\mathcal{T}_k$. Denote the first of them by $[t_{\flat 1}, t^{\flat 1}]$ and the second one by $[t_{\flat 2}, t^{\flat 2}]$. In the intervals $(t^z, t_{\flat 1})$, $(t^{\flat 1}, t_k)$, $(t^k, t_{\flat 2})$, and $(t^{\flat 2}, \hat{t})$, the point $x^{(1)}(t)$ is outside the set $S(1, \alpha, r, t) \bigcup S(2, \alpha, r, t)$. Therefore, in each interval, we can estimate the increment of the function $V^{(2)}$ using Lemma 13.4 assuming $\omega \leq \Delta$. Doing this and taking into account estimates (13.28), (13.29), and (13.34), we get

$$\begin{aligned}\mathrm{Var}\big(V^{(2)}, [t_0, \hat{t}]\big) \leq &\left(\left[\frac{\hat{t} - t_0}{\vartheta}\right] + 2\right) \cdot (6\lambda \Delta\sigma\mu + \lambda r)\\ &+ 4 \cdot 4\lambda \Delta\sigma\mu + 3\lambda\alpha + \lambda\chi(t_0, \hat{t}).\end{aligned} \tag{13.35}$$

E. Consider the case when for some $t \geq \hat{t}$ the point $x^{(1)}(t)$ is inside the set $O\big(\alpha, W_{\hat{c}}^{(2)}(t)\big)$. Since $Z(t) \subset W_{\hat{c}}^{(2)}(t)$, this includes, in particular, the case when $t^z \geq \hat{t}$.

At the instant $t^{\hat{c}}$, one has estimate (13.30). For $t \geq t^{\hat{c}}$, the point $x^{(1)}(t)$ is outside the set $O\big(\alpha, W_{\hat{c}}^{(2)}(t)\big)$. Since

$$Z(t) \subset W_{\hat{c}}^{(2)}(t), \quad K(t) \subset W_{\hat{c}}^{(2)}(t), \quad r \leq \alpha,$$

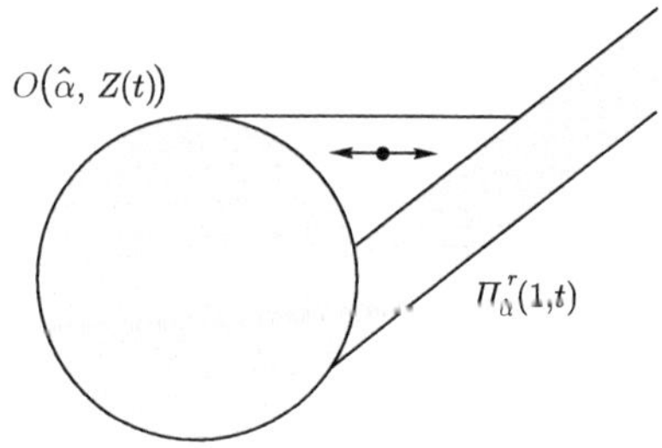

Fig. 13.10 Bad points near the switching set $S(1, \hat{\alpha}, r, t)$ at the place of conjunction of the sets $O(\hat{\alpha}, Z(t))$ and $\Pi^r_{\hat{\alpha}}(1, t)$

we get that the motion $x^{(1)}(\cdot)$ is outside the sets $O(\alpha, Z(t)), O(\alpha, K(t)), \Pi^r_\alpha(t)$, and, therefore, along the motion $x^{(1)}(\cdot)$ for $t \geq t^{\hat{c}}$, the "correct" first player's control works except, maybe, an interval $[t^{\hat{c}}, t^{\hat{c}} + \omega]$, where $\omega \leq \Delta$. So, using Lemma 13.4, for $t \in [t^{\hat{c}}, T]$, one gets estimate

$$V^{(2)}(t, x^{(1)}(t)) \leq \hat{c} + \lambda\alpha + 4\lambda\Delta\sigma\mu + \lambda\chi(t^{\hat{c}}, t). \tag{13.36}$$

Let for $t \geq \hat{t}$ the point $x^{(1)}(t)$ be outside the set $O(\alpha, W^{(2)}_{\hat{c}}(t))$. Then the motion is also outside the sets mentioned above, and in estimate of type (13.35) for $\mathrm{Var}(V^{(2)}, [t_0, t])$, only the last summand grows.

F. Thus, the final estimate at the instant is the maximum of two values $F(T)$ and $L(T)$:

$$\begin{aligned} V^{(2)}(T, x^{(1)}(T)) &\leq \max\{F(t), L(t)\}, \\ F(T) &= V^{(2)}(t_0, x_0) + \left(\left[\frac{\hat{t} - t_0}{\vartheta}\right] + 2\right) \cdot (6\lambda\Delta\sigma\mu + \lambda r) \\ &\qquad + 16\lambda\Delta\sigma\mu + 3\lambda\alpha + \lambda\chi(t_0, T), \\ L(T) &= \hat{c} + \lambda\alpha + 4\lambda\Delta\sigma\mu + \lambda\chi(t_0, T). \end{aligned} \tag{13.37}$$

Recall that the quantity $\hat{c}$ depends on $\hat{t}$: $\hat{c} = \tilde{c}(\hat{t})$.

Since $V^{(2)}(T, x^{(1)}(T)) = \varphi(x_1^{(1)}(T), x_2^{(1)}(T))$, inequality (13.37) is an estimate of the first player's guarantee when he uses the strategy **U** in a discrete scheme of control with a step Δ in system (13.9).

A problem with practical application of the strategy **U** is the following. At the instant $t \in [\bar{t}, T)$, there are "bad" points x located outside the set $S(1, \hat{\alpha}, r, t)$ ($S(2, \hat{\alpha}, r, t)$), for which the horizontal (vertical) direction to the line $\Pi(1, t)$ ($\Pi(2, t)$) cannot be determined as the horizontal (vertical) direction from the point x to the set $S(1, \hat{\alpha}, r, t)$ ($S(2, \hat{\alpha}, r, t)$) because the latter direction is not unique. This situation is shown schematically in Fig. 13.10. At the same time due to possible numeric inaccuracies, it is reasonable to think that the switching line for u_1 (u_2) obtained numerically is located in the set $S(1, \hat{\alpha}, r, t)$ ($S(2, \hat{\alpha}, r, t)$).

To exclude this problem for $t \in [\bar{t}, \hat{t}]$, one can do the following thing. Take into account that the line $\Pi(1,t)$ ($\Pi(2,t)$) for $t \in [\bar{t}, \hat{t}]$ crosses the horizontal lines outside the axis x_1 (x_2) with non-zero angle, and there is a lower estimate for this angle. Let us increase α up to some $\breve{\alpha} > \hat{\alpha}$ such that the set $O\big(\breve{\alpha}, Z(t)\big), t \in [\bar{t}, \hat{t}]$, covers that "bad" points in the horizontal (vertical) lines for the set $S(1, \hat{\alpha}, r, t)$ ($S(2, \hat{\alpha}, r, t)$), $r \in [0, \hat{r}]$. Then for each point $x \notin S(1, \breve{\alpha}, r, t)$ ($x \notin S(2, \breve{\alpha}, r, t)$), $t \in [\bar{t}, \hat{t}]$, there is no such a non-uniqueness of the direction to the set $S(1, \hat{\alpha}, r, t)$ ($S(2, \hat{\alpha}, r, t)$) for $r \in [0, \hat{r}]$. Estimate (13.37) holds, but we shall use it for $\alpha = \breve{\alpha}$ only.

For $t \in (\hat{t}, T)$, the choice of the control $u_i, i = 1, 2$, which takes into account the direction of the vector $D_i(t)u_i$ to the switching line $\Pi(i,t)$, is used for obtaining estimate (13.37) only for positions $x^{(1)}(t) \notin O\big(\alpha, W_{\hat{c}}^{(2)}(t)\big)$. If $\alpha \geq \hat{\alpha}, r \in [0, \hat{r}], r \leq \alpha$, and $t \in (\hat{t}, T)$, there is the inclusion $S(i, \alpha, r, t) \subset O\big(\alpha, W_{\hat{c}}^{(2)}(t)\big)$. Thus, the horizontal direction (for $i = 1$) from a point $x \notin O\big(\alpha, W_{\hat{c}}^{(2)}(t)\big)$ to the line $\Pi(1,t)$ coincides with the horizontal direction from this point to the set $S(1, \alpha, r, t)$. In the same way, the vertical direction from a point $x \notin O\big(\alpha, W_{\hat{c}}^{(2)}(t)\big)$ to the line $\Pi(2,t)$ coincides with the horizontal direction from this point to the set $S(2, \alpha, r, t)$.

Theorem 13.1. *Fix* $r \in [0, \hat{r}]$. *Let the multivalued strategy* $\mathbf{U}$ *defined in the interval* $[\bar{t}, T)$ *take the value* $\mathbf{U}_i(t,x) = \{u_i : |u_i| \leq \mu\}$ *in the set* $S(i, \breve{\alpha}, r, t), i = 1, 2$. *Let outside the set* $S(i, \breve{\alpha}, r, t)$ *the value* $\mathbf{U}_i(t,x)$ *equal either to* $+\mu$, *or to* $-\mu$ *be chosen in such a way that the vector* $D_i(t)\mathbf{U}_i(t,x)$ *is directed to the set* $S(i, \hat{\alpha}, r, t), i = 1, 2$. *Then for any initial position* $(t_0, x_0) \in Y$, *the strategy* $\mathbf{U}$ *in a discrete scheme of control with a step* $\Delta \leq \vartheta(\hat{t}, \hat{\alpha}, \hat{r})$ *guarantees in system* (13.9) *to the first player a result, which is described by formula* (13.37), *where* $\alpha = \breve{\alpha}$.

13.7.2 Stability of Suggested Control Method

Let $\hat{\xi}$ be the lower estimate for the angle between the line $\Pi(1,t)$ and horizontal lines outside the axis x_1, when $t \in [\bar{t}, \hat{t}]$. Define $\beta = r \sin \hat{\xi}$. Consider a neighborhood $O\big(\beta, \Pi(1,t)\big), t \in [\bar{t}, T)$. Take an arbitrary continuous line $\pi(1,t)$ in this neighborhood that will be used for constructing the component U_1^* of the strategy U^*. Let x be an arbitrary phase state at some instant t. Consider a ray with the beginning at this point and directing vector $D_1(t)$. If the ray crosses the line $\pi(1,t)$, then define $U_1^*(t,x) = +\mu$, otherwise $U_1^*(t,x) = -\mu$. In the same way, we can introduce a line $\pi(2,t)$ for constructing the component U_2^*. It is clear that the strategy U^* is a one-valued selector from the multivalued strategy $\mathbf{U}$.

Fix an arbitrary $\varepsilon > 0$. Choose an instant $\hat{t}$ such that $\hat{c} = \varepsilon/4$. Let the number $\breve{\alpha}$ obey the relation $3\lambda\breve{\alpha} = \varepsilon/2$. Choose numbers $\hat{\alpha} \in (0, \breve{\alpha})$ and $\hat{r} \in (0, \hat{\alpha}]$ such that there is an estimate $\vartheta(\hat{t}, \hat{\alpha}, \hat{r}) > 0$, which is less than the time of transfer of systems (13.9) and (13.10) from the set $\Pi_{\hat{\alpha}}^{\hat{r}}(1, \cdot)$ to the set $\Pi_{\hat{\alpha}}^{\hat{r}}(2, \cdot)$ and back in the interval $[\bar{t}, \hat{t}]$. Also, we demand that for chosen $\hat{\alpha}$, $\hat{r}$, and $\breve{\alpha}$, the property of absence of "bad" points x outside the sets

$$O\big(\breve{\alpha}, Z(t)\big) \bigcup O\big(\breve{\alpha}, K(i,t)\big) \bigcup \Pi^{r}_{\hat{\alpha}}(i,t), \quad i = 1,\, 2,$$

holds. Then the quantity

$$A(\hat{t}, \hat{\alpha}, \hat{r}) = \left[\frac{T - \bar{t}}{\vartheta(\hat{t}, \hat{\alpha}, \hat{r})} \right] + 2$$

is a fixed number. We choose $r^* \in (0, \hat{r}]$ and $\Delta^* \leq \vartheta(\hat{t}, \hat{\alpha}, \hat{r})$ such that

$$A(\hat{t}, \hat{\alpha}, \hat{r}) \cdot (6\lambda\Delta^*\sigma\mu + \lambda r^*) + 16\lambda\Delta^*\sigma\mu \leq \frac{\varepsilon}{2}.$$

For $\beta = r \sin \hat{\xi}$, one has

$$O\big(\beta, \Pi(i,t)\big) \subset S(i, \hat{\alpha}, r, t), \quad i = 1,\, 2,\ t \in [\bar{t}, \hat{t}].$$

From this due to (13.37), where $\alpha = \breve{\alpha}$, it follows that there exist such $\beta^* > 0$ and $\Delta^* > 0$ that for any $\beta \in [0, \beta^*]$ and $\Delta \in (0, \Delta^*]$, the following estimate is true:

$$\varphi\big(x_1^{(1)}(T), x_2^{(1)}(T)\big) \leq V^{(2)}(t_0, x_0) + \varepsilon + \lambda\chi(t_0, T). \tag{13.38}$$

Estimates (13.37) and (13.38) concern the case when at an instant t the first player knows the exact position $x^{(1)}(t)$ of system (13.9) while constructing its control. Now, let us consider the case of inexact measurements.

Assume that instead of the true position $x^{(1)}(t)$ at an instant t, the first player gets some measurement $\zeta(t)$ such that $\big\|\zeta(t) - x^{(1)}(t)\big\| \leq h$. He uses this measurement to produce the control $U^*\big(t, \zeta(t)\big)$. As a consequence from estimates (13.37) and (13.38), the next statement follows.

Corollary 13.1. *For any $\varepsilon > 0$, one can choose numbers $\gamma^* > 0, h^* > 0$, and $\Delta^* > 0$ such that if the strategy U^* in system* (13.9) *is built on the basis of the switching lines $\pi(1,t)$ and $\pi(2,t)$ located for each $t \in [\bar{t}, T)$ in the sets $O\big(\gamma^*, \Pi(1,t)\big)$ and $O\big(\gamma^*, \Pi(2,t)\big)$, respectively, the measurement inaccuracy is not greater than h^*, and the step $\Delta > 0$ of the discrete scheme of control obeys the inequality $\Delta \leq \Delta^*$, then for any initial position $(t_0, x_0) \in Y$ and for any realization $v(\cdot)$ of the second player's control, estimate* (13.38) *holds.*

To prove this statement, it is sufficient to take $\gamma^* \leq \beta^*/2, h^* \leq \beta^*/2$.

Remark 13.2. We talk about the strategy U^* as a quasioptimal one for system (13.9). The last summand in estimate (13.38) decreases as approximating system (13.10) gets closer to system (13.9). With that, the value $V^{(2)}(t_0, x_0)$ tends to the value $V(t_0, x_0)$ of the Value function for system (13.9). It is reasonable to investigate the limit of the switching lines $\Pi(1,t)$ and $\Pi(2,t)$. It is natural to try to prove that the limit lines define an optimal strategy of the first player in game (13.9). But in this work, we do not deal with such a study.

Remark 13.3. The ideal switching lines $\Pi(1,t)$ and $\Pi(2,t)$ for system (13.10) define an optimal strategy for all initial positions $(t_0, x_0) \in Y$ in this system. This strategy is stable with respect to small inaccuracies of numeric constructions and errors of measurement of the phase state of the system. This follows from estimate (13.38).

13.8 Simulation Results

Let the pursuers P_1, P_2, and the evader E move in the plane. This plane is called the *original geometric space*. At the initial instant t_0, velocities of all objects are parallel to the horizontal axis and sufficiently greater than the possible changes of the lateral velocity components. Velocity of each object has a constant component parallel to the horizontal axis. Magnitudes of these components are such that the horizontal crossings of the objects P_1 and E and the objects P_2 and E happen at the instant T. The dynamics of lateral motion is described by relations (13.1), (13.2); the resultant miss is given by formula (13.4).

Parameters of the game are taken as (13.11). The initial lateral velocities and accelerations are assumed to be zero:

$$\dot{z}^0_{P_1} = \dot{z}^0_{P_2} = \dot{z}^0_E = 0, \;\; a^0_{P_1} = a^0_{P_2} = a^0_E = 0.$$

The initial instant is $t_0 = 0$.

In the following figures, the horizontal axis is denoted by the symbol d. So, the coordinate d shows the longitudinal position of the objects. Controls of the objects are built on the basis of exact measurements of the players' positions and affect the vertical (lateral) coordinate.

In Fig. 13.11, trajectories of the objects are shown for the following values of the initial lateral deviations: $z_{p_1}(t_0) = -2, z_{p_2}(t_0) = 5$. The first player (who joins the pursuers) applies the quasioptimal control generated by the switching lines built in the framework of system (13.10) under quite fine grids on the parameter c and in time. The control of the second player (evader) is produced on the basis of its switching lines, which are also built in the framework of system (13.10). A typical picture of the switching lines for the second player's control is given in Fig. 13.12. There are six domains, in which the feedback control of the second player keeps one of the extreme values $+\nu$ and $-\nu$. The arrows show directions of the vector $E(t)\nu$ in different domains. The procedure for constructing switching lines for the second player is described in Ganebny et al. (2012a). The control of the second player defined by its switching lines is not justified theoretically yet. We consider it as an empirical quasioptimal one.

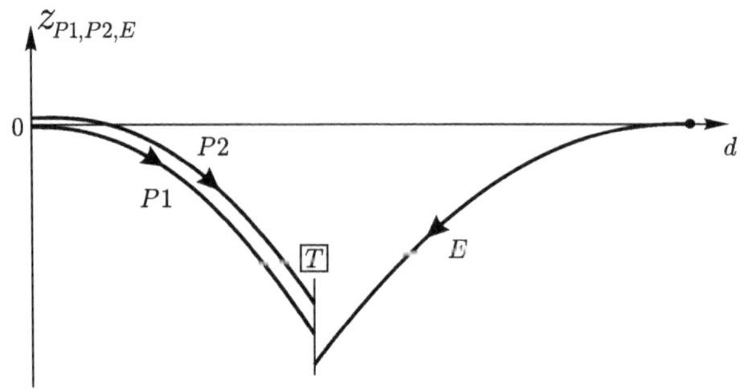

Fig. 13.11 Trajectories of the objects in the original geometric space for small initial deviations after application of the quasioptimal controls

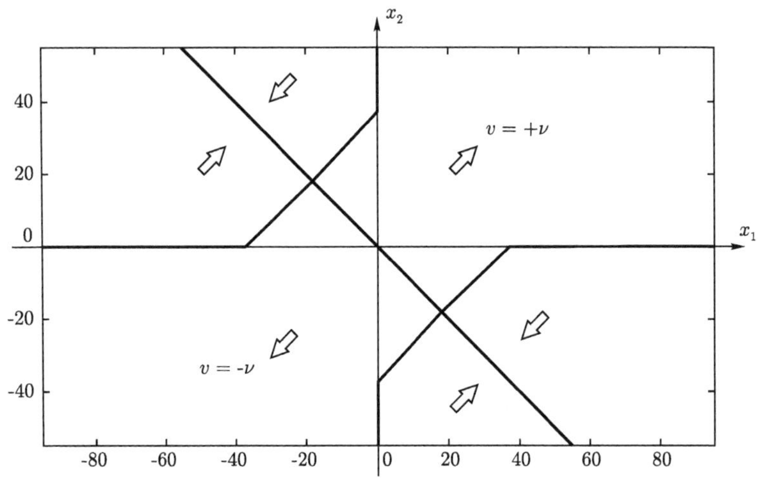

Fig. 13.12 Typical picture of the second player's switching lines

In Fig. 13.13, one can see the trajectories for the same initial lateral deviations, but under a random control of the second player (at each step of the discrete scheme a uniformly distributed value is taken from the interval $[-\nu, +\nu]$ and kept during this step). In comparison with the case of quasioptimal control of the second player, here, the situation of the exact capture is present.

Figure 13.14 shows trajectories for large initial lateral deviations: $z_{P_1}(t_0) = -120, z_{P_2}(t_0) = 150$. The first player uses its quasioptimal control based on the switching lines. The empirical quasioptimal control of the second one is produced by its switching lines.

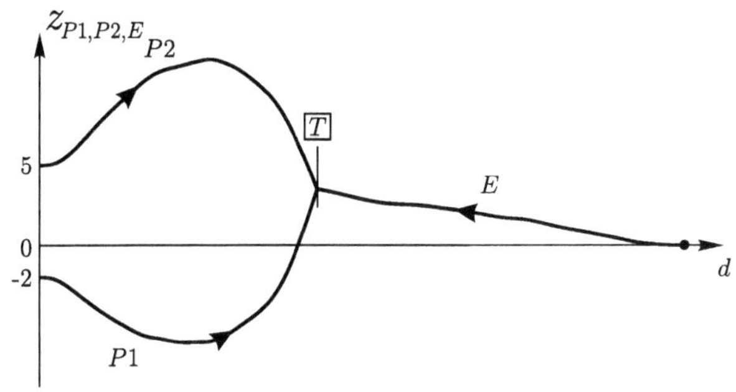

Fig. 13.13 Trajectories of the objects in the original geometric space for small initial deviations after application of the quasioptimal control of the first player and a random control of the second one

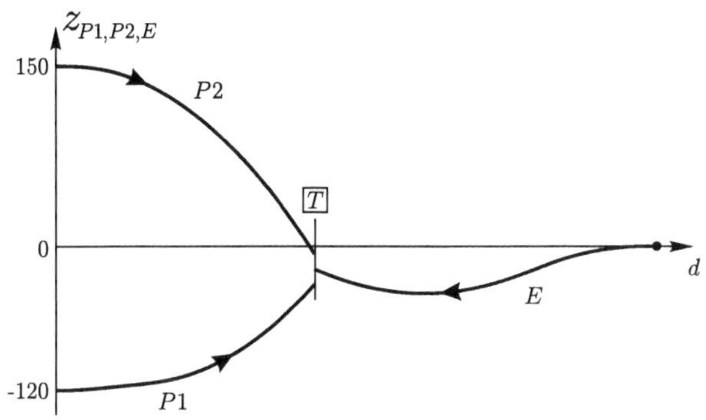

Fig. 13.14 Trajectories of the objects in the original geometric space for large initial deviations after application of the quasioptimal controls

13.9 Conclusion

The main result of the work is in description and justification of a quasioptimal feedback control of the first player in a zero-sum differential game with two "weak" equal pursuers and one evader. The approach is based on construction of two switching lines depending on time for two scalar controls of the first player in the approximating system. The control is stable with respect to inaccuracies of numeric constructions of the switching lines and errors of measurements of the current phase state of the system.

A specific property of the considered problem, which allowed to justify the suggested method of control, is that at the point of crossing the switching lines the value of the Value function of the approximating game decreases with decreasing time-to-go.

Estimates obtained during proof of the main theorem are quite simple, and the interest is in the general scheme of reasoning.

Acknowledgments We are grateful to the unknown reviewer for very helpful remarks.

This work was supported by Program of Presidium RAS "Dynamic Systems and Control Theory" under financial support of UrB RAS (project No.12-*П*-1-1002) and also by the Russian Foundation for Basic Research under grant no.12-01-00537.

References

Abramyantz TG, Maslov EP (2004) A differential game of pursuit of a group target. Izv Akad Nauk Teor Sist Upr 5:16–22 (in Russian)

Bardi M, Capuzzo-Dolcetta I (1997) Optimal control and viscosity solutions of Hamilton-Jacobi-Bellman equations. Birkhäuser, Boston

Blagodatskih AI, Petrov NN (2009) Conflict interaction controlled objects groups. Udmurt State University, Izhevsk (in Russian)

Botkin ND, Patsko VS (1983) Positional control in linear differential game. Eng Cybern 21(4):69–76

Botkin ND, Kein VM, Patsko VS (1984) The model problem of controlling the lateral motion of an aircraft during landing. J App Math Mech 48(4):395–400

Chikrii AA (1997) Conflict-controlled processes. Mathematics and its applications, vol 405. Kluwer Academic, Dordrecht

Ganebny SA, Kumkov SS, Le Ménec S, Patsko VS (2012a) Model problem in a line with two pursuers and one evader. Dyn Game Appl 2(2):228–257

Ganebny SA, Kumkov SS, Le Ménec S, Patsko VS (2012b) Study of linear game with two pursuers and one evader: Different strength of pursuers. In: Cardaliaguet P, Cressman R (eds) Advances in dynamic games. Annals of the international society of dynamic games, vol 12, chap 14. Birkhäuser, Boston, pp 269–292

Grigorenko NL (1991) The problem of pursuit by several objects. In: Differential games — developments in modelling and computation (Espoo, 1990). Lecture notes in control and information sciences, vol 156. Springer, Berlin, pp 71–80

Hagedorn P, Breakwell JV (1976) A differential game with two pursuers and one evader. J Optimiz Theory App 18(1):15–29

Isaacs R (1965) Differential games. Wiley, New York

Krasovskii NN (1985) Control of dynamic system. Nauka, Moscow (in Russian)

Krasovskii NN, Subbotin AI (1974) Positional differential games. Nauka, Moscow (in Russian)

Krasovskii NN, Subbotin AI (1988) Game-theoretical control problems. Springer, New York

Levchenkov AY, Pashkov AG (1990) Differential game of optimal approach of two inertial pursuers to a noninertial evader. J Optimiz Theory App 65:501–518

Le Ménec S (2011) Linear differential game with two pursuers and one evader. In: Breton M, Szajowski K (eds) Advances in dynamic games. Theory, applications, and numerical methods for differential and stochastic games. Annals of the international society of dynamic games, vol 11. Birkhäuser, Boston, pp 209–226

Patsko VS (2006) Switching surfaces in linear differential games. J Math Sci 139(5):6909–6953

Patsko VS, Botkin ND, Kein VM, Turova VL, Zarkh MA (1994) Control of an aircraft landing in windshear. J Optimiz Theory App 83(2):237–267

Petrosjan LA (1977) Differential games of pursuit. Leningrad University, Leningrad (in Russian)

Petrosjan LA, Tomski GV (1983) Geometry of simple pursuit. Nauka, Sibirsk. Otdcl., Novosibirsk (in Russian)

Pschenichnyi BN (1976) Simple pursuit by several objects. Kibernetika 3:145–146 (in Russian)

Rikhsiev BB (1989) Differential games with simple moves. Fan, Tashkent (in Russian)

Shima T, Shinar J (2002) Time-varying linear pursuit-evasion game models with bounded controls. J Guid Control Dynam 25(3):425–432

Shinar J, Shima T (2002) Non-orthodox guidance law development approach for intercepting maneuvering targets. J Guid Control Dynam 25(4):658–666

Stipanović DM, Melikyan AA, Hovakimyan N (2009) Some sufficient conditions for multi-player pursuit-evasion games with continuous and discrete observations. In: Bernhard P, Gaitsgory V, Pourtallier O (eds) Advances in dynamic games and applications. Annals of the international society of dynamic games, vol 10. Springer, Berlin, pp 133–145

Stipanović DM, Tomlin CJ, Leitmann G (2012) Monotone approximations of minimum and maximum functions and multi-objective problems. Appl Math Optim 66:455–473

Subbotin AI, Chentsov AG (1981) Guaranteed optimization in control problems. Nauka, Moscow (in Russian)

Tarasyev AM, Uspenskiy AA, Ushakov VN (1995) Approximation schemes and finite-difference operators for constructing generalized solutions of Hamilton-Jacobi equations. J Comput Syst Sci Int 33(6):127–139

Chapter 14
Collaborative Interception of Moving Re-locatable Target

Stéphane Le Ménec

Abstract This paper is dealing with a team of autonomous vehicles using on-board sensors for tracking and intercepting a moving target. The team of autonomous vehicles is composed of a pursuing vehicle and of several Unmanned Aircraft Vehicles (UAVs). The on-board sensors we talk about have limited capabilities in terms of range. Before acquiring the target, the pursuing vehicle relies on external discrete time information. The target is a slow moving target with respect to the pursuer velocity. The pursuer and the evader are both ground mobile vehicles. The pursuer is able to decide when receiving target re-location information coming from UAVs. UAVs are flying in a cooperative manner with the pursuing vehicle. This situation has been described in terms of a zero-sum two-player pursuit-evasion differential game with costly information. The pursuer minimizes the time to reach the target, while the target tries to evade and to maximize the capture time. After solving this pursuit-evasion game for simple kinematics, test and evaluation simulations with more realistic kinematics have been performed. We also discuss the 4D guidance law and the coordination algorithm we implemented for managing the UAVs.

Keywords Game theory • Differential games • Pursuit-Evasion games • Attainability sets • Minkowski's difference • Co-operative control

S. Le Ménec (✉)
EADS / MBDA, 1, Avenue Réaumur, 92 358 Le Plessis-Robinson Cedex, France
e-mail: stephane.le-menec@mbda-systems.com

V. Křivan and G. Zaccour (eds.), *Advances in Dynamic Games*, Annals of the International Society of Dynamic Games 13, DOI 10.1007/978-3-319-02690-9_14,

14.1 Introduction

Modern systems with local network capability, i.e. communication network between entities in the close vicinity, may allow the implementation of new guidance schemes involving several cooperating platforms; see Le Ménec et al. (2011). In the complex context of future applications as border control; fire fighting; convoy protection; area surveillance; disaster and crisis situation management; and many other applications; air (and ground) autonomous platforms will engage ground objectives with initial goal locations given at launch. We focus on one-on-one interceptions; i.e., one autonomous platform (herein called pursuer) reaching one target objective. In the following, we call target a generic objective. The ground target we consider is a generic vehicle moving slowly with respect to the kinematics of the pursuing platform. However, the pursuing vehicle considered so far is a Lock-On After Launch (LOAL) vehicle due to the sensor acquisition range being limited compared to the distances they must fly before interception. Receiving in-flight updated target designation information from radars covering the entire operation field would require long range network communications. The scenarios examined in this paper avoid using long range communication networks for data latency; operator availabilities, robustness and environment perturbation reasons. Moreover, re-location information based on large coverage radars may require having high altitude observation facilities dedicated to this task. In this manner, completing an interception path requires re-locating of the target due to the amount of time elapsed between target designation and in-flight autonomous target acquisition.

We propose a collaborative guidance scheme which relies on one intercepting vehicle plus on small Unmanned Aircraft Vehicles (UAVs) called observer vehicles flying over the target at prescribed times for target re-location (one unique target is assumed). The guidance scheme specifies when target re-locations are needed to guarantee interception whatever the behavior of the target is. In addition, the overall guidance scheme manages the coordination of observer vehicles to provide these re-locations. This algorithm implemented on-board the intercepting vehicle would guide the interceptor and manage the flight path of some small observer vehicles. Reachability techniques and pursuit-evasion game techniques detailed in Bernhard and Pourtallier (1994), Neveu et al. (1995), Olsder and Pourtallier (1995) and Isaacs (1965) have been used for computing re-location instants. The law applied for guiding the observer vehicles to the right place at the right instant is sometimes referenced as 4D guidance as described in Lee et al. (2007).

In the first part of this paper, using differential game concepts, we describe the one-on-one guidance law we use for moving the pursuing vehicle. Then, the second part of this paper is dedicated to the observer vehicle allocation plan management and the overall guidance algorithm. Examples performed using Matlab simulations are reported in the third section. The main contribution of this paper is related to computing information sets when the evader has heading angle constraints. In addition, using this guidance scheme based on information sets in a co-operative guidance framework involving several platforms is new according to the knowledge of the author.

14.2 Guidance Law Based on Discontinuous Information

We consider a two-player pursuit-evasion game between Pursuer P and Evader E. Maximum speed of P is a. P is not able to see target E beyond range R_0. When P asks information about target location, then a penalty term is applied; i.e. the interception is delayed of δ; i.e. P stops during time duration δ. In addition, E is not able to see P except when P re-localizes E. E has maximum speed b, which is smaller than a ($b < a$). Interception occurs when the range between E and P is less than R_0 (the game stops at $R = R_0$). The objective for P is to minimize the time to intercept E; meaning while E has the opposite objective (zero-sum pursuit-evasion game). The pursuer aims finding a compromise between the number of re-locations (penalty) and imprecise guidance (longer trajectory). The pursuit strategy is a piecewise open loop guidance law.

14.2.1 Information Sets

We start first defining some conventions. We call *stage* an open loop guidance period between two re-locations. Then, we introduce information sets C_n. $C_0 = B(0, R_0)$ is the set of positions where P perceives directly target E. C_0 is a ball centered on P described by distance R_0. Then, C_1 is the set of initial positions such that, whatever the evader does, evader E will be in C_0 at the next stage. In a recursive manner, we define C_n being the set of initial conditions so that P is able to bring E into C_{n-1} in one stage. C_n corresponds also to the set of initial conditions (initial positions of the target in the pursuer coordinates systems) such that the capture is guaranteed in n stages. In addition, we define τ_n as the optimal duration of a stage. During stage n, the position of the target in the pursuer coordinates system is described as follow:

$$x(\tau) = x_{n-1} + \int_0^{\tau} v(s)\, ds - \int_0^{\tau-\delta} u(s)\, ds \tag{14.1}$$

where v and u are, respectively, the evader and pursuer controls and δ is the time penalty. We note Q_τ and P_τ the respective sets of possible movements for E and P. Then,

$$x(\tau) = x_{n-1} + q - p \tag{14.2}$$

With $q \in Q_\tau$ and $p \in P_\tau$, Q_τ and P_τ being the possible moves (attainability sets), respectively, of E and P. Saying that at next stage the state of the game will be in C_{n-1} is equivalent to:

$$\exists\, \tau_n \mid x(\tau_n) \in C_{n-1} \tag{14.3}$$

$$\Longleftrightarrow \exists\, \tau_n \mid x_n + q - p \in C_{n-1} \tag{14.4}$$

$$\Longleftrightarrow \exists\, \tau_n \mid x_n \in (C_{n-1} - Q_\tau + P_\tau) \tag{14.5}$$

$$\Longleftrightarrow x_n \in \bigcup_\tau (C_{n-1} - Q_\tau + P_\tau) \tag{14.6}$$

First remark is that if $\tau < \delta$, then P does not move and as a consequence $\tau_n^{min} = \delta$. In addition, we notice from above equations that a control exists for P if and only if $C_{n-1} - Q_\tau \neq \emptyset$. This last remark provides us with an equation for computing τ_n^{max}.

$$C_n = \bigcup_{\tau_n^{min}}^{\tau_n^{max}} (C_{n-1} - Q_\tau + P_\tau) \tag{14.7}$$

For simplicity reasons, we consider that both players have no curvature constraint and can go in any direction. Therefore, Q_τ and P_τ are simple sets described by circles with radius related to the maximum speed of players.

$$Q_\tau = \int_0^\tau v(s)\, ds \text{ with } ||v|| \leq b = B(0, b\,\tau) \tag{14.8}$$

$$P_\tau = \int_0^{\tau-\delta} u(s)\, ds \text{ with } ||u|| \leq a = B(0, a\,(\tau - \delta)) \tag{14.9}$$

Then, using equation (14.7) and using properties of Minkowski's differences, we compute C_n as a ball of radius R_n and R_n as a function of R_{n-1}.

$$C_n = \bigcup_{\tau_n^{min}}^{\tau_n^{max}} B(0,\ R_{n-1}) - B(0,\ b\,\tau) + B(0,\ a\,(\tau - \delta)) \tag{14.10}$$

$$= B(0,\ a\,(\tau_n^{max} - \delta) - b\,\tau_n^{max} + R_{n-1}) \tag{14.11}$$

And,

$$C_{n-1} - Q_{\tau_n} \neq \emptyset \tag{14.12}$$

$$\Longleftrightarrow B(0,\ R_{n-1} - b\,\tau) \neq \emptyset \tag{14.13}$$

$$\Longleftrightarrow R_{n-1} \leq b\,\tau \tag{14.14}$$

$$\Longleftrightarrow \tau_n^{max} = \frac{R_{n-1}}{b} \tag{14.15}$$

Decision (14.15) can be made also on the basis of geometric reasoning (see Fig. 14.2). When $\delta \to 0$, then we say that the information is free. When the

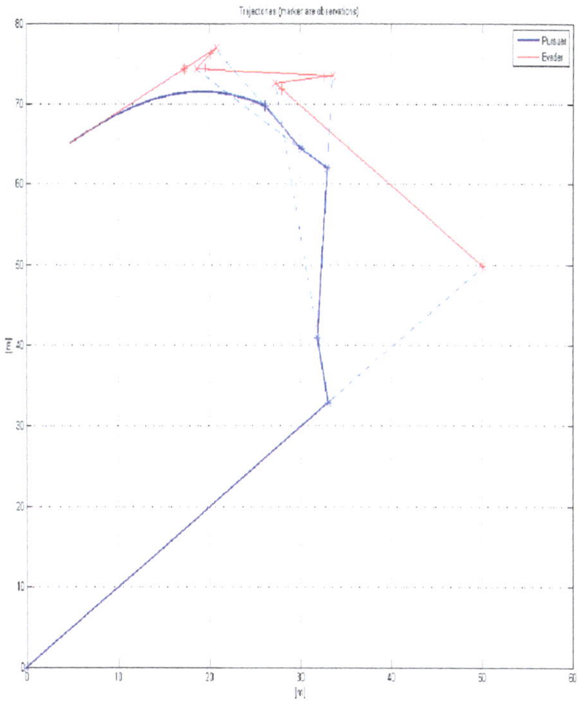

Fig. 14.1 Target trajectory in the pursuer coordinates system with P and P playing in an optimal manner

information is free, the strategies computed using the process detailed above are still piecewise open-loop strategies that minimize the number of relocations. In most of the following sections, we assume that the information is free which is more relevant to ground mobile vehicles. In the scenarios we consider in the following, it is not required for P to stop for receiving information (immediate reception).

14.2.2 Target Without Heading Angle Constraint

For the remaining sections (except for Sect. 14.4.2) we now only consider the case of $\delta = 0$. When P and E have both no curvature constraint, then during a stage the optimal behavior for P is to go towards the last observed position because P does not know if E is going left or right. In the meantime, E is going in the opposite direction in a way to maximize the capture time. Figure 14.1 explains what happens if P and E play optimally. The coordinates system of Fig. 14.1 is centered on P with fixed directions for the horizontal and vertical axes. According to the previous remark, we compute R_{n-1} as a function of R_n:

$$R_{n-1} = R_n + b\tau_n^{max} - a\tau_n^{max} \tag{14.16}$$

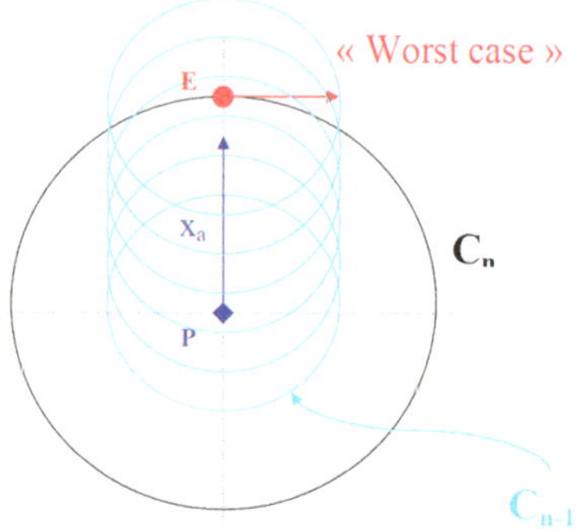

Fig. 14.2 Max duration τ_n^{max} of stage n

However, the min-max reasoning explained above has sense only and only if the game has finite time duration. If during a stage, E is going left or right following a 90° angle direction with respect to the pursuer velocity vector, then P can never see E if the stage duration is too long. P has to keep in mind that E can follow this kind of strategy for computing τ_n^{max}. τ_n^{max} is the longer duration such that P can bring the target in C_{n-1} at next stage. Figure 14.2 describes in a graphic manner Eq. (14.15). If $\tau_n > \tau_n^{max}$, then no control exists anymore; the target can be at more than R_{n-1}.

The duration of a stage can be less than τ_n^{max} if the stage does not start at the extreme boundary of an information set. Reasons why a stage does not start at the boundary of informations sets can be because it is the beginning of the game (first stage) or because at least one player did not play its optimal strategy at previous stage. For each stage, we need the distance from P to E at next stage less than or equal to R_{n-1}; i.e.:

$$||x|| \; - \; a\,\tau_n \; + \; b\,\tau_n \leq R_{n-1} \tag{14.17}$$

Then, the duration of next stage knowing E and P positions at the beginning of the stage is given by the following formula:

$$\tau_n^{opt} = \frac{||x|| \; - \; R_{n-1}}{a \; - \; b} \tag{14.18}$$

Applying the pursuer optimal strategy which is going towards the last observation of E and the stage duration τ_n^{opt} as described in Eq. (14.18) lead to trajectories described in Fig. 14.3 (trajectories in absolute referential) and Fig. 14.4 (trajectories in non-rotating coordinates frame attached to position P). If both players apply their optimal strategies; i.e. E going in the opposite direction to P; P going in direction to E; E and P using their maximum speed; then the end of each stage coincides exactly with the boundary of the next information step as plotted in Fig. 14.5.

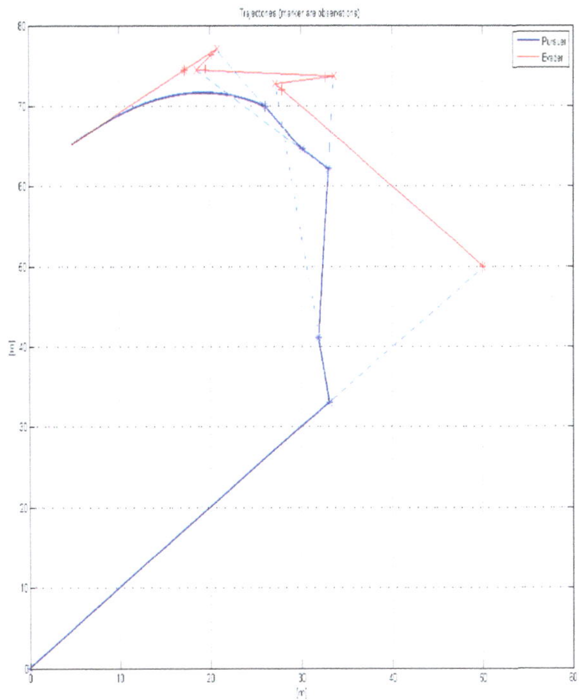

Fig. 14.3 Pursuer P goes towards the last observed position of target / Evader E during the prescribed time duration τ_n^{opt}; P starting at coordinates (0, 0) and E starting at (50, 40)

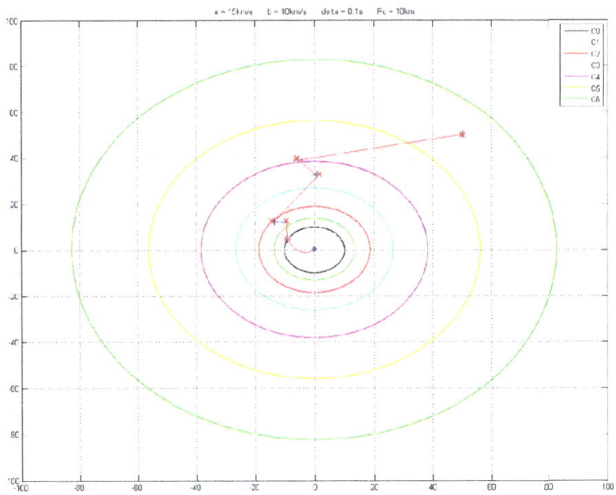

Fig. 14.4 Position of E in the pursuer axes system always crosses from C_n to C_{n-1}

14.2.3 *Target with Heading Angle Constraint*

To limit the possible movements of E and to be less pessimistic, we consider a constraint angle with respect to the pursuer evader line of sight (see Fig. 14.6 for definition of angle α). α is a direction constraint for E defined at relocation which

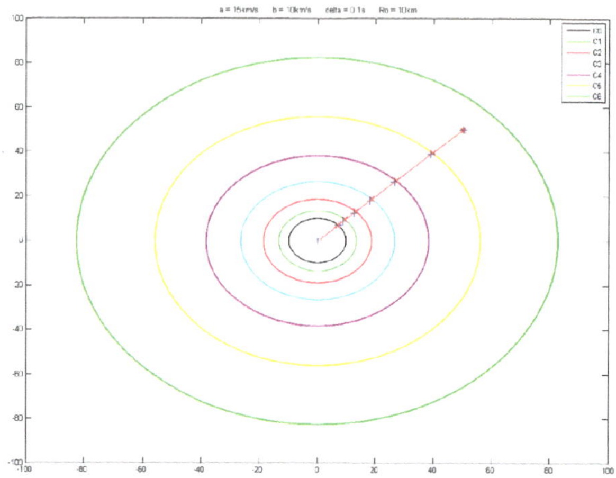

Fig. 14.5 Optimal trajectory in coordinates system centered on P

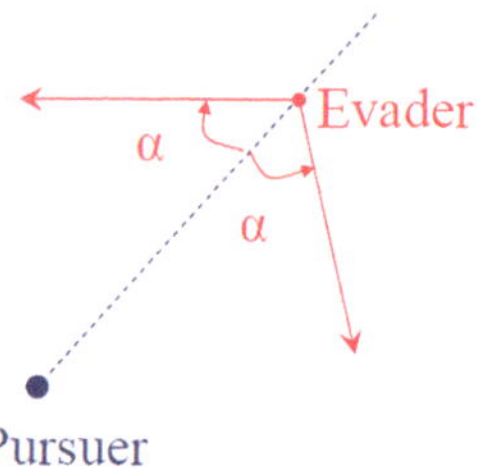

Fig. 14.6 Heading angle constraint on the Evader directions

remains constant all along the following stage. The direction corresponding to α is updated with respect to the line of sight between E and P when P observed the position of E. The meaning of this constraint is that E has to go not too far away from the direction to P. The α constraint can also be viewed as a design parameter that shapes the size of the information sets. Two scenarios have to be analyzed separately: $|\alpha| \geq 90°$ and $|\alpha| < 90°$; α being the maximum heading angle of E; Evader heading angle varying from 0 (towards P) to α. Situations corresponding to $\alpha \geq 90°$ are described in Fig. 14.7. Then, the optimal evasion strategy; i.e. the strategy which maximizes the evasion time corresponds to E going along direction $\pm\alpha$. P is always going towards the last observed position of E. E has always the tentative $\pm 90°$ strategy which limits the stage duration as in previous section (Sect. 14.2.2). Therefore, the maximum duration of a stage is still given by $\tau_n^{max} = \frac{R_{n-1}}{b}$. By doing some analytical computation and using Al Kashi theorem we obtain the following formulas:

$$\left(R_n - a\,\frac{R_{n-1}}{b}\right)^2 - 2\,R_n\,R_{n-1}\cos\alpha + 2a\,\frac{R_{n-1}^2}{b}\cos\alpha = 0 \quad (14.19)$$

And by considering the positive solution of Eq. (14.19) we still express R_n as a function of R_{n-1}. As in previous section (Sect. 14.2.2), we compute τ_n^{opt} which can be less than τ_n^{max}. At the end of a stage, distance d between P and E needs to be

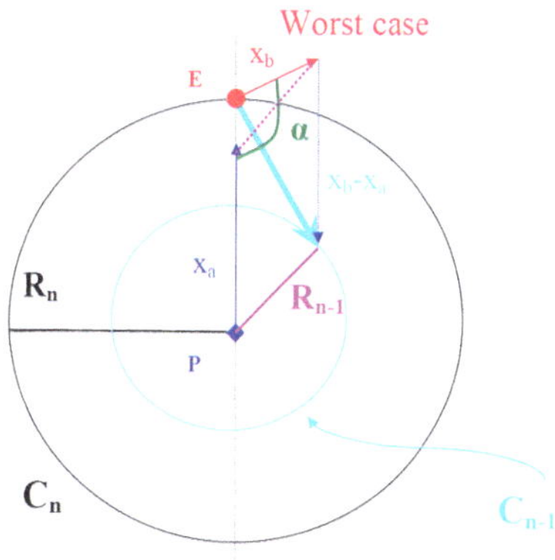

Fig. 14.7 Information set with Evader heading constraint $\alpha \geq 90°$

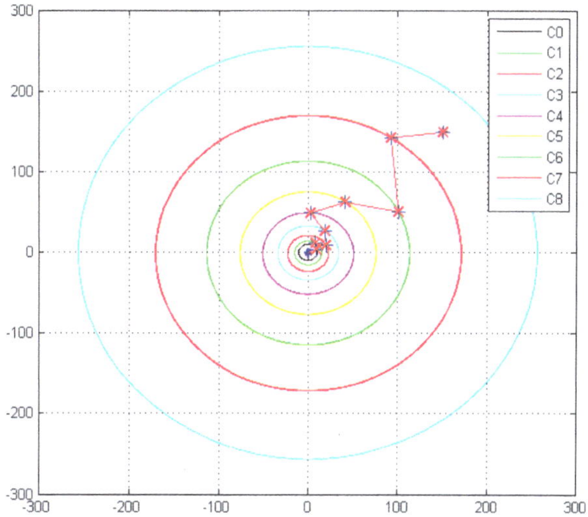

Fig. 14.8 Optimal strategies and heading angle constraint $\alpha = 150°$; trajectories in pursuer referential; at each re-location E changes from direction α to direction $-\alpha$

less than or equal to R_{n-1}. Using the geometry of Fig. 14.7, we are able to write the value of d and as a consequence the analytical value of τ_n^{opt}.

$$d = (||x|| - a\,\tau_n^{opt})^2 + (b\,\tau_n^{opt})^2 - 2\,(R_n - a\,\tau_n^{opt})\,b\,\tau_n^{opt}\,\cos\alpha \quad (14.20)$$

An example with $\alpha = 150°$ is reported in Figs. 14.8 and 14.9, respectively, in the pursuer coordinates frame and in inertial/absolute referential. Everything is going similarly if penalty terms δ is no more 0. For the second case $\alpha > 90°$, then there is no more distinction between the evader strategy which consists in maximizing the interception time and which consists in reducing as much as possible the stage duration (strategy called $\pm 90°$ strategy previously). Now, the optimal choice for E is always to follow direction $\pm\alpha$. We use numerical dichotomy for computing by simulation R_n and τ_n^{opt} when $\alpha > 90°$. Figure 14.10 shows how R_n evolves

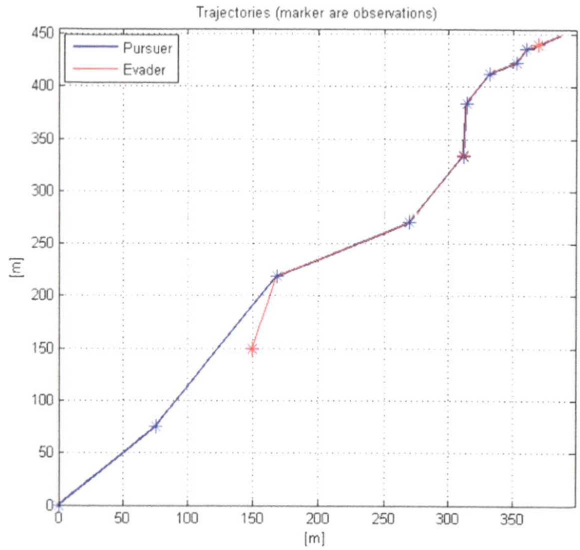

Fig. 14.9 Optimal strategies and heading angle constraint $\alpha = 150°$; trajectories in inertial / absolute referential; P starting at (0, 0) and E starting at (150, 150)

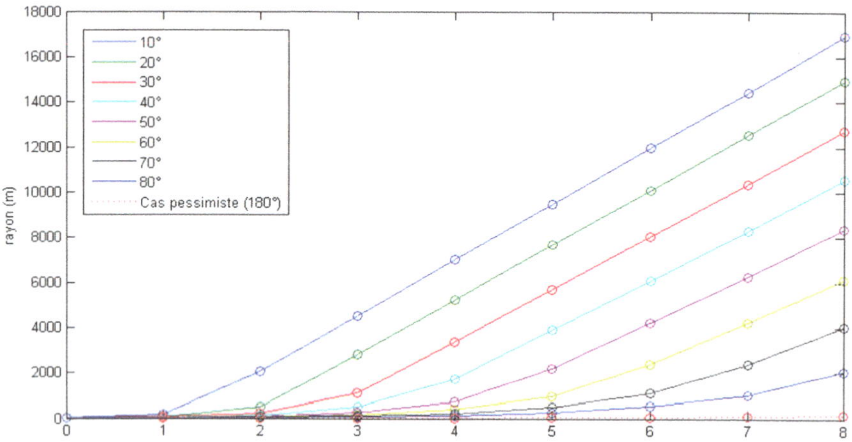

Fig. 14.10 Information set radius versus heading angle constraint

according to α. We confirm that R_n increases when α decreases; it is easier (less re-location required) to capture targets with less maneuverability.

14.3 Observer 4D Guidance Law

We move now to the definition of a co-operative guidance scheme in charge of intercepting a ground mobile vehicle so-called evader by a ground mobile vehicle (the pursuer). In addition, UAVs (fix body, flying at constant speed, no hovering capability) so-called observers are in charge of providing information on request to

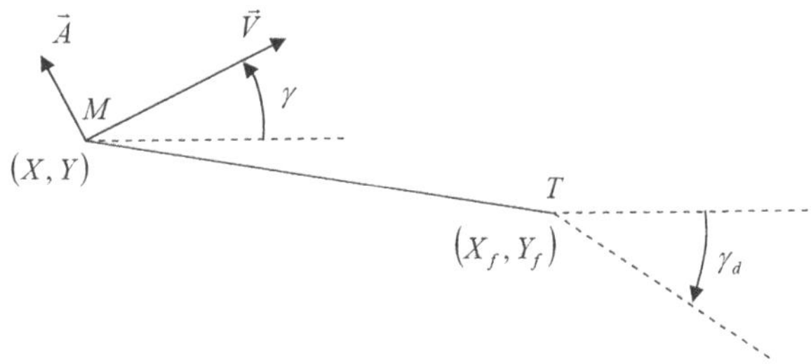

Fig. 14.11 Biased proportional navigation

the pursuer. The pursuer and the UAVs are in the same team; have same objective which is having the pursuer intercepting the evader as fast as possible. Nobody has a global view of the situation; just an initial designation at start of the game. The pursuer and the observers have limited sensing range. Therefore, observers need to be at the right place, at the right time in a way to support the pursuer vehicle, i.e. in a way to precisely re-localize the evader position. Ideally, the pursuing vehicle provides the observers with a list of future tentative (rough) evader positions $((x_1, y_1, t_1), (x_2, y_2, t_2) \ldots)$, where (x_i, y_i) are Cartesian coordinates and t_i are times. The one-on-one guidance law studied in Sect. 14.2 only provides the next immediate re-location instant τ_n. Moreover, the following episode (value of τ_{n-1}) depends on what happens in stage n; i.e. it is related to the fact that the pursuer and the evader apply optimal controls or not. For estimating the future target positions we assume that the target moves in straight line and at constant speed. According to the number of UAVs we have, we use UAVs to cover different assumptions about the target heading direction (see Sect. 14.4 about simulation examples for more explanations). We update these sequences with the new information received after each observation.

For controlling UAVs trajectories we implemented a guidance law called Impact Time Control (ITC) which is an extension of a classical guidance law called Biased Proportional Navigation (BPN). BPN is Proportional Navigation (PN) plus impact angle constraint. If talking about M reaching T, PN is a well-known guidance law which applies a gain to the turning rate of the line of sight between M and T to ensure that M is in collision course with respect to T trajectory; i.e. to nullify the line of sight turning rate. BPN adds an extra term to PN for M being able to satisfy a terminal slope constraint γ_d (see Fig. 14.11). PN and BPN are guidance laws for moving targets that can also be used against fix targets as in the case of UAVs we guide towards non-moving observation points. By controlling the terminal slope of BPN and by shaping the UAV trajectory, ITC is able to control the impact time. For more details, please refer to Lee et al. (2007).

There exist plenty of other guidance laws based also on close loop controls (Glizer 1996; Jeon et al. 2006) and on way-point-based approaches. The problem of optimal control of Dubins' car is extremely difficult (see, for example, Patsko and Turova 2011). General solutions enabling time arrival control are preprogrammed mid-course strategies in which the vehicles approach to the objectives (observation points) via predefined way-points. However, since this approach is difficult to cope with unpredicted changes in the engagement conditions such as the target motion, this study only focuses on feedback guidance laws. We obtain reasonable results with ITC guidance laws as soon as the required corrections on impact times are not too large. However, if the UAV is too far or too close to the aimed point, it cannot have the desired impact time. Choosing an observer which has the ability to reach the right place at the specified time is mandatory, otherwise the algorithm has to be restarted from the beginning (initial target designation; re-compute information sets and so on; for more details, see the end of the algorithm provided in Sect. 14.3.1) By the way, we need to remember that the impact times we talk about are rough designations only.

14.3.1 Co-operative Guidance Algorithm

We implemented an auction algorithm for allocating UAVs to 4D positions (location plus time). The planning algorithm we use runs as follow:

Vehicles: Pursuer (P), Evader (E) and Observers (Os)
Initial position of P and E known from each other
$n = 2$
While $n > 1$
    Computation of information sets C_1 to C_n
        Flag_information_sets ← "valid"
        n is the number of observations required as defined in Sect. 14.2
        n such that E is inside C_n and outside C_{n-1}
        n and E position $\Rightarrow \tau_n^{opt}$
        If $n = 0$ then go directly to step **"Perfect information homing"**
        If $n = 1$ then go directly to step **"Pursuer trajectory"**
    Computation of m assumptions about the Evader future position:
        m are the heading angles we assume for the Evader path
        m heading angles are chosen between $\pm\alpha$
        (α being the angular constraint on the Evader direction
        respect to the Pursuer Evader line of sight as defined in Sect. 14.2.3)
        m is the number of UAVs we have
        Evader position is extrapolated during τ_n^{opt}
    While *Distance Pursuer to Evader > R_0*
            And *Flag_information_sets = "valid"*
        If *Last simulation step was a successful information step*
            about E position
            Then *Update n: $n \leftarrow (n - 1)$*
                If $n = 1$ then go directly to step **"Pursuer trajectory"**
                Computation of new τ_n^{opt} (next observation time;
                    computed by the pursuer and broad-casted to all the UAVs)
                P is also in charge of computing the m new observation points

at new τ_n^{opt} *(and send these objective points to all the UAVs)*
Allocation of the m observation points to the m UAVs
using an auction method for allocating the best UAV
in a decentralized manner without over allocations
(best UAV is related to how close an UAV can reach
an observation point satisfying the required impact time)
End if
If $t < \tau_n^{opt}$
Then "STEP Pursuer trajectory" *Pursuer vehicle goes straight*
to the Evader last observed position
The m UAVs go to their positions using ITC (Sect. 14.3)
When they arrive at destination (or have no objective) the UAVs
continue straight until they receive new observation points
Else *(*$t \geq \tau_n^{opt}$*)*
If *There is an UAV able to detect the evader*
Then *Forward the Evader position to the Pursuer*
Else *(An observation instant has been missed)*
Flag_information_sets ← "not_valid"
Interception has failed
A first information about E position is required to restart
Information sets C_1 *to* C_n *need to be recomputed;*
As the value of n, τ_n^{opt} *and the reallocation of the UAVs;*
n can be larger; extra information sets can be required;
End if
End While
$t \leftarrow t + 1$ *(t is time step simulation)*
End While
Perform the **"STEP Perfect information homing"**

The co-operative guidance scheme described above aims providing the pursuer with evader positions as required in Sect. 14.2. Even if the UAVs are able to track more the evader positions, this information is not used to improve the pursuer trajectory. We claim that sending data through data links is bandwidth consuming and controlling the amount of data we exchange through data networks is also something important.

14.4 Simulation Examples

14.4.1 Comparison to Standard Strategies

Figure 14.12 shows how the co-operative guidance algorithm described in Sect. 14.3.1 works when there is only one re-location step involving three UAVs for

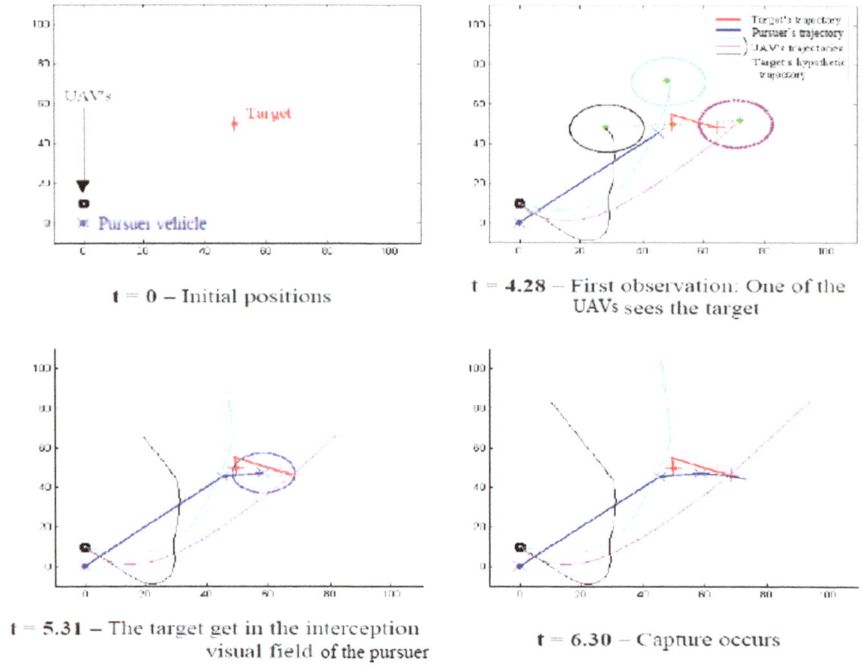

Fig. 14.12 Co-operative guidance scheme with 3 UAVs and one step re-location; sensor detection ranges have been plotted around the UAVs and around the pursuer (R_0) at target detection

observation purpose. In addition, for illustration purpose, we compare the results we obtain when applying three different guidance schemes (see Fig. 14.13):

- The pursuer applies a collision course strategy; straight line extrapolation of the evader trajectory plus impact point computation under this assumption;
- The pursuer applies pure pursuit strategy; goes towards the last observed point without re-location;
- The pursuer applies the guidance algorithm defined in Sect. 14.3.1; three UAVs; and one step re-location.

Figure 14.14 shows what happens in the pursuer coordinates frame; i.e. the trajectories of Fig. 14.13 are superposed on the information sets.

14.4.2 Realistic 3D Simulation

Then, the co-operative guidance algorithm has been implemented in a 3D simulation with more realistic models. The simulation parameters are as follows:

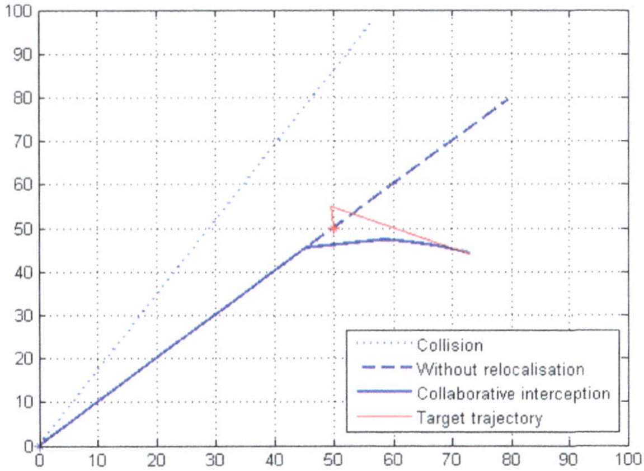

Fig. 14.13 Co-operative guidance algorithm versus collision course and versus pure pursuit without UAV re-location

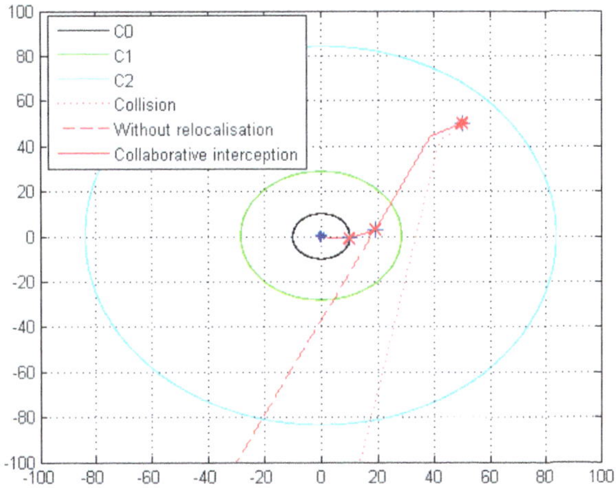

Fig. 14.14 Trajectories of Fig. 14.13 in pursuer coordinates frame

$$R_{0\,pursuer} = 800\,m$$

$$R_{0\,UAV} = 1200\,m$$

$$\delta = 4\,s \text{ (setting parameter even if the pursuer does not stop)}$$

$$a = 160\,m/s \text{ (pursuer and UAV velocity)}$$

$$b = 70\,m/s \text{ (evader and UAV velocity)}$$

$$x_{E0} = 3200\,m$$

$$y_{E0} = 3200\,m$$

$$x_{P0} = 10{,}000\,m$$

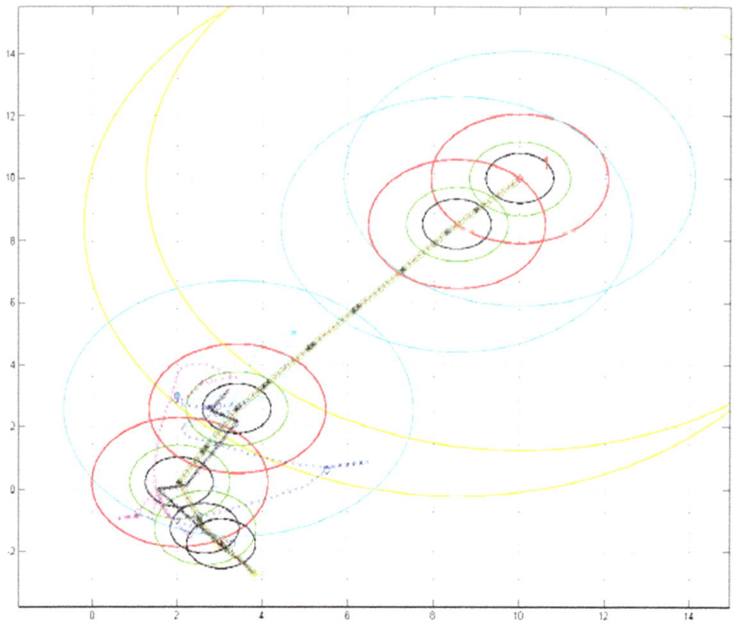

Fig. 14.15 Evader trajectory with 4 bends; trajectories in absolute referential; trajectories plus information sets

$$y_{P0} = 10{,}000\,m$$
$$\alpha = \pm 180^0 \text{ (no constraint angle)}$$

The trajectory of the evader has been defined in a random manner. First, we fix a number of bends and then we decide randomly what are the evader heading angles and the duration of each bend. The trajectories corresponding to a case with the evader performing five segments (four bends) are drawn in Figs. 14.15 and 14.16 (trajectories in absolute referential). The complete scenario plus the information sets have been plotted in Fig. 14.15 (information sets are moving with the pursuer position). Figure 14.16 is a zoom of the bottom left part of Fig. 14.15. The sensor detection ranges are plotted in Figs. 14.16 in place of the information sets at re-location instants for the UAVs and at final detection for the pursuer.

Two-hundred runs have been performed in Monte Carlo mode for evaluating the co-operative guidance algorithm with the parameters listed at the beginning of Sect. 14.4.2. Target trajectories with 0 up to 4 bends occur during these 200 runs and the ratio of success over target miss is 95.9%. We consider a run as a fail as soon as UAVs are not able to re-localize the evader at prescribed time. Then, the time interval $[t_{min}, t_{max}]$ we use for choosing the duration of each segment has been adjusted in a way to have 250 random evader trajectories with 1 up to 6 bends. Then, the ratio of success is equal to 82.2%.

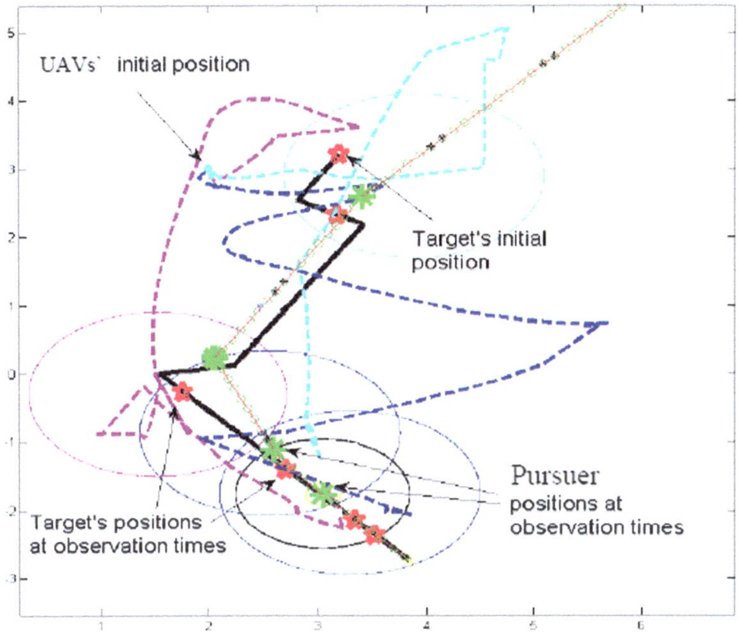

Fig. 14.16 Trajectories of Fig. 14.15; sensor detection ranges plotted at target re-location

14.5 Conclusion

A co-operative guidance algorithm involving one pursuer, one evader, and several UAVs has been proposed. This algorithm is based on a one-on-one pursuit evasion differential game with discontinuous and costly information. The pursuer and the UAVs belong to the same team. The pursuer tries to minimize the time to be at range R_0 from the evader; meaning while the evader aims to maximize this time. The sensing capabilities of the vehicles are limited. The sensor detection range of the pursuer is R_0. UAVs which have similar sensor detection limitation are in charge of providing target re-locations at discontinuous time. The co-operative guidance algorithm proposed has been evaluated using Monte Carlo runs. Encouraging results have been obtained. Larger parametric studies could be performed to better evaluate the potential of this kind of approach which mixes:

- Strategies based on a pursuit evasion differential game;
- Close loop 4D guidance laws;
- An auction algorithm for coordinating UAVs in a decentralized manner;
- And co-operative guidance concepts.

The one-on-one pursuit evasion differential game has been studied considering vehicles without curvature constraints. Computing information sets with more realistic kinematics would require implementing numerical algorithms for calculating

Minkowski's differences. The expression of C_n will be no more analytical. As controlling the UAVs is the central part of the application, other guidance logics could be used to improve the overall algorithm performances. Among other features, the 4D guidance law (Impact Time Control guidance law) that we use for shaping the UAVs trajectories could be updated. What happens if the number of pursuers we consider is larger than one is also an open problem.

References

Bernhard P, Pourtallier O (1994) Pursuit evasion game with costly information. Dyn Control 4:365–382

Glizer VY (1996) Optimal planar interception with fixed end conditions: Closed-form solution. J Optimiz Theory App 88(3):503–539

Isaacs R (1965) Differential games, a mathematical theory with applications to warfare and pursuit, control and optimization. Wiley, New York

Jeon IS, Lee JI, Tahk MJ (2006) Impact-time-control guidance law for anti-ship missiles. IEEE T Contr Syst T 14(2):260–266

Le Ménec S, Markham K, Tsourdos A, Shin HS, Piet-Lahanier H (2011) Cooperative allocation and guidance for air defence application. In: 18th IFAC World Congress, Milano, Italy, pp 3897–3902, session dedicated to Missile Guidance, Control and Estimation

Lee JI, Jeon IS, Tahk MJ (2007) Guidance law to control impact time and angle. IEEE T Aero Elec Syst 43(1):301–310

Neveu D, Pignon J, Raimondo A, Nicolas J, Pourtallier O (1995) Pursuit games with costly information: Application to the ASW helicopter versus submarine game. In: New trends in dynamic games and applications, Annals of the ISDG, vol 3. Birkhäuser, Boston, pp 247–257

Olsder GJ, Pourtallier O (1995) Optimal selection of observation times in a costly information game. In: New trends in dynamic games and applications, Annals of the ISDG, vol 3. Birkhäuser, Boston, pp 227–246

Patsko VS, Turova VL (2011) Homicidal chauffeur game: History and modern studies. In: Advances in dynamic games; theory, applications, and numerical methods for differential and stochastic games, Annals of the ISDG, vol 11. Birkhäuser, Boston, pp 227–251

Chapter 15
The Effect of Pursuer Dynamics on the Value of Linear Pursuit-Evasion Games with Bounded Controls

Josef Shinar, Valery Y. Glizer, and Vladimir Turetsky

Abstract Linear pursuit-evasion games with bounded controls are considered. The cases of an ideal, a first-order, and a second-order pursuer against an ideal and a first-order evader are analyzed. For these cases, the values of the games are compared with each other, indicating the effect of the pursuer dynamics. It is shown that replacing the second-order pursuer by a first-order approximation underestimates the value of the game (the guaranteed miss distance).

Keywords Pursuit-evasion games • Bounded control • Linearized geometry • Dynamics order • Scalarizing transformation • Game value

15.1 Introduction

The optimal performance of modern interceptor missiles against maneuverable targets can be analyzed by using the mathematical model of pursuit-evasion games with bounded controls. In general, the endgame geometry of the interception can be linearized with respect to a nominal collision course, allowing planar analysis.

J. Shinar
Faculty of Aerospace Engineering, Technion - Israel Institute of Technology, Haifa 32000, Israel
e-mail: aer4301@aerodyne.technion.ac.il

V.Y. Glizer (✉) • V. Turetsky
Department of Applied Mathematics, Ort Braude College, P.O.B. 78, Karmiel 21982, Israel
e-mail: valery48@braude.ac.il; turetsky1@braude.ac.il

V. Křivan and G. Zaccour (eds.), *Advances in Dynamic Games*, Annals of the International Society of Dynamic Games 13, DOI 10.1007/978-3-319-02690-9_15,

The complete solution of planar linear pursuit-evasion games with bounded controls is well known (Gutman 1979; Gutman and Leitmann 1976; Shima and Shinar 2002; Shinar 1981). The solution involves the decomposition of the game space into a regular and a singular region. The regular region of the game space is covered by the family of candidate optimal trajectories. In this region the optimal strategies of both players are "bang-bang" type and the value of the game depends on the initial conditions. The solution also indicates the existence of a singular region, including the large majority of expected initial conditions of the endgame. In the singular region the optimal strategies of both players are arbitrary and the game value is constant. This value can be interpreted as the guaranteed minimal miss distance against an optimally evading target.

The guaranteed minimal miss distance is an important parameter in the design of interceptor missiles, determining the size of the warhead needed to destroy the target. Even if in the presence of measurement noise the actual miss distance distribution is certainly larger than guaranteed minimal miss distance, obtained by assuming perfect information, the value of the guaranteed minimal miss distance serves as a reference of the interceptor efficiency.

The guaranteed miss distance depends on the maneuverability ratio of the players and their respective dynamics. So far only pursuit-evasion games with simple (ideal and first-order) dynamics were analyzed in the literature. If the dynamics of both players are ideal, point capture (zero miss distance) can be achieved from the singular region if the pursuer has a superior maneuverability (Gutman and Leitmann 1976).

If the pursuer has first-order dynamics, but the evader dynamics is ideal (the worst possible case for the pursuer) point capture is not possible (Gutman 1979). In this case the guaranteed miss distance can be minimized by increasing the maneuverability ratio of the pursuer. If the dynamics of the evader is also of the first order, achieving point capture requires in addition to superior maneuverability also superior pursuer agility (acceleration rate).

In this paper the yet unpublished results on a linear pursuit-evasion game with bounded controls and with second-order pursuer dynamics against an evader of ideal and first-order dynamics are presented. In this case point capture is of course not possible, but very small miss distance can be guaranteed by sufficient maneuverability ratio and fast dynamics of the pursuer. Comparison with the earlier results is also presented. The structure of the paper is the following: in the next section an interception scenario is formulated as a linear pursuit-evasion game with bounded controls, which includes also the commonly used modeling assumptions. It is followed by the general solution of such pursuit-evasion game with bounded controls and previously published results are summarized. The new results presented in this paper start with an analytical investigation of the game solution with second-order pursuer dynamics followed by details of the game solution with comparison to a model of first-order dynamics. The concluding section points out the importance of using the more realistic model of second-order pursuer dynamics in guaranteed miss distance assessment.

15.2 Problem Formulation

15.2.1 Solution Outline

An (aerial) interception scenario belongs to the family of *pursuit-evasion* problems. The objective of the interceptor missile (called in the sequel the *pursuer*) is destroying the target (called in the sequel the *evader*). Target destruction can be achieved either by a direct hit or (if a hit cannot be achieved) by detonating an explosive warhead in its vicinity. Therefore, the natural cost function of an interception is the distance of closest approach (called the *miss distance*), to be minimized by the *pursuer*.

The *evader*'s acceleration strategy can be either known or unknown to the *pursuer*. Only if the *evader*'s acceleration strategy or its future acceleration profile is known to the *pursuer*, the interception can be formulated as an optimal control problem. Otherwise, the *evader*'s trajectory is not predictable and the optimal control formulation is conceptually inappropriate. In such a case, assuming that the *evader*'s acceleration bounds are known, a *robust* control formulation, requiring successful interception against any feasible (or admissible) target maneuver, can be used.

Since in most aerial interception scenarios the *evader*'s acceleration is independently controlled, another relevant formulation of the problem is in the context of *zero-sum differential games*. In such a game the *pursuer* and the *evader* wish to optimize (minimize/maximize, respectively) simultaneously the same cost function by using their respective optimal strategies. If the *minmax* and *maxmin* processes lead to the same solution, the game has a *saddle-point* and the corresponding optimized cost is the *value* of the game. Thus, the solution of the game with a *saddle-point* is a triplet composed of the *optimal strategies* of the *pursuer* and the *evader* and the *value* of the game, all expressed as functions of the state and time.

Based on such game solution, the best interceptor's guidance law (the realization of the *optimal pursuer strategy*) and the best evasive maneuver (the realization of the *optimal evader strategy*) can be found. If both *players* use their optimal strategies the outcome of the interception, e.g. the *guaranteed miss distance,* will be the *value* of the game. The pursuer cannot achieve a smaller *miss distance* and the *evader* cannot generate a larger one, as long as the opponent uses its optimal strategy.

The formulation of an aerial interception as a *zero-sum differential game* was first suggested by Isaacs (1965) and since then it was used in a great number of research papers and publications. Due to the nonlinear nature of the scenario only very few reduced dimensional pursuit-evasion games, based on oversimplified assumptions, could be solved.

15.2.2 Modeling Assumptions

In order to obtain some kind of generalized (hopefully closed form) solutions, in all analytical studies simplified models, relating to the scenario and to the dynamics of the *players*, were used. When the analytical solution of such model is obtained, it becomes necessary to verify the validity of each simplifying assumption in the context of the solution. The most commonly used assumptions are reviewed in the sequel.

A great part of interception analysis has been carried out in a deterministic mind set, assuming that all state variables and parameters are known to both participants. This means that all state variables of the problem can be (and are) measured with high accuracy. This *perfect information* assumption is unfortunately not valid. Some variables, such as the acceleration of the opponent, are not measurable, so they have to be reconstructed from measured data by an observer. Moreover, all measurements are imprecise. This fact is expressed by saying that an actual measurement is the sum of the actual value plus an additive error, modeled as a noise of a given family. Having a large sequence of measurements, the noise can be filtered and the unmeasured state variables can be obtained by an estimator. One should remember that the outcome of a realistic noise corrupted scenario will not be identical to the outcome predicted by a *perfect information* analysis.

Another frequently used assumption is that the flying vehicles can be represented by their center of gravity, where the mass is concentrated. Such an assumption, neglecting the angular motions, called the *point-mass* approximation, is very useful for trajectory computations and for miss distances that are either negligibly small or very large. If the miss distance is of the order of the interceptor and/or the target dimensions, a lethality analysis with more details is needed.

In many studies interceptor and target velocities are assumed to be constant or known as function of time. In the case of a maneuvering aerial vehicle this assumption is simply not physical due to the maneuver-dependent induced aerodynamic drag force. Different velocity profiles lead to different flight times and different miss distances.

The maneuvering dynamics of a flying vehicle has a rather complex (not necessarily linear) structure, while in many studies ideal (instantaneous) dynamics or first-order linear dynamics are assumed. While the assumption of ideal interceptor dynamics can lead to totally unrealistic results, the representation of first-order dynamics preserves, at least qualitatively, a realistic behavior. In any case, the value of an *equivalent* time constant has to be selected carefully for approximating the true dynamics.

15.2.3 Linearized Interception Model

In spite of adopting some (or even all) of the above-mentioned simplifying assumptions, interception kinematics remains generally nonlinear. There is no need to emphasize the difficulties of analyzing nonlinear problems, particularly when optimization is involved. Therefore, much effort has been devoted to create linear interception models in order to obtain closed form optimal solutions. The linearization is based on assuming that the relative interception trajectory is sufficiently close to the initial *collision course* trajectory, to be used as a reference.

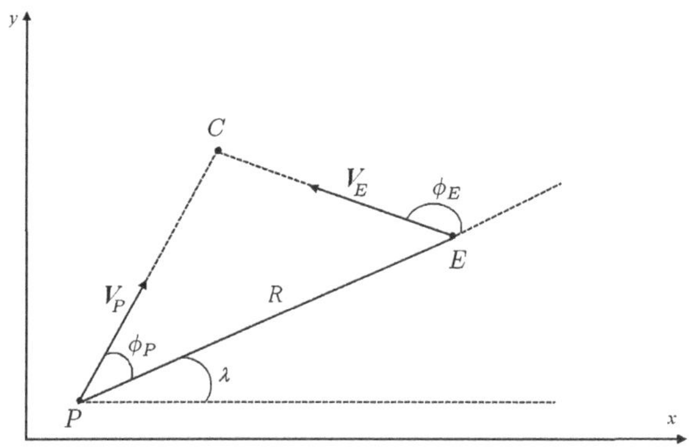

Fig. 15.1 Collision course geometry

The notion of *collision course* comes from an ancient naval background for intercepting a vessel by another. It relates therefore to a planar constant speed scenario. The *collision* plane is defined by the line of sight vector R and the velocity vector of the *evader* (target). Assuming that the target moves on a straight line and the *pursuer* (interceptor) speed is larger ($V_P > V_E$), there exists a unique direction in the *interception* plane for the *pursuer* reaching the *evader* in a finite time, as illustrated in Fig. 15.1.

Assuming constant speeds ($\dot{V}_P = \dot{V}_E = 0$), the two conditions for *collision* can be written as

$$V_P \sin\phi_P - V_E \sin\phi_E = 0, \tag{15.1}$$

$$V_P \cos\phi_P - V_E \cos\phi_E = -\dot{R} = V_{c_0} > 0 \tag{15.2}$$

where $\phi_P \in [0, \pi]$, $\phi_E \in [0, \pi]$ are the respective *aspect* angles, and V_c is the constant closing velocity. Equation (15.1) indicates that the line of sight angle remains fixed: $\lambda(t) \equiv \lambda_0$. This equation determines the required direction of the interceptor missile with respect to the non-rotating line of sight:

$$(\phi_P)_{\text{col}} = \arcsin\left[V_E \sin\phi_E / V_P\right]. \tag{15.3}$$

This direction, called the *collision course*, is constant ($\dot{\phi}_P = 0$) as long as the target does not maneuver ($\dot{\phi}_E = 0$) and the velocities are fixed. For a given angle ϕ_E, system (15.1)–(15.2) can have a unique solution ϕ_P only if $V_P > V_E$, as in a missile/aircraft interception. If $V_P < V_E$, as in the case of anti-ballistic missile defense, there may be either two solutions or none.

Based on (15.2), the relative motion in the x-direction becomes predictable as a function of time.

$$x(t) = R(t) = R_0 - V_c t = V_c(t_f - t), \tag{15.4}$$

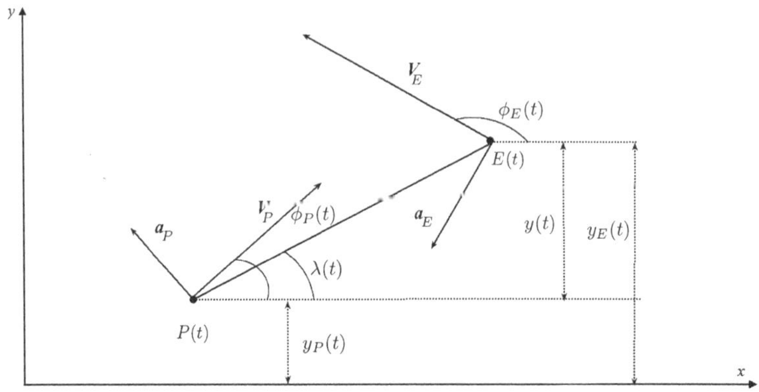

Fig. 15.2 Planar interception geometry

where $t_f = R_0/V_c$ is the predicted (fixed) final time of the engagement (collision).

If during the interception engagement the difference between the actual flight conditions and an ideal collision course remains small, then by setting $\lambda_0 = 0$ the line of sight angle $\lambda(t)$ remains also small. In this case one can write two sets of identical *linear* equations of motion normal to the reference line of sight in two perpendicular planes. Thus, a valid trajectory linearization, based on the smallness of angular deviations from the *collision course*, leads also to the decoupling of the original three-dimensional motion to two planar motions in perpendicular planes (Isaacs 1965). The validity of the linearization is preserved, even if the velocities V_P and V_E are not constant, but known as functions of time. In an aerial interception scenario, where both the interceptor and the target are equally affected by gravity, the respective term can be left out of the equations and the direction of the two perpendicular planes is immaterial.

This decoupling property is the reason for the common practice of designing interceptor missiles in a cruciform configuration, having two identical guidance channels acting in perpendicular planes. Based on the decoupling property of the linearized motion, this paper concentrates on linearized planar interception models. In such linearized planar interception model the x-axis of the (inertial) coordinate system is aligned with the initial line of sight (LOS), i.e. $R_0 = x_0$.

In a planar scenario the only state variable of interest is the relative position "y" between the interceptor missile and the target normal to the initial reference LOS as seen in Fig. 15.2.

$$y(t) \triangleq y_E(t) - y_P(t). \tag{15.5}$$

The basic equations of linearized motion normal to the initial LOS and the respective initial conditions are

$$\dot{y}(t) = V_E \sin\phi_E(t) - V_P \sin\phi_P(t); \;\; y(0) = 0, \tag{15.6}$$

$$\ddot{y}(t) = (a_E)_\perp - (a_P)_\perp; \;\; \dot{y}(0) = V_E \sin\phi_{E_0} - V_P \sin\phi_{P_0} \tag{15.7}$$

where $(a_E)_\perp = a_E \cos\phi_E(t)$ and $(a_P)_\perp = a_P \cos\phi_P(t)$ are the respective acceleration components normal to the initial LOS ($\lambda(0) = 0$). The relationship between the actual accelerations and the respective acceleration commands $(a_E^c)_\perp$ and $(a_P^c)_\perp$ are expressed in most cases by a linear transfer function.

So the equations of a linearized planar pursuit-evasion game can be written in the following form:

$$\dot{X} = AX + Bu + Cv, \tag{15.8}$$

where $X \in R^n$ is the state vector

$$X^T = [y, \dot{y}, (a_P)_\perp, (a_E)_\perp, \ldots], \tag{15.9}$$

the dimension n depends on the dynamics of the pursuer and the evader; A is an $n\times n$ matrix, B and C are n-dimensional vectors; u and v are the normalized (scalar) control variables $(a_P^c)_\perp = ua_P^{\max}$; $(a_E^c)_\perp = va_E^{\max}$, where $a_P^{\max}$ and $a_E^{\max}$ are the maximal values of acceleration commands in the y-direction, which depend on the initial aspect angles ϕ_{P_0} and ϕ_{E_0}, respectively.

Thus, the normalized controls satisfy the constraints

$$|u| \leq 1, \;\; |v| \leq 1. \tag{15.10}$$

By the definition of the coordinate system (see Fig. 15.2),

$$X_1(0) = 0, \;\; X_2(0) = V_E \sin\phi_{E_0} - V_P \sin\phi_{P_0}. \tag{15.11}$$

The notation (15.9) assumes that the maneuvering dynamics of the pursuer and the evader may be not ideal. In this case, both lateral accelerations are state variables. Moreover, it is assumed that

$$X_3(0) = 0, \;\; X_4(0) = 0. \tag{15.12}$$

The aerial interception engagement of an independently controlled maneuverable evader can be formulated as a zero-sum differential game of pursuit-evasion. It is a two-person zero-sum game that sometimes is considered as a "two-sided" optimal control problem, but its solution is generally more complex. Although the necessary conditions of game optimality look similar to those of an optimal control problem, the sufficiency conditions are very different and are more difficult to verify.

The natural cost function of an interception engagement with bounded controls (15.10) is the *miss distance* (the distance of closest approach), which is the absolute value of the final lateral separation between the *players*

$$J = |y(t_f)|. \tag{15.13}$$

In this paper, similarly to many other studies, *perfect information* is assumed, which means that both *players* have perfect knowledge of the state variables and the parameters of the engagement.

The solution of a two-person zero-sum differential game consists in general of four elements: the *optimal strategies* of the two *players* and (possibly) two outcomes, namely the *upper value* and the *lower value* of the game. The *players' strategies* (U for the *minimizer* and V for the *maximizer*) are mappings from the sets of information available for each *player* to the respective set of admissible controls. Since the *players* optimize the cost function independently, it is important whether the minimization or the maximization occurs first. The *upper value* of the game J_{up} is defined as

$$J_{\text{up}} = \min\max\{J(U,V)\} = \max\left\{J\left(U^*,V\right)\right\}, \tag{15.14}$$

while the *lower value* of the game J_{low} is

$$J_{\text{low}} = \max\min\{J(U,V)\} = \min\left\{J\left(U,V^*\right)\right\}. \tag{15.15}$$

where U^* and V^* are the respective *optimal strategies.* These two outcomes are generally different. Obviously

$$J_{\text{up}} \geq J_{\text{low}}. \tag{15.16}$$

If both outcomes are equal (min max = max min), one says that the game has a *value* J^* and the respective *optimal strategies* U^* and V^* are *saddle-point strategies* (or in other words the game has a *saddle point*). If the information available for each *player* is the state vector of the game, the realizations of the optimal strategies are feedback controls. Note that satisfaction of the necessary conditions of game optimality provides only *candidate optimal strategies.* One way to verify that the sufficiency conditions are also satisfied is to fill the entire game space with *candidate optimal* trajectories (Shima and Shinar 2002).

Pursuit-evasion games with separated dynamics admit a *saddle point* and have a *value.* The game solution provides the optimal guidance law of the interceptor missile (*optimal pursuer strategy*), the optimal missile avoidance strategy (*optimal evader strategy*), and the corresponding guaranteed outcome (*value*) of the game.

15.2.4 Terminal Projection Transformation

In this subsection, a useful methodology facilitating the solution of linear games is presented. If the state variables are not involved in the "running" cost, as in (15.13), the vector differential equation of a planar interception (15.8) can be reduced to a scalar one by using the transformation (Bryson and Ho 1975; Krasovskii and Subbotin 1988), called here as the *terminal projection*,

$$Z(t) = D\Phi(t_f,t)X(t), \tag{15.17}$$

where $D = [1,0,\ldots,0]$ and $\Phi(t_f,t)$ is the transition matrix of the original homogeneous system $\dot{X} = AX$. The new state variable, denoted by $Z(t)$, is the *zero-effort miss distance*, the miss distance that created if none of the *players* use any control until the final time of the interception. The notion of the *zero-effort miss distance* has a central role in modern missile guidance theory.

Based on this scalarization, the cost function of the interception game (15.13) can be rewritten as

$$J = |Z(t_f)|. \tag{15.18}$$

The time derivative of $Z(t)$ becomes, using the property of the transition matrix $\dot{\Phi}(t_f,t) = -\Phi(t_f,t)A$,

$$\dot{Z}(t) = \tilde{B}\left(t_f,t\right)u(t) + \tilde{C}\left(t_f,t\right)v(t), \tag{15.19}$$

where

$$\tilde{B}\left(t_f,t\right) = D\Phi(t_f,t)B;\;\; \tilde{C}\left(t_f,t\right) = D\Phi\left(t_f,t\right)C. \tag{15.20}$$

Note that due to (15.19) the evolution of $Z(t)$ depends only on the controls. Thus, integrating (15.19),

$$Z\left(t_f\right) = Z(t) + \int_t^{t_f}\left\{\tilde{B}\left(t_f,s\right)u(s) + \tilde{C}\left(t_f,s\right)v(s)\right\}ds. \tag{15.21}$$

In reality every interceptor missile and airborne target have inherent physical limitations on the maximal value of the admissible lateral accelerations. Such saturation phenomenon creates nonlinear dynamics with the difficulties of obtaining closed form solutions. One approach to circumvent the effect of such nonlinearities is limiting the acceleration commands to the value of the admissible lateral accelerations as indicated by (15.10). In this case the cost function of the form (15.13) is appropriate. Such "hard" control constraint guarantees that the actual lateral accelerations respect the admissible physical limits (at least as long as the physical limits are non-decreasing during the interception).

The realizations of the *candidate optimal strategies* using bounded controls can become discontinuous (of the "bang-bang" type). The resulting eventual chattering control creates an unnecessary excessive control effort and may create also inconveniences of implementation. Nevertheless, the chattering phenomenon can be eliminated by several feasible modifications in the guidance law (Glizer et al. 2012; Turetsky and Glizer 2005).

15.3 Scalarized Game Solution

In this section several dynamic models of planar linearized pursuit-evasion games with bounded control are presented. The presentations use the scalar (reduced) state variable Z, the *zero-effort miss distance*. The cost function used in this formulation is (15.18) to be minimized by the *pursuer* and maximized by the *evader* subject to the dynamics (15.19) and the constraints (15.10).

The Hamiltonian of the game is

$$H = \lambda_Z[\tilde{B}\left(t_f, t\right) u + \tilde{C}\left(t_f, t\right) v], \tag{15.22}$$

where λ_Z is the costate variable satisfying

$$\dot{\lambda}_Z = -\partial H/\partial Z = 0, \tag{15.23}$$

$$\lambda_Z(t_f) = \partial J/\partial Z|_{t_f} = \text{sign}\{Z(t_f)\};\ \ Z(t_f) \neq 0, \tag{15.24}$$

leading to

$$\lambda_Z(t) = \text{sign}\{Z(t_f)\};\ \ Z(t_f) \neq 0, \tag{15.25}$$

as long as $\lambda_Z(t)$ is continuous. This allows determining the optimal strategies as

$$u^*(t) = -\,\text{sign}\{Z(t_f)\tilde{B}\left(t_f, t\right)\}, \tag{15.26}$$

$$v^*(t) = \text{sign}\{Z(t_f)\tilde{C}\left(t_f, t\right)\}. \tag{15.27}$$

Substituting (15.26) and (15.27) into (15.21) yields

$$Z\left(t_f\right) = Z\left(t\right) - \text{sign}\left\{Z\left(t_f\right)\right\} \int_t^{t_f} \left\{\left|\tilde{B}\left(t_f, s\right)\right| - \left|\tilde{C}\left(t_f, s\right)\right|\right\} ds. \tag{15.28}$$

Assuming that $Z(t)$ does not change sign, a *candidate optimal* trajectory that terminates with the miss distance $Z(t_f)$ can be constructed by backward integration using (15.28) and one can test whether the family of such *optimal* trajectories fills the entire game space. Regions that are left empty by such construction are *singular* and within them another pair of *optimal strategies* has to be found. This procedure will be carried out in the following subsections for different dynamic game models.

15.3.1 Ideal Pursuer and Evader Dynamics

This is the simplest game model, where the directly controlled normalized lateral accelerations are the control variables ($a_P^c = (a_P)_\perp = ua_P^{\max}$, $a_E^c = (a_E)_\perp = va_E^{\max}$) and the state vector has only two components ($n = 2$):

$$X^T = [x_1, x_2] = [y, \dot{y}]. \tag{15.29}$$

In this model,

$$A = \begin{bmatrix} 0 & 1 \\ 0 & 0 \end{bmatrix}, \quad B = \begin{bmatrix} 0 \\ -a_p^{\max} \end{bmatrix}, \quad C = \begin{bmatrix} 0 \\ a_e^{\max} \end{bmatrix}. \tag{15.30}$$

The components of the transition matrix $\Phi(t_f, t)$ involved in this case are

$$\varphi_{11} = 1; \quad \varphi_{12} = t_{\text{go}}, \tag{15.31}$$

yielding the *zero-effort miss distance*:

$$Z(t_{\text{go}}) = y + \dot{y}t_{\text{go}}, \tag{15.32}$$

and the coefficients in (15.19) as

$$\tilde{B}\left(t_f, t\right) = \tilde{B}(t_{\text{go}}) = -t_{\text{go}}a_P^{\max}; \quad \tilde{C}\left(t_f, t\right) = \tilde{C}(t_{\text{go}}) = t_{\text{go}}a_E^{\max}, \tag{15.33}$$

as the functions of the time-to-go

$$t_{\text{go}} \triangleq t_f - t. \tag{15.34}$$

Remark 15.1. In order to not overload the paper with multiple notations, we keep here and in what follows the same notation for the zero-effort miss distance Z and the coefficients $\tilde{B}$ and $\tilde{C}$ in (15.19), although the independent variables are different in different models.

Due to the form of Z, $\tilde{B}$, and $\tilde{C}$, the time-to-go (15.34) is chosen as a new independent variable, leading to the scalar system

$$\frac{dZ}{dt_{\text{go}}} = t_{\text{go}}a_P^{\max}u - t_{\text{go}}a_E^{\max}v, \tag{15.35}$$

and the cost function

$$J = |Z(0)|. \tag{15.36}$$

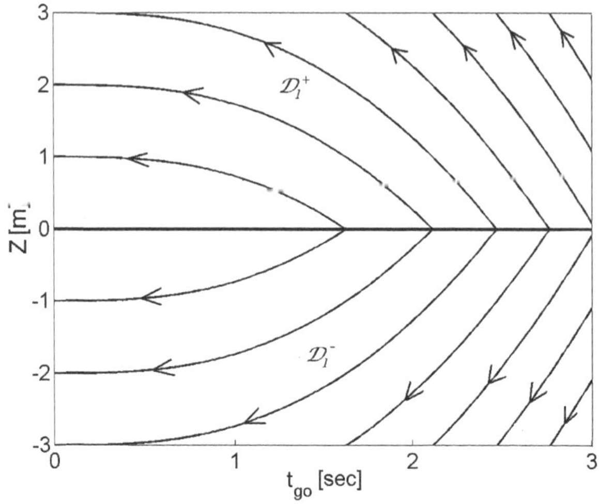

Fig. 15.3 DGL/I game space decomposition for $\mu < 1$

We denote the game with the dynamics (15.35), the cost function (15.36) and the control constraints (15.10) as DGL/I (differential game law/ideal).

By defining the *pursuer/evader* maneuver ratio

$$\mu \triangleq a_P^{\max}/a_E^{\max}, \tag{15.37}$$

due to (15.26)–(15.28), one obtains for a fixed $t_{\text{go}} > 0$

$$Z(0) = Z(t_{\text{go}}) - \frac{1}{2}t_{\text{go}}^2\,(\mu - 1)\,a_E^{\max}\,\text{sign}\left\{Z(t_{\text{go}})\right\}. \tag{15.38}$$

From (15.38) it is clear that the game value $J^* = |Z(0)|$ depends on the value of μ. If $\mu < 1$, zero miss distance cannot be achieved from any initial condition. Optimal trajectories originate from the t_{go}-axis (serving as a *dispersal line*) and fill the entire game space, as seen in Fig. 15.3.

If $\mu = 1$, the optimal trajectories are parallel lines to the t_{go}-axis filling the game space, as shown in Fig. 15.4.

If $\mu > 1$, the reduced game space is decomposed in two parts: a *singular* region, denoted as $\mathcal{D}_0$, and a *regular* one, denoted as $\mathcal{D}_1$, as shown in Fig. 15.5. The boundaries that separate the two regions are symmetrical parabolas described by the equation

$$Z_\pm^*\left(t_{\text{go}}\right) = \pm\frac{1}{2}t_{\text{go}}^2\,(\mu - 1)\,a_E^{\max} \tag{15.39}$$

Within this *singular* region the *optimal strategies* are *arbitrary* and the *value* of the game is constant (zero). Outside the boundaries (15.39) there is the *regular* region, denoted as $\mathcal{D}_1$, where the *optimal strategies* are given by (15.26) and (15.27),

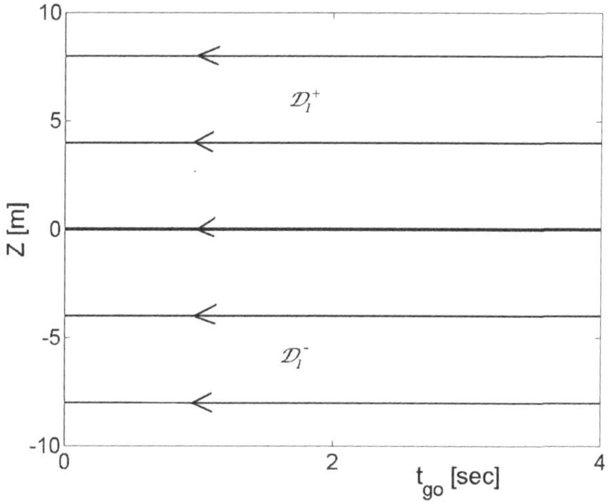

Fig. 15.4 DGL/I game space decomposition for $\mu = 1$

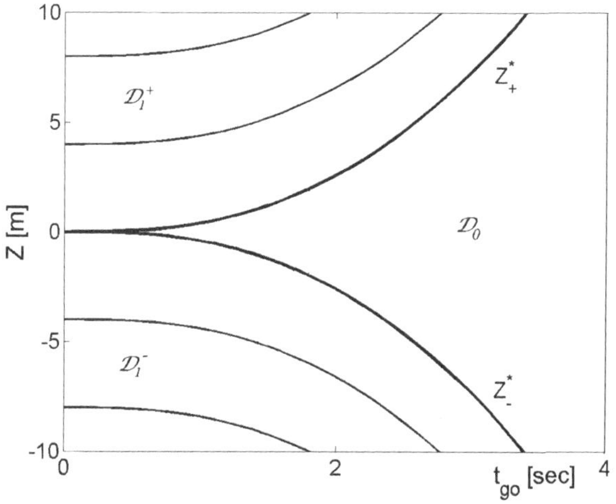

Fig. 15.5 DGL/I game space decomposition for $\mu > 1$

while the *value* of the game depends on the initial conditions, according to (15.28). This game solution was published by Gutman and Leitmann (1976).

The implementation of an interceptor guidance law based on this game solution, denoted DGL/I, is not unique, due to the existence of the *singular* region for $\mu > 1$, which is the case of practical interest. One option is to use the "bang-bang" guidance law of (15.26) and (15.27) everywhere. Another option suggested in Gutman and Leitmann (1976) is to use in $\mathcal{D}_0$ a linear guidance law in such a way that on the boundaries of the region the maximal admissible acceleration is reached. This guidance law turns out to be Proportional Navigation (PN) (Adler 1956). In a linearized scenario the small line of sight angle $\lambda(t) << 1$ can be written as

$$\lambda(t) \cong \tan\lambda(t) = y(t)/x(t). \tag{15.40}$$

Using this approximation the line of sight rate becomes

$$\dot{\lambda} = \frac{d}{dt}\left(\frac{y}{x}\right) = \frac{\dot{y}x - y\dot{x}}{x^2} = \frac{\dot{y}}{x} - \frac{\dot{x}}{x^2}y. \tag{15.41}$$

Since for the linearized geometry $x(t) = V_c t_{\text{go}}$,

$$\dot{\lambda} = \frac{\dot{y}}{V_c t_{\text{go}}} + \frac{y}{V_c t_{\text{go}}^2} = \frac{1}{V_c t_{\text{go}}^2}[y + \dot{y}t_{\text{go}}]. \tag{15.42}$$

The linear approximation of the final separation is $y(t_f)$. Therefore, the expression for the linearized predicted *zero-effort miss distance* for PN becomes

$$Z_{\text{PN}}(t_{\text{go}}) = y + \dot{y}t_{\text{go}}, \tag{15.43}$$

leading to conclude that

$$\dot{\lambda} = \frac{Z_{\text{PN}}}{V_c t_{\text{go}}^2}. \tag{15.44}$$

The classical form of PN is

$$(a_P)_\perp = N' V_c \dot{\lambda}. \tag{15.45}$$

In order to obtain maximal admissible acceleration on the boundaries of D_0 the effective navigation ratio N' depends on μ. By using (15.39) with (15.43)–(15.45) the appropriate value of N' is

$$N' = 2\mu/(\mu - 1). \tag{15.46}$$

By accepting this suggestion, the interceptor guidance law can be expressed in the entire reduced game space (Z, t_{go}) by using the saturation operator ($\text{sat}\{f(t)\} = f(t)$ if $|f(t)| < 1$ and $\text{sat}\{f(t)\} = \text{sign}\, f(t)$ if $|f(t)| \geq 1$),

$$(a_P)_\perp = a_P^{\max}\,\text{sat}\left\{\frac{2\mu}{\mu - 1}\frac{Z_{\text{PN}}(t_{\text{go}})}{t_{\text{go}}^2 a_P^{\max}}\right\}. \tag{15.47}$$

Since the model of ideal dynamics, being far from reality, cannot provide a reliable element in guided missile design, for such a purpose more realistic models are needed.

15.3.2 Ideal Evader and First-Order Pursuer Dynamics

This model, used first in Gutman (1979) acknowledges the strong effect of interceptor dynamics on the homing performance and approximates it by a first-order transfer function with time constant τ_P. The assumption of ideal *evader* dynamics (although this is not realistic) provides the "worst case" for the *pursuer* and therefore it is on the "safe side" for guided missile design.

Using such a model, the state vector of the game is three-dimensional ($n = 3$):

$$X^T = [x_1, x_2, x_3] = [y, \dot{y}, (a_P)_\perp]. \tag{15.48}$$

In this model,

$$A = \begin{bmatrix} 0 & 1 & 0 \\ 0 & 0 & -1 \\ 0 & 0 & -\dfrac{1}{\tau_P} \end{bmatrix}, \quad B = \begin{bmatrix} 0 \\ 0 \\ \dfrac{a_P^{\max}}{\tau_P} \end{bmatrix}, \quad C = \begin{bmatrix} 0 \\ a_E^{\max} \\ 0 \end{bmatrix}. \tag{15.49}$$

The components of the transition matrix $\Phi(t_f, t)$ involved in this case are

$$\varphi_{11} = 1; \;\; \varphi_{12} = \tau_P \theta; \;\; \varphi_{13} = -\tau_P^2 \psi(\theta), \tag{15.50}$$

yielding the *zero-effort miss distance*

$$Z(\theta) = y + \dot{y}\tau_P\theta - (a_P)_\perp \tau_P^2 \psi(\theta), \tag{15.51}$$

and the coefficients in (15.19) as

$$\tilde{B}\left(t_f, t\right) = \tilde{B}(\theta) = -a_P^{\max}\tau_P\psi(\theta); \;\; \tilde{C}\left(t_f, t\right) = \tilde{C}(\theta) = \tau_P\theta a_E^{\max}, \tag{15.52}$$

where

$$\theta \triangleq t_{go}/\tau_P, \tag{15.53}$$

$$\psi(\theta) \triangleq e^{-\theta} + \theta - 1. \tag{15.54}$$

By using θ as a new independent variable and the *normalized zero-effort miss distance*, defined by

$$z(\theta) \triangleq \frac{Z(\theta)}{\tau_P^2 a_E^{\max}}, \tag{15.55}$$

as a new state variable, the scalar system (15.19) is transformed to

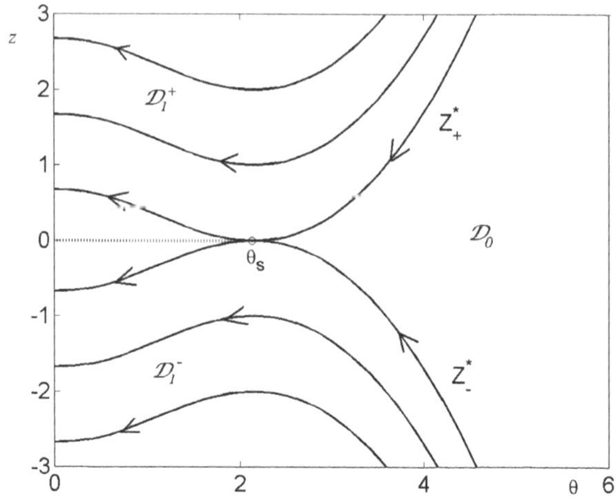

Fig. 15.6 DGL/0 game space decomposition for $\mu > 1$

$$\frac{dz}{d\theta} = \mu\psi(\theta)u - \theta v. \tag{15.56}$$

The cost function becomes

$$J = |z(0)|. \tag{15.57}$$

We denote the game with the dynamics (15.56), the cost function (15.57) and the control constraints (15.10) as DGL/0.

Due to (15.26)–(15.28), one obtains for a fixed $\theta > 0$,

$$z(0) = z(\theta) - \operatorname{sign}\{z(\theta)\} \int_0^\theta h(\theta) d\theta, \tag{15.58}$$

where

$$h(\theta) \triangleq \mu\psi(\theta) - \theta. \tag{15.59}$$

One can easily see that for $\mu > 1$ the integral in (15.58) has a minimum attained for $\theta = \theta_s$, where θ_s is the nonzero solution of the equation

$$\mu\psi(\theta) - \theta = 0. \tag{15.60}$$

For small values of θ ($\theta < \theta_s$) the integrand is negative, which means that zero miss distance ($z(0) = 0$) can never be achieved. The decomposition of the reduced (z, θ) game space for $\mu > 1$ is shown in Fig. 15.6.

The two *limiting* trajectories (Z_+^*, Z_-^*), satisfying the condition that $z(\theta)$ does not change sign, reach the θ-axis tangentially at $\theta = \theta_s$. The reduced game space is decomposed into a *singular* region $\mathscr{D}_0$, which is between these trajectories for $\theta > \theta_s$ and the *regular* region $\mathscr{D}_1$. In $\mathscr{D}_1$ the *optimal strategies* are given by (15.26) and (15.27) while the nonzero *value* of the game depends on the initial conditions.

In the *singular* region the *optimal strategies* are *arbitrary* and the *value* of the game is a nonzero constant J_s. This *singular* game *value*, which is the smallest guaranteed miss distance that an optimally playing *pursuer* can achieve against an optimally playing *evader*, depends on the physical parameter μ. Once θ_s is found from the solution of (15.60), J_s can be computed from (15.58) by setting $Z(\theta_s) = 0$ and by direct integration between θ_s and zero. The larger is μ, the smaller is θ_s and consequently also J_s is smaller. For a sufficiently large value of μ the guaranteed miss distance J_s is very small.

Every trajectory starting in $\mathscr{D}_0$ must go through the *throat* $[Z(\theta_s) = 0]$. This is a *dispersal point* for the *evader* to decide on the maneuver direction for $\theta < \theta_s$. The *pursuer* must wait at that point (for an infinitely short time) in order to follow the direction selected by the *evader* in order to obtain an outcome not larger than J_s.

In the (practically unimportant) case of $\mu < 1$, there is no *singular* region and the game space decomposition looks very similar to Fig. 15.3.

Similarly to DGL/I, the implementation of the guidance law, based on the DGL/0 solution, is not unique, due to the existence of the singular region. One option is (similarly to DGL/I) to use the "bang-bang" guidance law of (15.26) and (15.27) everywhere. In Gutman (1979) it is suggested to use in $\mathscr{D}_0$ a linear guidance law in such a way that on the boundaries of the region the maximal admissible acceleration is reached. Another interesting option is to use in $\mathscr{D}_0$ a linear-quadratic game solution that guarantees reaching the *throat* with minimal control effort.

The interceptor guidance law DGL/0 has an important advantage. Its implementation requires only (see (15.42) and (15.51)) the knowledge of the line of sight rate and own acceleration, but not the target acceleration. Although it cannot guarantee zero miss distance even in an ideal situation, the guaranteed miss distance can be made very small.

15.3.3 First-Order Evader and Pursuer Dynamics

If there is sufficient information on the *evader* dynamics, approximating it by a first-order transfer function provides a more realistic and balanced game model (Shinar 1981). In this game the state vector is four-dimensional ($n = 4$):

$$X^T = [x_1, x_2, x_3, x_4] = [y, \dot{y}, (a_E)_\perp, (a_P)_\perp]. \tag{15.61}$$

In this model,

$$A = \begin{bmatrix} 0 & 1 & 0 & 0 \\ 0 & 0 & 1 & -1 \\ 0 & 0 & -\dfrac{1}{\tau_E} & 0 \\ 0 & 0 & 0 & \dfrac{1}{\tau_P} \end{bmatrix}, \quad B = \begin{bmatrix} 0 \\ 0 \\ 0 \\ \dfrac{a_P^{\max}}{\tau_P} \end{bmatrix}, \quad C = \begin{bmatrix} 0 \\ 0 \\ \dfrac{a_E^{\max}}{\tau_E} \\ 0 \end{bmatrix}. \tag{15.62}$$

The components of the transition matrix $\Phi(t_f, t)$ involved in this case are

$$\varphi_{11} = 1; \;\; \varphi_{12} = t_{go}; \;\; \varphi_{13} = \tau_E^2 \psi(\theta/\varepsilon); \;\; \varphi_{14} = -\tau_P^2 \psi(\theta), \tag{15.63}$$

yielding the *zero-effort miss distance*

$$Z(\theta) = y + \dot{y}\tau_P\theta + (a_E)_\perp \tau_E^2 \psi(\theta/\varepsilon) - (a_P)_\perp \tau_P^2 \psi(\theta), \tag{15.64}$$

and the coefficients in (15.19) as

$$\tilde{B}(t_f, t) = \tilde{B}(\theta) = -a_P^{\max} \tau_P \psi(\theta); \;\; \tilde{C}(t_f, t) = \tilde{C}(\theta) = a_E^{\max} \tau_E \psi(\theta/\varepsilon). \tag{15.65}$$

where θ is defined by (15.53),

$$\varepsilon \triangleq \tau_E/\tau_P. \tag{15.66}$$

By using θ as a new independent variable and the *normalized zero-effort miss distance*, defined by (15.55), (15.64), as a new state variable, the scalar system (15.19) is transformed to

$$\frac{dz}{d\theta} = \mu\psi(\theta)u - \varepsilon\psi(\theta/\varepsilon)v. \tag{15.67}$$

We denote the game with the dynamics (15.67), the cost function (15.57) and the control constraints (15.10) as DGL/1. Due to (15.26)–(15.28), for a fixed $\theta > 0$, $z(0)$ is given by (15.58), where now

$$h(\theta) = \mu\psi(\theta) - \varepsilon\psi(\theta/\varepsilon). \tag{15.68}$$

Depending on the values of the parameters μ and ε, $h(\theta)$ can be either positive or negative. If $\mu > 1$ and $\mu\varepsilon < 1$, the function $\int_0^\theta h(\xi)d\xi$ has a minimum at $\theta = \theta_s$, where θ_s is the positive solution of the equation

$$\mu\psi(\theta) - \varepsilon\psi(\theta/\varepsilon) = 0, \tag{15.69}$$

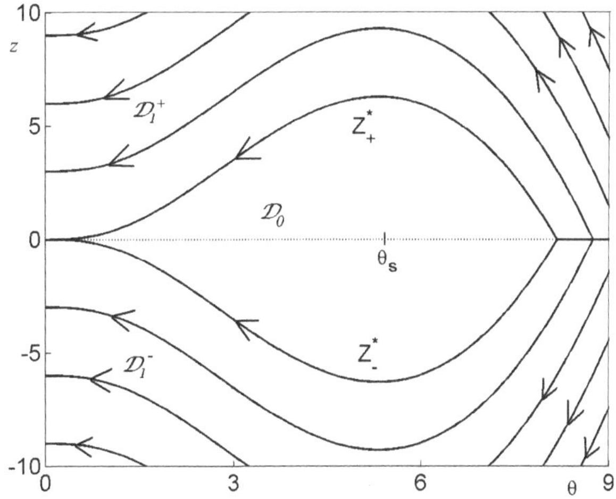

Fig. 15.7 DGL/1 game space decomposition for $\mu < 1$ and $\mu\varepsilon \geq 1$

Table 15.1 Conditions for various game solution structures

	$\mu < 1$	$\mu = 1$	$\mu > 1$
DGL/I	Fig. 15.3	Fig. 15.4	Fig. 15.5
DGL/0			
$\mu\varepsilon = 0$	Fig. 15.3	Fig. 15.4	Fig. 15.6
DGL/1			
$\mu\varepsilon < 1$	Fig. 15.3	Fig. 15.4	Fig. 15.5
$\mu\varepsilon \geq 1$	Fig. 15.7		Fig. 15.5

and the game space decomposition is similar to Fig. 15.6 (the one of DGL/0 for $\mu > 1$).

For $\mu > 1$ and $\mu\varepsilon \geq 1$, the only nonnegative solution of (15.68) is $\theta = 0$; $h(\theta) > 0$ for $\theta > 0$, and the game space decomposition is similar to Fig. 15.5. In this case from any initial position inside the *singular* region $\mathcal{D}_0$ zero miss distance is guaranteed by using *arbitrary* strategies. For the case of $\mu < 1$ and $\mu\varepsilon \geq 1$ there is a bounded *singular* region, where zero miss distance can be achieved, as seen in Fig. 15.7. This game space decomposition was first presented by Shima and Shinar (2002).

Due to the existence of the *singular* region, the implementation of the interceptor guidance law based on this game solution, denoted as DGL/1, is also not unique and options similar to those of DGL/0 and DGL/I can be adopted.

It should also be noted that the implementation of DGL/1 requires the knowledge of the current *evader* maneuver as a component of the state vector, which cannot be measured from another platform. It has to be reconstructed from available measurements by an observer for a noise free case or by an estimator if the available measurements are corrupted by noise.

In Table 15.1 the conditions for various game solution structures are summarized.

15.4 Second-Order Pursuer Dynamics

In general, the autopilot dynamics of a well-designed interceptor missile is rather complex. It contains several nonlinear elements and its linearized version is of high order. The approximation of such high order dynamics by a simple first-order model is very difficult. The approximation by a non oscillatory second-order dynamic model seems more suitable and (having two independent parameters, the frequency ω_P and the damping factor ζ_P) is also easier. In this section, results of a not yet published analytical investigation of the game solution with second-order pursuer dynamics is presented. The maneuvering dynamics of the pursuer is described by a linear second-order transfer function with the damping coefficient $\zeta_P > 0$ and the frequency ω_P

$$H_P(s) = \frac{1}{1 + 2\zeta_P s/\omega_P + s^2/\omega_P^2}. \tag{15.70}$$

In the pursuit-evasion differential game with the second-order pursuer dynamics (15.70) and the first-order evader dynamics, the state vector is of the dimension $n = 5$:

$$X^T = [x_1, x_2, x_3, x_4, x_5] = [y, \dot{y}, (a_E)_\perp, (a_P)_\perp, (\dot{a}_P)_\perp]. \tag{15.71}$$

In this model,

$$A = \begin{bmatrix} 0 & 1 & 0 & 0 & 0 \\ 0 & 0 & 1 & -1 & 0 \\ 0 & 0 & -1/\tau_E & 0 & 0 \\ 0 & 0 & 0 & 0 & 1 \\ 0 & 0 & 0 & -\omega_P^2 & -2\zeta_P\omega_P \end{bmatrix}, \quad B = \begin{bmatrix} 0 \\ 0 \\ 0 \\ 0 \\ \omega_P^2 a_P^{\max} \end{bmatrix}, \quad C = \begin{bmatrix} 0 \\ 0 \\ a_E^{\max}/\tau_E \\ 0 \\ 0 \end{bmatrix}, \tag{15.72}$$

and, in addition to (15.11)–(15.12), it is assumed that

$$X_5(0) = 0. \tag{15.73}$$

15.4.1 Scalarized Game Solution

The components of the transition matrix $\Phi(t_f, t)$ involved in this case are

$$\varphi_{11} = 1; \;\; \varphi_{12} = \theta_2/\omega_P; \;\; \varphi_{13} = \tau_E^2 \psi(\theta_2/\varepsilon_2), \tag{15.74}$$

$$\varphi_{14} = \frac{\theta_2^2}{2\omega_P^2} + \omega_P \int_0^{\theta_2} \varphi_{15}(\xi) d\xi, \tag{15.75}$$

$$\varphi_{15} = \begin{cases} -\exp(-\zeta_P\theta_2)\left(C_1 \cosh(\sqrt{\zeta_P^2 - 1}\theta_2) + C_2 \sinh(\sqrt{\zeta_P^2 - 1}\theta_2)\right) \\ -\dfrac{\theta_2 - 2\zeta_P}{\omega_P^3}, & \zeta_P > 1, \\ \dfrac{1}{\omega_P^3}\left(-(\theta_2 + 2)\exp(-\theta_2) - \theta_2 + 2\right), & \zeta_P = 1, \\ -\exp(-\zeta_P\theta_2)\left(C_1 \cos(\sqrt{1 - \zeta_P^2}\theta_2) + C_2 \sin(\sqrt{1 - \zeta_P^2}\theta_2)\right) \\ -\dfrac{\theta_2 - 2\zeta_P}{\omega_P^3}, & \zeta_P < 1, \end{cases} \tag{15.76}$$

where

$$\theta_2 \triangleq \omega_P(t_f - t), \tag{15.77}$$

$$\varepsilon_2 \triangleq \omega_P \tau_E, \tag{15.78}$$

$$C_1 = \frac{2\zeta_P}{\omega_P^3}, \quad C_2 = \frac{2\zeta_P^2 - 1}{\omega_P^3\sqrt{|\zeta_P^2 - 1|}}. \tag{15.79}$$

The pursuit-evasion game with the dynamics (15.8), (15.72) can be scalarized in non-dimensional form by choosing θ_2 as a new independent variable and the normalized zero-effort miss distance

$$z_2 \triangleq \frac{\omega_P^2}{a_E^{\max}} Z, \tag{15.80}$$

as a new state variable. The scalar state variable z_2 satisfies the differential equation

$$\frac{dz_2}{d\theta_2} = h_P(\theta_2)u - h_E(\theta_2)v, \tag{15.81}$$

and the initial condition

$$z_2(\theta_{20}) = z_{20}, \tag{15.82}$$

Moreover,

$$h_P(\theta_2) = \mu\tilde{h}_P(\theta_2), \tag{15.83}$$

where μ is defined by (15.37) and

$$\tilde{h}_P(\theta_2) =$$

$$\begin{cases} \omega_P^3 \exp(-\zeta_P\theta_2)\left(C_1 \cosh\left(\sqrt{\zeta_P^2 - 1}\theta_2\right) + C_2 \sinh\left(\sqrt{\zeta_P^2 - 1}\theta_2\right)\right) + \\ \theta_2 \quad 2\zeta_P, & \zeta_P > 1 \\ (\theta_2 + 2)\exp(-\theta) + \theta_2 - 2, & \zeta_P = 1, \\ \omega_P^3 \exp(-\zeta_P\theta_2)\left(C_1 \cos\left(\sqrt{1 - \zeta_P^2}\theta_2\right) + C_2 \sin\left(\sqrt{1 - \zeta_P^2}\theta_2\right)\right) + \\ \theta_2 - 2\zeta_P, & \zeta_P < 1. \end{cases} \tag{15.84}$$

For a first-order or ideal ($\tau_E = 0$) evader

$$h_E(\theta_2) = \begin{cases} \varepsilon_2\left(\exp(-\theta_2/\varepsilon_2) + \dfrac{\theta_2}{\varepsilon_2} - 1\right), & \varepsilon_2 > 0, \\ \theta_2, & \varepsilon_2 = 0. \end{cases} \tag{15.85}$$

Due to (15.11)–(15.12) and (15.73), the initial conditions are

$$\theta_{20} = \omega_P t_f, \quad z_{20} = \frac{\omega_P^2 t_f X_2(0)}{a_E^{\max}}. \tag{15.86}$$

Note that u and v in the equation (15.81) are in fact $u = u(t_f - \theta_2/\omega_P)$ and $v = v(t_f - \theta_2/\omega_P)$ and they satisfy the constraints (15.10). Since $z_2(0) = \omega_P^2 X_1(t_f)/a_E^{\max}$, the cost functional (15.13) becomes

$$J = |z_2(0)|. \tag{15.87}$$

Thus, the original differential game (15.8), (15.10), (15.13) is reduced to the scalar differential game with the dynamics (15.81), the constraints (15.10), and the cost functional (15.87).

In Glizer and Turetsky (2008) it is shown that a original differential game has a saddle point, consisting of the optimal strategies

$$u^0(t, X) = u^*\left(\omega_P(t_f - t), \frac{\omega_P^2}{a_E^{\max}} D\Phi(t_f, t)X\right),$$

$$v^0(t, X) = v^*\left(\omega_P(t_f - t), \frac{\omega_P^2}{a_E^{\max}} D\Phi(t_f, t)X\right) \tag{15.88}$$

where $D = [1, 0, 0, 0, 0]$, $u^*(\theta_2, z_2)$ and $v^*(\theta_2, z_2)$ are the optimal strategies in the scalar differential game (15.81), (15.10), (15.87). The latter have the form

$$u^*(\theta_2, z_2) = \begin{cases} \mathrm{sign} h_P(\theta_2) \mathrm{sign} z_2, & (\theta_2, z_2) \in \mathscr{D}_1, \\ \text{arbitrary subject to (9)}, & (\theta_2, z_2) \in \mathscr{D}_0, \end{cases} \tag{15.89}$$

$$v^*(\theta_2, z_2) = \begin{cases} \mathrm{sign} h_E(\theta_2) \mathrm{sign} z_2, & (\theta_2, z_2) \in \mathscr{D}_1, \\ \text{arbitrary subject to (9)}, & (\theta_2, z_2) \in \mathscr{D}_0, \end{cases} \tag{15.90}$$

where $\mathscr{D}_1$ and $\mathscr{D}_0$ are the *regular* and *singular* regions, $\mathscr{D}_0 = R^2 \backslash \mathscr{D}_1$ and the *regular* region $\mathscr{D}_1$ is closed.

The structure of the *singular* and *regular* regions depends on the behavior of the determining function

$$\Gamma(\theta_2) = |h_P(\theta_2)| - |h_E(\theta_2)|, \quad \theta_2 \geq 0. \tag{15.91}$$

In the next subsection, the behavior of the function $\Gamma(\theta_2)$ and the structure of the regions $\mathscr{D}_0$ and $\mathscr{D}_1$ are analyzed in detail.

15.4.2 Analysis of $\Gamma(\theta_2)$

15.4.2.1 Positiveness of $h_P(\theta_2)$ and $h_E(\theta_2)$

It is known (Shinar 1981) that $h_E(\theta_2) > 0$ for $\theta_2 > 0$. The following lemma establishes the same property of the function $h_P(\theta_2)$.

Lemma 15.1. *For all $\theta_2 > 0$, the function $h_P(\theta_2)$ is positive.*

Proof. Based on (15.83)–(15.84), for all $\zeta_P > 0$,

$$h_P(0) = h'_P(0) = 0. \tag{15.92}$$

Now, let us start with the case $\zeta_P > 1$. The second derivative of $h_P(\theta_2)$ is

$$h''_P(\theta_2) = \frac{\mu \exp(-\zeta_P \theta_2) \sinh\left(\sqrt{\zeta_P^2 - 1}\theta_2\right)}{\sqrt{\zeta_P^2 - 1}}, \tag{15.93}$$

directly yielding

$$h''_P(\theta_2) > 0, \quad \theta_2 > 0. \tag{15.94}$$

Equation (15.92) and the inequality (15.94) imply that $h'_P(\theta_2) > 0$ for $\theta_2 > 0$, and, consequently, $h_P(\theta_2) > 0$ for $\theta_2 > 0$.

Proceed to the case $\zeta_P = 1$. In this case, the second derivative of $h_P(\theta_2)$ is

$$h''_P(\theta_2) = \mu\theta_2 \exp(-\theta_2), \tag{15.95}$$

yielding (15.94). Therefore, similarly to the case $\zeta_P > 1$, the function $h_P(\theta_2)$ is positive for $\theta_2 > 0$.

Finally, consider the case $\zeta_P < 1$. In this case, the first and the second derivatives of $h_P(\theta_2)$ are

$$h'_P(\theta_2) =$$

$$\mu\left[1 - \exp(-\zeta_P\theta_2)\left(\cos\left(\sqrt{1-\zeta_P^2}\theta_2\right) + \frac{\zeta_P}{\sqrt{1-\zeta_P^2}}\sin\left(\sqrt{1-\zeta_P^2}\theta_2\right)\right)\right], \tag{15.96}$$

$$h''_P(\theta_2) = \frac{\mu\exp(-\zeta_P\theta_2)\sin\left(\sqrt{1-\zeta_P^2}\theta_2\right)}{\sqrt{1-\zeta_P^2}}. \tag{15.97}$$

By virtue of (15.97), for $\theta_2 > 0$, the local minimum of $h'_P(\theta_2)$ is attained for $\theta_{2n} = 2\pi n/\sqrt{1-\zeta_P^2}$, $n = 1, 2, \ldots$, and the local minima of $h'(\theta_2)$ are

$$h'_P(\theta_{2n}) = \mu(1 - \exp(-\zeta_P\theta_{2n})) > 0, \quad n = 1, 2, \ldots. \tag{15.98}$$

Moreover,

$$\lim_{\theta_2\to\infty} h'_P(\theta_2) = \mu > 0. \tag{15.99}$$

Equation (15.92) and inequalities (15.98)–(15.99) yield $h'_P(\theta_2) > 0$ for $\theta_2 > 0$, and, consequently, $h_P(\theta_2) > 0$ for $\theta_2 > 0$. This completes the proof of the lemma. □

Lemma 15.1 is illustrated by Fig. 15.8, where the function $h_P(\theta_2)$ is depicted for $\mu = 2$ and three different values of ζ_P.

Lemma 15.1 allows to rewrite the function $\Gamma(\theta_2)$, given by (15.91), as

$$\Gamma(\theta_2) = h_P(\theta_2) - h_E(\theta_2), \quad \theta_2 \geq 0. \tag{15.100}$$

Fig. 15.8 Function $h_P(\theta_2)$

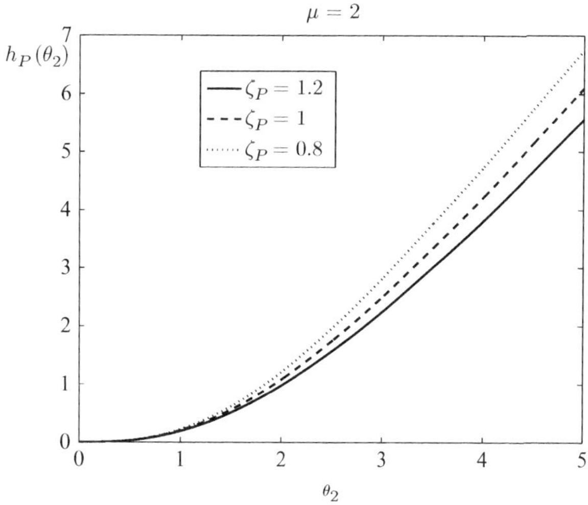

15.4.2.2 Zeros of $\Gamma(\theta_2)$

Lemma 15.2. *For any $\zeta_P > 0$, $\mu > 1$ and $\varepsilon_2 \geq 0$, the function $\Gamma(\theta_2)$ has at least one positive zero.*

Proof. If $\varepsilon_2 > 0$,

$$\Gamma(0) = \Gamma'(0) = 0, \quad \Gamma''(0) = -\frac{1}{\varepsilon_2} < 0. \tag{15.101}$$

For $\varepsilon_2 = 0$,

$$\Gamma(0) = 0, \Gamma'(0) = -1. \tag{15.102}$$

Both (15.101) and (15.102) imply that the function $\Gamma(\theta_2)$ decreases in the vicinity of $\theta_2 = 0$, i.e. there exists $\delta > 0$ such that

$$\Gamma(\theta_2) < 0, \quad \theta_2 \in (0, \delta). \tag{15.103}$$

Moreover, for $\theta_2 \to +\infty$,

$$\Gamma(\theta_2) \sim (\mu - 1)\theta_2 - 2\mu\zeta_P + \varepsilon_2. \tag{15.104}$$

This means that if

$$\mu > 1, \tag{15.105}$$

then there exists $M > 0$ such that

$$\Gamma(\theta_2) > 0, \quad \theta_2 > M. \tag{15.106}$$

The conditions (15.103) and (15.106) yield that the equation $\Gamma(\theta_2) = 0$ must have at least one positive zero. □

Remark 15.2. From the proof of Lemma 15.2, it directly follows that the equation $\Gamma(\theta_2) = 0$ can have only an odd number of positive zeros, taking into account their multiplicities.

Lemma 15.3. *For any $\zeta_P \geq 1$, $\mu > 1$ and $\varepsilon_2 \geq 0$, the function $\Gamma(\theta_2)$ has a single zero $\theta_{2s} > 0$.*

Proof. Due to Lemma 15.2, there exists at least one zero of $\Gamma(\theta_2)$. Let us prove that $\Gamma(\theta_2)$ has exactly one positive zero.

Case 1. $\varepsilon_2 > 0$. For this, it is sufficient to show that the second derivative $\Gamma''(\theta_2)$ has no more than two distinct positive zeros.

First, consider the case $\zeta_P > 1$. In this case,

$$\Gamma''(\theta_2) = \frac{\mu}{\sqrt{\zeta_P^2 - 1}} \exp(-\zeta_P \theta_2) \sinh\left(\theta_2 \sqrt{\zeta_P^2 - 1}\right) - \frac{1}{\varepsilon_2} \exp(-\theta_2/\varepsilon_2), \tag{15.107}$$

and the equation $\Gamma''(\theta_2) = 0$ is equivalent to

$$f_1(\theta_2) \triangleq \exp((1/\varepsilon_2 - \zeta_P)\theta_2) \sinh\left(\theta_2 \sqrt{\zeta_P^2 - 1}\right) = \frac{\sqrt{\zeta_P^2 - 1}}{\mu \varepsilon_2}. \tag{15.108}$$

The derivative of the function $f_1(\theta_2)$, defined in (15.108),

$$f_1'(\theta_2) = \exp((1/\varepsilon_2 - \zeta_P)\theta_2) \times$$

$$\left[(1/\varepsilon_2 - \zeta_P) \sinh\left(\theta_2 \sqrt{\zeta_P^2 - 1}\right) + \sqrt{\zeta_P^2 - 1} \cosh\left(\theta_2 \sqrt{\zeta_P^2 - 1}\right)\right]. \tag{15.109}$$

Thus, the equation $f_1'(\theta_2) = 0$ is equivalent to the equation

$$(1/\varepsilon_2 - \zeta_P)(w - 1/w) + \sqrt{\zeta_P^2 - 1}(w + 1/w) = 0, \tag{15.110}$$

where

$$w \triangleq \exp\left(\theta_2 \sqrt{\zeta_P^2 - 1}\right) > 1. \tag{15.111}$$

Equation (15.110) can be rewritten as

$$w^2 = \frac{\zeta_P - 1/\varepsilon_2 + \sqrt{\zeta_P^2 - 1}}{\zeta_P - 1/\varepsilon_2 - \sqrt{\zeta_P^2 - 1}}. \tag{15.112}$$

If

$$\zeta_P - 1/\varepsilon_2 - \sqrt{\zeta_P^2 - 1} > 0, \tag{15.113}$$

the Eq. (15.112) has the single positive root

$$\tilde{w}_1 = \sqrt{\frac{\zeta_P - 1/\varepsilon_2 + \sqrt{\zeta_P^2 - 1}}{\zeta_P - 1/\varepsilon_2 - \sqrt{\zeta_P^2 - 1}}} > 1, \tag{15.114}$$

yielding by simple algebra the single positive zero of the equation $f_1'(\theta_2) = 0$:

$$\tilde{\theta}_2 = \frac{1}{2\sqrt{\zeta_P^2 - 1}} \ln \frac{\zeta_P - 1/\varepsilon + \sqrt{\zeta_P^2 - 1}}{\zeta_P - 1/\varepsilon - \sqrt{\zeta_P^2 - 1}} > 0. \tag{15.115}$$

Thus, if (15.113) is valid, then $f_1'(\tilde{\theta}_2) = 0$, while $f_1'(\theta_2) > 0$ for $\theta_2 \in [0, \tilde{\theta}_2)$ and $f_1'(\theta_2) < 0$ for $\theta_2 > \tilde{\theta}_2$. Therefore, the function $f_1(\theta)$ increases monotonically for $\theta_2 \in [0, \tilde{\theta}_2]$ from $f_1(0) = 0$ to $f_1(\tilde{\theta}_2)$ and decreases monotonically for $\theta_2 > \tilde{\theta}_2$. This means that the Eq. (15.108) has no more than two roots, depending on the value of $\frac{\sqrt{\zeta_P^2-1}}{\mu\varepsilon_2}$. If the condition (15.113) is not valid, then $f_1'(\theta_2) > 0$ for all $\theta_2 > 0$, i.e. $f_1(\theta_2)$ increases monotonically for all $\theta_2 > 0$. In this case, the Eq. (15.108) has exactly one root. It means that $\Gamma''(\theta_2) = 0$ has no more than two distinct positive roots, which completes the proof of the lemma in the case $\zeta_P > 1$.

Proceed to the case $\zeta_P = 1$. In this case, the equation $\Gamma''(\theta) = 0$, similarly to (15.108) is equivalent to the equation

$$f_2(\theta_2) \triangleq \exp((1/\varepsilon_2 - 1)\theta_2)\theta_2 = \frac{1}{\mu\varepsilon_2}. \tag{15.116}$$

The derivative of the function f_2, defined in (15.116), is

$$f_2'(\theta_2) = \frac{1}{\varepsilon_2} \exp((1/\varepsilon_2 - 1)\theta_2)[(1 - \varepsilon_2)\theta_2 + \varepsilon_2]. \tag{15.117}$$

If

$$\varepsilon_2 > 1, \tag{15.118}$$

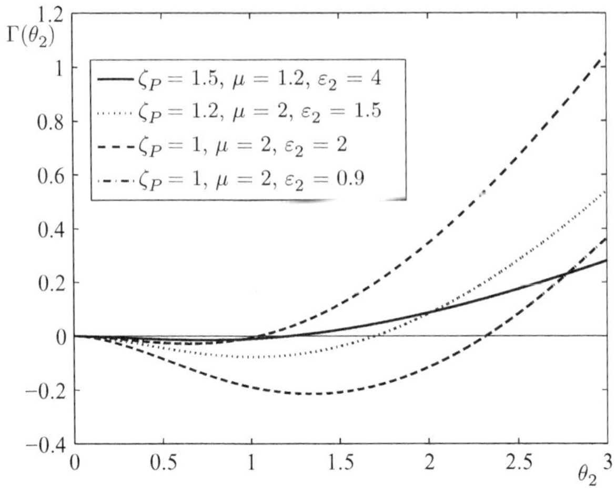

Fig. 15.9 Function $\Gamma(\theta_2)$: $\zeta_P \geq 1$

then the equation $f_2'(\theta_2) = 0$ has the single root

$$\tilde{\theta}_2 = \frac{\varepsilon_2}{\varepsilon_2 - 1} > 0. \tag{15.119}$$

In this case, the function $f_2(\theta_2)$ increases monotonically for $\theta_2 \in [0, \tilde{\theta}_2]$ from $f_2(0) = 0$ to $f_2(\tilde{\theta}_2)$ and decreases monotonically for $\theta_2 > \tilde{\theta}_2$. This means that the Eq. (15.116) has no more than two roots, depending on the value of $\frac{1}{\mu\varepsilon_2}$. If the condition (15.118) is not valid, i.e. $\varepsilon_2 \leq 1$, then $f_2'(\theta_2) > 0$ for all $\theta_2 > 0$, and $f_2(\theta_2)$ increases monotonically for all $\theta_2 > 0$. In this case, the Eq. (15.116) has exactly one root. It is shown that $\Gamma''(\theta)$ has no more than two distinct positive zeros, which completes the proof of the lemma in this case.

Case 2. $\varepsilon_2 = 0$. The proof is similar to Case 1. □

In Fig. 15.9, the function $\Gamma(\theta_2)$ is depicted for different parameters. It is seen that in all cases $\Gamma(\theta_2)$ has the single positive zero.

Let us introduce the function

$$\bar{\Gamma}(\theta_2) = \mu\left(-\frac{\exp(\ \zeta_P\theta_2)}{\sqrt{1-\zeta_P^2}} + \theta_2 - 2\zeta_P\right) - \varepsilon_2\left(\exp(-\theta_2/\varepsilon_2) + \frac{\theta_2}{\varepsilon_2} - 1\right). \tag{15.120}$$

Lemma 15.4. *If the parameters* $\zeta_P < 1$, $\mu > 1$ *and* $\varepsilon_2 > 0$ *satisfy the condition*

$$\bar{\Gamma}\left(\frac{2\pi}{\sqrt{1-\zeta_P^2}}\right) \geq 0, \tag{15.121}$$

then the function $\Gamma(\theta_2)$ has a single zero $\theta_{2s} > 0$.

Proof. Due to Lemma 15.2, there exists at least one zero of $\Gamma(\theta_2)$. Let us prove that subject to the condition (15.121), the equation $\Gamma(\theta_2) = 0$ has exactly one positive root.

Note that for $\zeta_P < 1$, the function $h_P(\theta_2)$ has the form (see (15.83)–(15.84))

$$h_P(\theta_2) =$$
$$\mu\left[\exp(-\zeta_P\theta_2)\left(A\cos\left(\sqrt{1-\zeta_P^2}\theta_2\right) + B\sin\left(\sqrt{1-\zeta_P^2}\theta_2\right)\right) + \theta_2 - 2\zeta_P\right], \tag{15.122}$$

where $A = 2\zeta_P$, $B = (2\zeta_P^2 - 1)/\sqrt{1-\zeta_P^2}$. Thus,

$$A^2 + B^2 = 4\zeta_P^2 + \frac{4\zeta_P^4 - 4\zeta_P^2 + 1}{1-\zeta_P^2} = \frac{1}{1-\zeta_P^2}. \tag{15.123}$$

Due to (15.122)–(15.123), the function $h_P(\theta_2)$ can be rewritten as

$$h_P(\theta_2) = \mu\left[\frac{\exp(-\zeta_P\theta_2)}{\sqrt{1-\zeta_P^2}}\sin\left(\sqrt{1-\zeta_P^2}\theta_2 + \varphi_P\right) + \theta_2 - 2\zeta_P\right], \tag{15.124}$$

where $\sin\varphi_P = 2\zeta_P\sqrt{1-\zeta_P^2}$, $\cos\varphi_P = 2\zeta_P^2 - 1$. Equation (15.124) leads to the inequality

$$h_P(\theta_2) \geq \bar{h}_P(\theta_2) \triangleq \mu\left(-\frac{\exp(-\zeta_P\theta_2)}{\sqrt{1-\zeta_P^2}} + \theta_2 - 2\zeta_P\right), \quad \theta_2 \geq 0, \tag{15.125}$$

which by (15.100) and (15.120) yields

$$\Gamma(\theta_2) \geq \bar{\Gamma}(\theta_2), \quad \theta_2 \geq 0. \tag{15.126}$$

Now, let us show that the equation $\bar{\Gamma}(\theta_2) = 0$ has a single positive root. Due to (15.120),

$$\bar{\Gamma}(0) = -\mu/\sqrt{1-\zeta_P^2} - 2\mu\zeta_P < 0, \tag{15.127}$$

and for $\theta_2 \to \infty$,

$$\bar{\Gamma}(\theta_2) \sim (\mu - 1)\theta_2 - 2\mu\zeta_P + \varepsilon_2. \tag{15.128}$$

Since $\mu > 1$, there exists $\bar{M} > 0$ such that $\bar{\Gamma}(\theta_2) > 0$ for $\theta_2 > \bar{M}$, which, along with (15.127), means that the function $\bar{\Gamma}(\theta_2)$ has at least one positive zero. The derivative of $\bar{\Gamma}(\theta_2)$ is

$$\bar{\Gamma}'(\theta_2) = \frac{\mu\zeta_P \exp(-\zeta_P\theta_2)}{\sqrt{1-\zeta_P^2}} + \mu + \exp(-\theta_2/\varepsilon_2) - 1 > 0, \quad \theta_2 > 0. \tag{15.129}$$

Therefore, the function $\bar{\Gamma}(\theta_2)$ increases monotonically for $\theta_2 > 0$, having one zero $\bar{\theta}_2 > 0$. Thus, the function $\bar{\Gamma}(\theta_2)$ has no zeros for $\theta_2 > \bar{\theta}_2$:

$$\bar{\Gamma}(\theta_2) > 0, \quad \theta_2 > \bar{\theta}_2. \tag{15.130}$$

Due to the monotonicity of $\bar{\Gamma}(\theta_2)$, the condition (15.121) guarantees that

$$\bar{\theta}_2 \le \theta_2^{**} = \frac{2\pi}{\sqrt{1-\zeta_P^2}}. \tag{15.131}$$

By virtue of the inequalities (15.125)–(15.126),

$$\Gamma(\theta_2) > 0, \quad \theta_2 > \theta_2^{**}. \tag{15.132}$$

Now, let us show that $\Gamma(\theta_2)$ has no more than one (and, consequently, exactly one) zero on the interval $(0, \theta_2^{**}]$. For this purpose, similarly to the proof of Lemma 15.3, it is sufficient to prove that the second derivative $\Gamma''(\theta_2)$ has no more than two distinct positive zeros on the interval $(0, \theta_2^{**}]$. For $\zeta_P < 1$, the equation $\Gamma''(\theta_2) = 0$ is equivalent to

$$f_{31}(\theta_2) \triangleq \sin\left(\sqrt{1-\zeta_P^2}\,\theta_2\right) = \frac{\sqrt{1-\zeta_P^2}}{\mu\varepsilon_2} \exp(-(1/\varepsilon_2 - \zeta_P)\theta_2) \triangleq f_{32}(\theta_2). \tag{15.133}$$

Depending on the sign of $1/\varepsilon_2 - \zeta_P$ and the value of the coefficient $\frac{\sqrt{1-\zeta_P^2}}{\mu\varepsilon_2}$, the Eq. (15.133) has no more than two distinct positive roots on the interval $(0, \theta_2^{**}]$. This is illustrated in Fig. 15.10, where the case of two roots is depicted for $1/\varepsilon_2 > \zeta_P$, for $1/\varepsilon_2 = \zeta_P$ and for $1/\varepsilon_2 < \zeta_P$. In these cases, the value of $\bar{\Gamma}(2\pi/\sqrt{1-\zeta_P^2})$ is 8.37, 4.14 and 3.20, respectively. This completes the proof of the lemma. □

In Fig. 15.11, a function $\Gamma(\theta_2)$ is depicted for the parameters satisfying the condition (15.121). It is seen that $\Gamma(\theta_2)$ has a single positive zero.

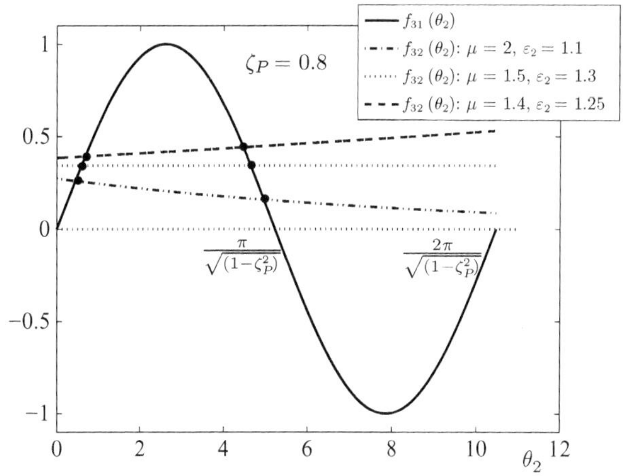

Fig. 15.10 Functions $f_{31}(\theta_2)$, $f_{32}(\theta_2)$

Fig. 15.11 Function $\Gamma(\theta_2)$: $\zeta_P < 1$

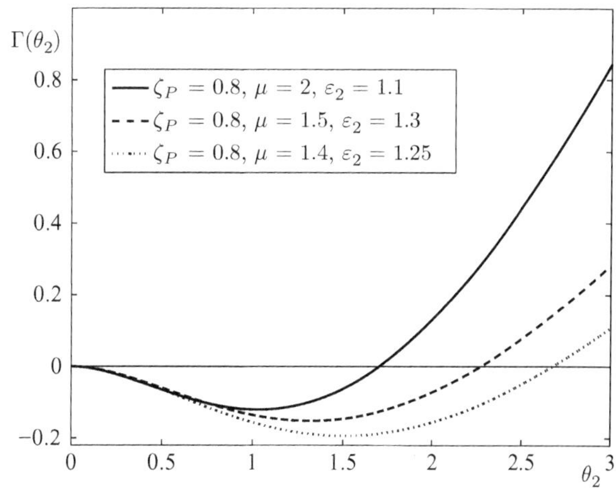

Now, let us find the domain of the parameters μ, ε_2, for which the condition (15.121) is satisfied for all $\zeta_P \in (0, 1)$. Due to (15.120), this domain is described as

$$\mu \geq \mu^{**}(\varepsilon_2) \triangleq \sup_{\zeta_P \in (0,1)} \frac{\varepsilon_2 \sqrt{1-\zeta_P^2}\left(\exp(-\theta_2^{**}(\zeta_P)/\varepsilon_2) + \theta_2^{**}(\zeta_P)/\varepsilon_2 - 1\right)}{(\theta_2^{**}(\zeta_P) - 2\zeta_P)\sqrt{1-\zeta_P^2} - \exp(-\zeta_P \theta_2^{**}(\zeta_P))}, \tag{15.134}$$

where

$$\theta_2^{**}(\zeta_P) \triangleq \frac{2\pi}{\sqrt{1-\zeta_P^2}}. \tag{15.135}$$

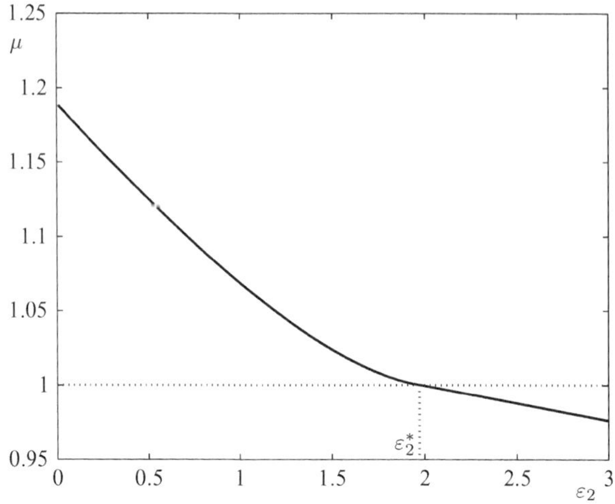

Fig. 15.12 Curve $\mu = \mu^{**}(\varepsilon_2)$

For any $\varepsilon_2 > 0$ and $\mu \geq \mu^{**}(\varepsilon_2)$, the equation $\Gamma(\theta_2) = 0$ has a single positive root for all $\zeta_P \in (0, 1)$. In Fig. 15.12, the curve $\mu = \mu^{**}(\varepsilon_2)$ is depicted. It is seen that for $\varepsilon_2 \geq \varepsilon_2^* \approx 1.98$, $\mu^{**}(\varepsilon_2) \leq 1$, i.e. for all $\varepsilon_2 \geq \varepsilon_2^*$, the equation $\Gamma(\theta_2) = 0$ has a single positive root for all $\mu > 1$ and all $\zeta_P \in (0, 1)$.

15.5 Game Solution

15.5.1 Case of a Single Positive Root of the Equation $\Gamma(\theta_2) = 0$

In this subsection, we consider the following set of parameters μ, ε_2, and ζ_P:

$$\Theta = \{(\mu > 1, \varepsilon_2 > 0, \zeta_P > 0) : (\zeta_P \geq 1) \vee [(0 < \zeta_P < 1) \wedge (\mu \geq \mu^{**}(\varepsilon_2))]\}. \tag{15.136}$$

For any $(\mu, \varepsilon_2, \zeta_P) \in \Theta$, either Lemma 15.3 or Lemma 15.4 is valid. Based on the analysis of Sect. 15.4 and the results of Glizer and Turetsky (2008), the following theorem can be formulated.

Theorem 15.1. *Let* $(\mu, \varepsilon_2, \zeta_P) \in \Theta$. *Then the optimal strategies in the differential game* (15.81), (15.10), (15.87) *are*

$$u^*(\theta_2, z_2) = \begin{cases} \mathrm{sign} z_2, & (\theta_2, z_2) \in \mathscr{D}_1, \\ \text{arbitrary subject to } (15.10), & (\theta_2, z_2) \in \mathscr{D}_0, \end{cases} \tag{15.137}$$

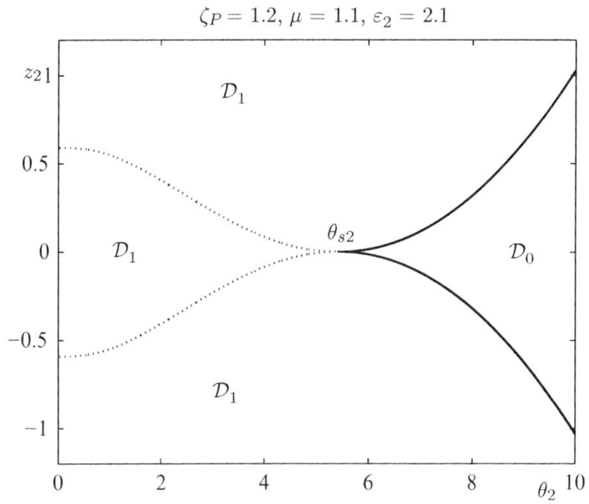

Fig. 15.13 Game space decomposition, $\zeta_P > 1$

$$v^*(\theta_2, z_2) = \begin{cases} \text{sign} z_2, & (\theta_2, z_2) \in \mathscr{D}_1, \\ \text{arbitrary subject to (15.10)}, & (\theta_2, z_2) \in \mathscr{D}_0, \end{cases} \tag{15.138}$$

where the singular region $\mathscr{D}_0$ *is*

$$\mathscr{D}_0 = \left\{ (\theta_2, z_2) : \ \theta_2 > \theta_{2s}, |z_2| < \int_{\theta_{2s}}^{\theta_2} \Gamma(\xi) d\xi \right\}, \tag{15.139}$$

$\mathscr{D}_1 = R^2 \backslash \mathscr{D}_0$, $\theta_{2s} > 0$ *is the single positive root of the equation* $\Gamma(\theta_2) = 0$. *Moreover, the game value is*

$$J_{2s}^* = J_{2s}^*(\theta_{20}, z_{20}) = \begin{cases} J_{s2_0}^* = \displaystyle\int_{\theta_{2s}}^{0} \Gamma(\xi) d\xi > 0, \quad (\theta_{20}, z_{20}) \in \mathscr{D}_0, \\ J_{s2_1}^* = |z_{20}| + \displaystyle\int_{\theta_{20}}^{0} \Gamma(\xi) d\xi, \ (\theta_{20}, z_{20}) \in \mathscr{D}_1. \end{cases} \tag{15.140}$$

In Figs. 15.13–15.15, the game space decomposition is shown for $\mu = 1.1$, $\varepsilon_2 = 2.1$, and three values of ζ_P: $\zeta_P = 1.2$, $\zeta_P = 1$, and $\zeta_P = 0.8$, respectively. For $\zeta_P = 0.8$, $\bar{\Gamma}(\theta_2^{**}(\zeta_P)) = 9.26 > 0$, i.e. the condition (15.121) is satisfied. Moreover, since $\mu = 2.1 > \mu^{**}(1.1) = 1.06$, the condition (15.121) with $\mu = 2.1$, $\varepsilon_2 = 1.1$

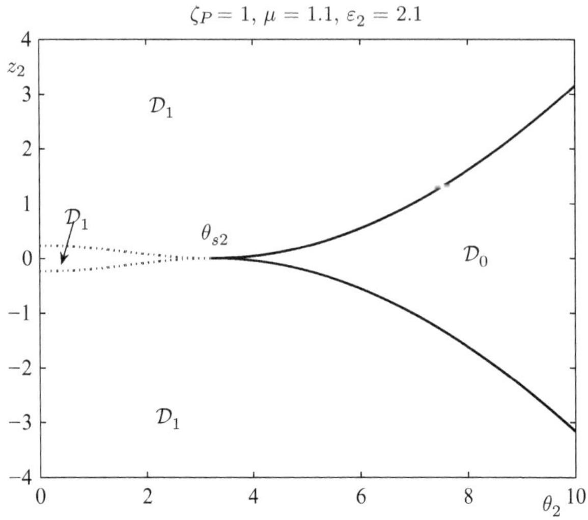

Fig. 15.14 Game space decomposition, $\zeta_P = 1$

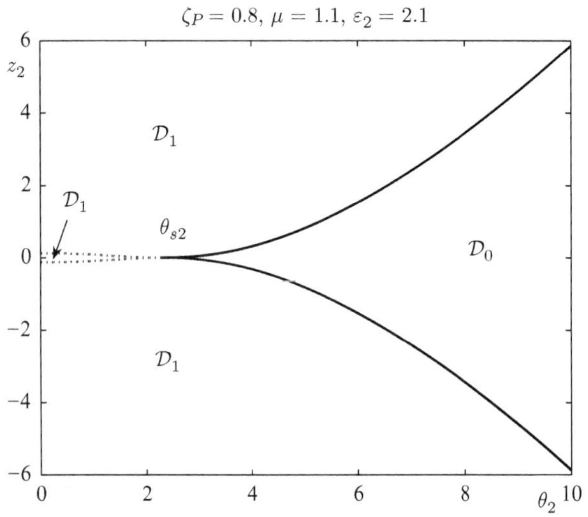

Fig. 15.15 Game space decomposition, $\zeta_P < 1$

is satisfied for all $\zeta_P \in (0, 1)$. For $\mu = 2.1$, $\varepsilon_2 = 1.1$: $\theta_{2s} = 5.42$ and $J^*_{2s} = 0.59$ for $\zeta_P = 1.2$; $\theta_{2s} = 3.23$ and $J^*_{2s} = 0.24$ for $\zeta_P = 1$; $\theta_{2s} = 2.30$ and $J^*_{2s} = 0.12$ for $\zeta_P = 0.8$, i.e. θ_{2s} and J^*_{2s} decrease monotonically for decreasing ζ_P.

It can be seen that the regular zone for $\theta_2 < \theta_{2s}$ between the boundaries becomes smaller as ζ_P is smaller.

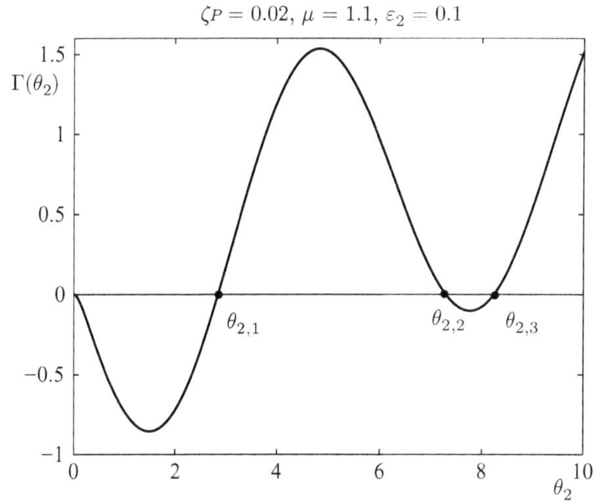

Fig. 15.16 Equation $\Gamma(\theta_2) = 0$ has three positive roots

15.5.2 Case of the Multiple Positive Roots of $\Gamma(\theta_2) = 0$

The condition (15.121) of Lemma 15.4 is sufficient for the uniqueness of a positive root of the equation $\Gamma(\theta_2) = 0$. Therefore, if this condition is violated, we can expect the appearance of at least two more positive roots of $\Gamma(\theta_2) = 0$. In Fig. 15.16, the case of three positive roots is depicted. For these parameters the roots are $\theta_{2,1} = 2.85$, $\theta_{2,2} = 7.30$, and $\theta_{2,3} = 8.24$.

The game space decomposition, corresponding to the parameters of Fig. 15.16, is shown in Fig. 15.17. The zero $\theta_{2,1}$ plays the same role as θ_{2s} in the case of a single positive root of $\Gamma(\theta_2) = 0$. The presence of the two additional zeros $\theta_{2,2}$ and $\theta_{2,3}$ makes the behavior of the boundary of $\mathscr{D}_0$ non-monotonic: namely, its upper part has a local maximum at $\theta = \theta_{2,2}$ and local minimum at $\theta = \theta_{2,3}$.

15.5.3 Comparison with Pursuer Having First-Order Dynamics

The analysis of a linear pursuit-evasion game with bounded controls and second-order dynamics detailed in the previous sections clearly demonstrated the complexity created by approximating the real dynamics of an interceptor missile by a second-order transfer function. In order to reduce such complexity, the efforts of replacing the second-order dynamics by some first-order dynamics, based on apparent similarities in their step response were made, as illustrated in Fig. 15.18. Since the normalized state and independent variables are different for the different

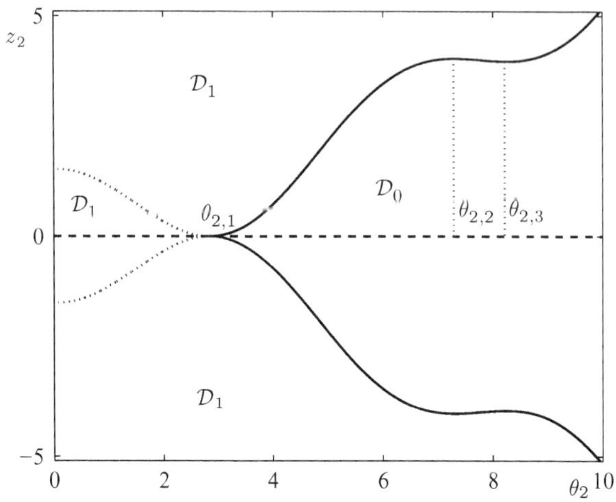

Fig. 15.17 Game space decomposition for three positive roots of $\Gamma(\theta_2) = 0$

Fig. 15.18 Step response comparison

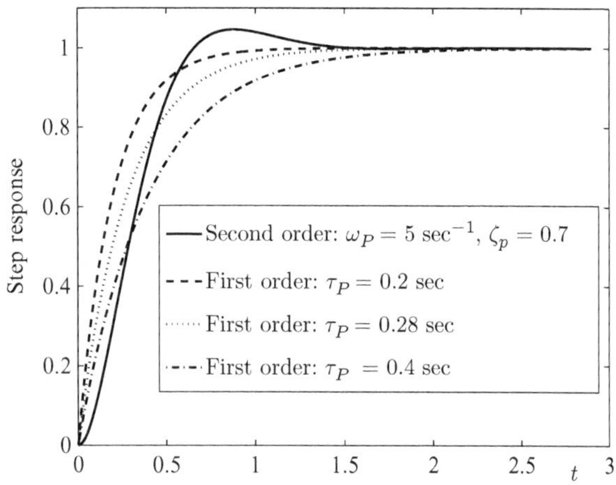

order of dynamics, the comparison is made using dimensional variables. The second-order dynamics selected for this illustration had $\omega_P = 5\,\mathrm{s}^{-1}$ and $\zeta_P = 0.7$. The set of compared first-order dynamics had time constants of $\tau_P = 0.2, 0.28$, and 0.4 s, respectively. The comparison of the guaranteed miss distances in a 4 s scenario against an ideal evader with $a_E^{\max} = 50\,\mathrm{m/s}^2$ obtained by the game solution, as the function of the pursuer maneuverability advantage μ, shown in Fig. 15.19, clearly demonstrates that using the first-order dynamic model strongly underestimates the guaranteed miss distances. Since these results should represent a pessimistic "worst case" situation, underestimating the guaranteed miss distance can be critical in the preliminary design phase.

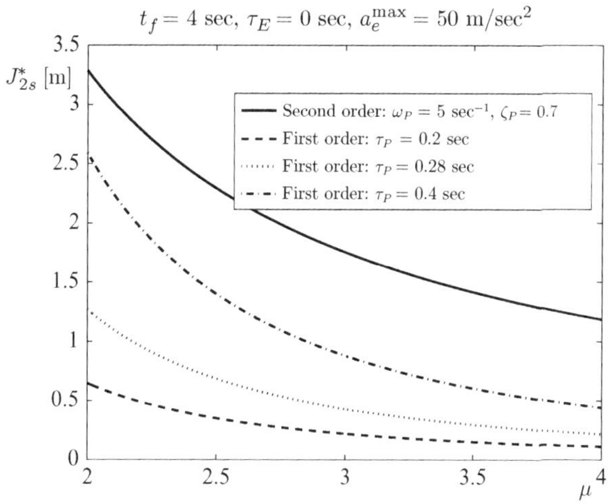

Fig. 15.19 Guaranteed miss distance comparison

15.6 Conclusions

This paper investigated the effect of the pursuer dynamics on the homing performance of an interceptor missile, modeled as the pursuer in a linear pursuit-evasion game with bounded controls. The homing performance is characterized by the guaranteed miss distance against an optimally evading target. The elaborate mathematical analysis, detailed in the paper, allowed to draw the following conclusions. The basic structure of a linear pursuit-evasion game with second-order pursuer dynamics is similar to the one with first-order pursuer dynamics. The reduced order (normalized zero effort miss distance, normalized time-to-go) game space is decomposed into two regions, a regular and a singular ones. The regular region is filled by "bang-bang" strategies controlled optimal trajectories. Such optimal trajectories do not fill the entire game space, indicating the existence of a singular region, where the optimal strategies are arbitrary and game value is a nonzero constant. This singular region has a major importance, because (being an unbounded region) it includes the large majority of the initial conditions of practical interest. The constant nonzero game value represents the guaranteed miss distance that can be achieved in the interception of an optimally evading target. The investigation of the analytical solution yields that for small values of the time-to-go the determining function of the game is negative, indicating that capture (zero miss distance) is not possible. Moreover, the analysis determined that for $\zeta_P \geq 1$ the determining function has a single zero. For $\zeta_P < 1$ another condition has to be satisfied for guaranteeing the existence of a single zero. If this condition is violated, odd numbers of additional roots may exist. The comparison of the homing performance, characterized by the guaranteed miss distance against an optimally evading target, shows that replacing the genuine second-order approximation of the actual dynamics by a simpler first-order dynamic model strongly underestimates the guaranteed miss distances. The guaranteed miss distance can be reduced by increasing the pursuer maneuverability and the frequency of the pursuer dynamics, as well as by reducing the damping factor.

References

Adler FP (1956) Missile guidance by three-dimensional proportional navigation. J Appl Phys 27:500–507
Bryson A, Ho Y (1975) Applied optimal control. Hemisphere, New York
Glizer VY, Turetsky V (2008) Complete solution of a differential game with linear dynamics and bounded controls. Appl Math Res Express. article ID: abm012, vol 2008, 49 pages
Glizer VY, Turetsky V, Fridman L, Shinar J (2012) History-dependent modified sliding mode interception strategies with maximal capture zone. J Frank Inst 349:638–657
Gutman S (1979) On optimal guidance for homing missiles. J Guid Control 2:296–300
Gutman S, Leitmann G (1976) Optimal strategies in the neighborhood of a collision course. AIAA J 14:1210–1212
Isaacs R (1965) Differential games. Wiley, New York
Krasovskii N, Subbotin A (1988) Game-theoretical control problems. Springer, New York
Shima T, Shinar J (2002) Time-varying linear pursuit-evasion game models with bounded controls. J Guid Control Dyn 25:425–432
Shinar J (1981) Solution techniques for realistic pursuit-evasion games. In: Leondes C (ed) Advances in control and dynamic systems, vol 17. Academic, NewYork, pp 63–124
Turetsky V, Glizer VY (2005) Continuous feedback control strategy with maximal capture zone in a class of pursuit games. Int Game Theory Rev 7:1–24